Conformational Analysis of Medium-Sized Heterocycles

Conformational Analysis of Medium-Sized Heterocycles

Edited by

Richard S. Glass

Richard S. Glass
Dept. of Chemistry
The University of Arizona
Tucson, AZ 85721

Library of Congress Cataloging-in-Publication Data

Conformational analysis of medium-sized heterocycles.

Bibliography: p.
Includes index.
1. Heterocyclic compounds. 2. Conformational analysis. I. Glass, Richard S.
QD400.C66 1988 547'.59 88-17210
ISBN 0-89573-283-1

Printed in the United States of America.

Distributed in North America by:

VCH Publishers, Inc.
220 East 23rd Street, Suite 909
New York, New York 10010

Distributed Worldwide by:

VCH Verlagsgesellschaft mbH
P.O. Box 1260/1280
D-6940 Weinheim
Federal Republic of Germany

Contributors

Roger W. Alder School of Chemistry, University of Bristol, Bristol BS8 1TS England

Frank A. L. Anet Department of Chemistry and Biochemistry, University of California, Los Angeles, CA 90024–1569

David A. Case Department of Molecular Biology, Research Institute of Scripps Clinic, La Jolla, CA 92037

Richard S. Glass Department of Chemistry, The University of Arizona, Tucson, AZ 85721

Patricia S. Hill Department of Chemistry, Millersville University, Millersville, PA 17551

Victor S. Hruby Department of Chemistry, The University of Arizona, Tucson, Arizona 85721

Louis D. Quin Department of Chemistry, The University of Massachusetts, Amherst, MA 01003

William N. Setzer Department of Chemistry, The University of Alabama in Huntsville, Huntsville, AL 35899

Jonathan M. White School of Chemistry, University of Bristol, Bristol BS8 1TS England

Preface

There have been a few books published that treat conformational analysis of heterocycles. Among these are the following general texts. *The Conformational Analysis of Heterocyclic Compounds* by F. G. Riddell is an excellent, brief introduction to the area, aimed at those with an undergraduate degree in chemistry. *Comprehensive Heterocyclic Chemistry,* edited by A. R. Katritzky and C. W. Rees, also presents aspects of conformational analysis at a more advanced level. *Stereochemistry of Heterocyclic Compounds* by W. L. F. Armarego consists of two volumes and extensively treats over 4,500 references up to 1974 and part of 1975. However, one is struck by the relatively minor attention paid to conformational analysis of medium-sized heterocycles in these already published books. That is, conformational analysis of six-membered heterocycles has been extensively considered, particularly from the point of view of the effect of the presence of a ring heteroatom in place of a carbon atom on the conformational properties of cyclohexane, which itself was so well developed by Hassel, Barton, and others. Heterocycles with small rings (three- and four-membered), five-, and, to some extent, seven-membered rings have been reviewed, as have large rings (twelve and more ring members). These latter systems have been investigated largely because of the interest in macrolide and polyether antibiotics as well as crown ethers, cryptands, and related compounds initiated by Pedersen, Cram, and Lehn. But there remains a gap in the review of eight- to eleven-membered heterocycles, which are generally referred to as medium-sized rings. This inattention does not reflect the outstanding advances that have been made in this area and of which the general reader is unaware. To be sure, there are a number of specialized series on heterocycles that review the advances in this area on a yearly basis; but these series are aimed at the specialist, not the general reader. This volume aims to fill this gap.

The contributors to this volume critically review conformational analysis of medium-sized heterocycles. Seven- and, in some cases, twelve- and even fourteen-membered heterocycles are included, at the discretion of the authors of each chapter, to contrast or extend the conformational anomalies or trends discerned for the eight- to eleven-membered heterocycles. The presentation conveys, for the general reader, the exciting advances made in this area. There is a growing literature in this area that is largely inaccessible to the general reader because much of it is included in research publications with emphasis in other areas. Consequently, the titles of these papers do not reflect their content of conformational information. Although most of this material appears in the specialized series already referred to, it is difficult for the nonexpert to develop a reasonable perspective of the field from these annual reviews. The authors of the chapters in this volume are all on the cutting edge of research in these areas and, therefore, they are familiar with all of the relevant data and are able to provide this information as well as critical insight for the nonspecialist. As will be evident in reading this volume, these experts provide a balanced and critical perspective that will enable all readers to fully appreciate the present status of the field.

Although much further work needs to be done in the conformational analysis of medium-sized heterocycles, there have been many important advances in this field that have been largely unappreciated. For example, it is generally believed that medium-sized ring compounds have a relatively large number of conformations of comparable energy with relatively low barriers for interconversion. That is, medium-sized rings are "floppy" and, consequently, are an unpromising area for conformational analysis. Although this perception is true in some cases, a variety of conformational constraints, such as endocyclic *cis* double bonds, aromatic ring fusions, and heteroatom-heteroatom bonds, dramatically change this picture as illustrated extensively in this book. The conformational possibilities for medium-sized cycloalkanes are remarkably restricted by substituents, and, as discussed in this book, in many cases by substituting heteroatoms for ring carbon atoms. Indeed, a major focus of this book is the effect caused by substituting heteroatoms (O, N, S, and P) for ring carbon atoms on the conformational properties of medium-sized cycloalkanes. Toward this end, an overview of the conformations of cycloalkanes with seven- to twelve-membered rings is presented in Chapter 2 and these conformations are compared and contrasted with the corresponding heterocycles throughout the text.

The series of which this volume is a part emphasizes the use of modern instrumental methods in stereochemical analysis. Such an emphasis is especially appropriate for this volume because the advances made in the conformational analysis of medium-sized heterocycles could not have been done without the progress in structural methods. Of particular importance are X-ray crystallographic analysis, nuclear magnetic resonance spectroscopy (especially use of FT instruments, 2D and nuclear Overhauser enhancement methods), photoelectron spectroscopy, and theoretical and computational methods. This last method is of such importance in this area that it is critically evaluated in Chapter 1. The theoretical methods that have been used most often up to now are semiempirical quantum mechanical techniques and force field (molecular mechanics) calculations. The present status and future development of these areas, as well as consideration of other theoretical methods that may supplant these, are discussed in Chapter 1. Application of these methods are discussed throughout the text. Although the use of photoelectron spectroscopy for conformational analysis is relatively new, it is of proven importance and its potential utility so great that an overview is presented in Chapter 4 and exemplified throughout the volume.

Medium-sized rings first attracted interest because of their high strain energies compared with five- to seven-membered rings on the one hand and large (twelve-membered and larger) rings on the other. They were recognized to suffer from angle, torsional, and nonbonded strain. The novel nonbonded strain due to transannular interactions drew particular interest. Chemical consequences of transannular interactions were extensively investigated by Prelog, Cope, Leonard, and others. Such studies are also important today as exemplified by the unprecedented redox behavior of certain bicyclic diamines and 1,5-dithiocane and the stabilization of their corresponding radical cations by transannular bonding. The unusual molecular architectures of medium-sized heterocycles are currently receiving extensive scrutiny owing to their propensity to coordinate metal ions and convey unusual properties to these complexes. Such studies are exemplified by those using 1,4,7-triazonane and 1,4,7-trithionane as ligands. Another important recent application of conformational analysis of medium-sized heterocycles involves the mapping of peptide hormone receptors. Many important peptide hormones (e.g., enkephalins) are small, linear, flexible molecules whose conformational space also depends on solvent and state (solution versus solid) effects. Ring formation restricts their conformations and some cyclic analogues are conformationally well-defined. Many of these cyclic analogues are medium-sized heterocycles and are biologically active. Insight into their conformations provides information on receptor binding and transmittal of biological information (transduction). This is of sufficient import and future applications are so promising that a separate chapter is devoted to a general review of cyclic peptides incorporating medium-sized heterocycles (Chapter 6). An understanding of the results of conformational analysis of medium-sized heterocycles is also important in designing stereochemical control in the reactions of these compounds. Such control is crucial in the use of such reactions in the synthesis of complex natural products. Important advances have already been made in using medium-sized carbocycles in complex synthesis and there are already some important results with medium-sized heterocycles using the concept of local conformational control. Such future applications are very promising. The intent in pointing out these applications of conformational analysis of medium-sized heterocycles is twofold. Those interested in these and related applications are encouraged to carefully read this volume and become familiar with the results and methods reviewed here. In addition, medium-sized heterocycles provide a unique geometry that features transannular juxtapositions. Such systems should be exploited more widely by the nonspecialist for studies that capitalize on these unique features. It is also hoped that this volume will stimulate more research in this rapidly developing area, which has already yielded important surprises.

The organization of the book is as follows. As already pointed out, Chapter 1 consists of a critical review of theoretical and computational methods that delineates their limitations, calls attention to presently unsolved problems, and outlines possibilities for future advances that will have an important impact on heterocyclic analysis. Chapters 2 through 5 critically review the results of conformational analysis of oxygen-, nitrogen-, sulfur-, and phosphorus-containing medium-sized heterocycles in that order, and a chapter is devoted to each of these heteroatoms. Mixed heterocyclic systems are discussed within these major chapters as follows:

N, O systems in Chapter 3; S, O, and/or N systems in Chapter 4; P, O, N, and/or S systems in Chapter 5; Chapter 6 treats peptides with medium-sized rings containing N, O, and/or S. As already mentioned, Chapter 6 deals exclusively with peptides. There is also some relevant discussion of peptides in other chapters. Other natural products or their derivatives and pharmaceuticals containing medium-sized heterocycles are presented throughout the book. In each chapter, the heterocycles are treated sequentially in order of increasing ring size beginning with seven-membered rings. Bicyclic compounds composed of medium-sized heterocycles are also featured in Chapter 3 on Nitrogen Heterocycles because there have been substantial and important studies in this area. The instrumental methods used in the conformational analyses presented in this book are highlighted and the extensive index is replete with references to specific methods, as well as listings under the specific compounds in which the method was used.

RICHARD S. GLASS

Tucson, Arizona

Contents

1. Theoretical and Computational Methods of Conformational Analysis in Heterocyclic Rings

David A. Case

2. Medium-sized Oxygen Heterocycles

Frank A. L. Anet

3. Nitrogen Heterocycles

Roger W. Alder and Jonathan M. White

1

Theoretical and Computational Methods of Conformational Analysis in Heterocyclic Rings

David A. Case
Research Institute of Scripps Clinic
La Jolla, CA

1.1. Introduction

The use of theoretical methods and computer calculations in conformational analysis has a long history. In some sense it is fruitless to separate "experiment" and "theory," since any interpretation of experimental results relies upon some "theoretical" picture of the nature of the conformational states available to the molecular system. Thus the classic work of Hassel[1] and Barton[2] on conformational preferences in six-membered rings contributed much to our understanding of the nature of the potential energy surface for these systems. For cyclic systems, the emergence of "theoretical" studies, particularly using computers, really began with Hendrickson's analyses of cycloalkanes in the early 1960s.[3] Since that time, computer methods have become powerful enough, and specialized enough, to warrant a separate chapter in a volume such as this one.

There are two logically distinct problems to be faced in theoretical analyses of molecular conformations. The first task (discussed in section 1.2) is to compute the molecular potential energy surface, that is, the internal energy as a function of the nuclear coordinates. This is generally only possible to do on a local or point-by-point basis, and it can be a formidable task to obtain *global* information (e.g., the location of all minima on the potential energy surface) for systems of the size discussed in this text. Once an estimate of the potential energy surface is available, a second step (discussed in section 1.3) is required to relate this surface to the observable properties of the molecule, such as the equilibrium populations or rate constants for interconversion of conformers. This can also be a difficult task, especially for condensed systems.

This chapter does not attempt to provide a comprehensive summary of all calculations on medium-sized heterocyclic rings. This would be an impossible task in any event, and would largely overlap the contents of later chapters that discuss particular systems. Rather, the focus here is on the *methods* involved in such calculations, with a particular emphasis on the reliability one can expect from various classes of computation. As computers become more powerful and less expensive, new types of calculation become feasible, and it is certain that the flavor of theoretical approaches to medium-sized rings will change in a dramatic way over the next decade. For this reason, approaches will be discussed whose primary focus to date has been on smaller systems, but which are likely to be applied to seven- to 12-membered rings in the near future.

1.1.1. Some Definitions

Terminology is not uniform in this field, but it is usually possible to avoid misunderstandings if the context is made clear. Those whose primary interest is in computing molecular potential energy surfaces often use the term "conformation" or "conformer" to describe *points* on those surfaces, that is, particular nuclear configurations. In this spirit, for example, Burkert and Allinger[4] use the term "conformation" to refer to virtually any arrangement of the atoms in a molecule and reserve the term "conformer" for those arrangements that correspond to (local) energy minima. Authors whose background is more closely tied to statistical mechanics or thermodynamics generally use these terms to refer to certain types of *stereoisomers*, that is, to assemblies of molecules at or near potential energy minima. Riddell, for example, defines conformations as "stereoisomers that can be interconverted either by rotation about bonds of order approximately one, with any concomitant small distortions of bond lengths and angles, or by inversion at a three-coordinate centre in the molecule, or by pseudorotation on phosphorus."[5]

These two definitions are functionally equivalent when the potential energy surface consists of relatively narrow minima separated by substantial barriers. For surfaces that allow large-amplitude motions or have low barriers between distinct potential energy minima, it is important to avoid confusion. A classic example arises in considering the twist and boat arrangements of cyclohexane. Strauss' language is typical: "The difference in energy between the boat and twist is rather small and the boat and twist conformers must be considered as a set of interconverting forms rather than as distinct chemical entities. It is appropriate to call the resulting conformation a twist-boat conformation."[6] The attempt in this chapter is to be as clear as possible, with the terms "arrangement," "configuration," "local minimum," and "transition state" used to describe *points* on a surface and the terms "conformer" and "conformation" reserved to describe (potentially) observable *collections* of molecules whose configurations are near particular energy minima.

1.2. Computing Molecular Potential Energy Surfaces

Computational methods for estimating molecular potential energy surfaces fall broadly into two classes. The most fundamental, and potentially the most useful, calculations are based on quantum mechanical principles that compute the electronic energy as a function of nuclear positions. These methods are of great generality and in many cases can proceed directly from "first principles," that is, without any recourse to experimental data for calibration. Other quantum methods (called semiempirical) use experimental data as input for simplified calculations. In addition to providing numerical information about molecular energies, such calculations can also provide valuable models for understanding the ways in which electronic interactions affect conformations; such models, for example, explain the effects of electronegativity or lone pairs on conformational preferences. These calculations are surveyed in section 1.2.1.

A second type of potential extends the scope of conventional vibrational force fields: it uses a similar classical model of additive interactions (such as terms representing bond stretches or bond angle deformations) and adds torsional and nonbonded terms in an attempt to describe the potential energy surface for arbitrary configurations. These are most commonly called "molecular mechanics" potentials because of the analogy to a mechanical set of springs. Unlike vibrational force fields, the molecular mechanics potentials do not require a minimum-energy geometry as input information; rather, local minima are determined as points on the surface at which all of the derivatives of the energy vanish. These empirical potentials require either experimental or quantum mechanical data for calibration, and are described in section 1.2.2.

All of these approaches are increasingly interdependent: quantum mechanical calculations are being used to calibrate molecular mechanics potentials and to refine vibrational force fields. In turn, molecular mechanics potentials are often used to provide geometries for the (more expensive) quantum calculations. Some conclusions and discussion of prospects for future calculations are given in section 1.2.3.

1.2.1. Quantum Mechanical Calculations

A variety of quantum mechanical methods are used to study conformational properties of organic molecules, including cyclic systems. The most common invoke the Hartree-Fock molecular orbital approximation, which in favorable cases can give excellent results for predictions of molecular geometry. The difference between the Hartree-Fock results and the true (nonrelativistic) electronic energy is defined as the correlation contribution; correlation effects can be of considerable importance for accurate studies of barrier heights and relative energies of various isomers. These calculations can also be divided (somewhat arbitrarily) into *ab initio* methods, in which no recourse to experimental data is used, and semiempirical methods, in which certain "test" molecules are used to provide parameters used in the theory. In the following, these various methods are outlined and comparisons of special relevance to conformational analyses of rings are offered.

1.2.1.1. *Ab initio* Molecular Orbital Methods. The nonrelativistic electronic Hamiltonian for a polyatomic molecule is (in atomic units):

$$H_{el} = -(\tfrac{1}{2})\sum_i \nabla_i^2 - \sum_i \sum_\alpha \frac{Z_\alpha}{r_{i\alpha}} + \sum_i \sum_{j>i} \frac{1}{r_{ij}} \qquad (1\text{-}1)$$

If the interelectronic repulsions (the final term above) were to be neglected, the wavefunction would have the form of any antisymmetrized product of one-electron functions (molecular orbitals). Using the complete Hamiltonian, but restricting the wavefunctions to those that have the form of an antisymmetrized product of molecular orbitals, yields the Hartree-Fock approximation. This level of theory can give results of good quality for a variety of chemical problems, particularly for predictions of molecular geometries. It has the additional advantage that the orbitals themselves are easy to visualize, so that a variety of qualitative models can be made to aid interpretation.

Basis Sets. In practice, the molecular orbitals themselves are rarely solved for but are expanded in a *basis set* of functions that resemble atomic orbitals. For polyatomic calculations, the almost universal choice uses basis functions that have a Gaussian form:

$$Nr^l e^{-\zeta r^2} Y_l^m(\theta, \phi) \qquad (1\text{-}2)$$

Here (r, θ, ϕ) give the coordinates of the electron relative to a particular nucleus, N is a normalization function, and Y_l^m is a real spherical harmonic. Although these functions are not themselves solutions to any atomic problem, integrals involving them are (relatively) easy to compute, and expansions in Gaussian orbitals are often referred to by the letters "LCAO" (for linear combination of atomic orbitals).

A device widely used with Gaussian basis functions is that of a *contracted* set. Here not all of the coefficients are independently varied; rather, the ratios among certain coefficients are kept fixed at predetermined values. This gives substantial savings in computation time with little loss in accuracy. Systematic studies of contraction schemes and their effects on molecular calculations have been carried out under the leadership of Pople, and basis sets proposed by his group are by far the most commonly used in organic chemistry, primarily because of the extensive literature with which new results can be compared.[7] Three of the most common basis sets are called STO-3G, 4-31G, and 6-31G*.

The STO-3G basis is a minimal set, in which first-row atoms have six Gaussian functions of the *s* type and three of the *p* type on first-row atoms, along with three Gaussian *s* functions on each hydrogen atom. Contraction reduces the number of variable coefficients of two *s*'s and one *p* on first-row atoms and one *s* function on hydrogens. (The letters "STO" derive from the fact that the contracted orbitals approximate Slater-type orbitals, which have an exponential dependence on *r*.) In a standard nomenclature, this basis set is denoted by (6*s* 3*p*/3*s*) contracted to [2*s* 1*p*/1*s*].

The 4-31G basis set is a "split valence" set in which the valence electrons can occupy two atomiclike orbitals, one made from a linear contraction of three Gaussian functions and the other consisting of a single Gaussian with exponent ζ different from those employed in the first contraction. This extra flexibility allows the valence electrons in a molecule to change their radial distributions from those found in atoms to different distributions that are more appropriate for the molecular environment. For superminicomputers of the general capabilities of the DEC VAX 11/780, this is about the largest basis set that is practical for direct calculations on molecules of the size discussed in this text (i.e., having at least seven nonhydrogen atoms). As more powerful computers and specialized array processors or parallel machines become more widely available, this may be expected to change.

A third popular basis set is 6-31G*, which improves upon the 4-31G set by using a linear combination of six Gaussians for each inner shell 1*s* orbitals and by adding a single set of *d*-type Gaussian "polarization" functions on each first-row atom. This extra valence flexibility also allows atoms in a molecule to adjust more fully to their environment, and for many types of calculations a 6-31G* basis set gives results quite close to the Hartree-Fock limit. Even more flexible basis sets have some general use; those such as 6-31G** and *p*-type polarization functions to all hydrogen atoms, generally at a great cost in computation speed.

For a molecule with about 40 basis functions in a minimal set (such as B_6H_{10}, which is slightly smaller than the rings discussed herein), a 4-31G calculation is approximately six times as expensive in terms of central processing unit (CPU) time as an STO-3G calculation, and 6-31G* is about 30 times as expensive as STO-3G. (For comparison, a MINDO/3 calculation would be approximately 100 times cheaper than STO-3G[8,9]). Disk-space considerations are also important for calculations on large molecules, since the number of integrals to be stored and manipulated rises as the fourth power of the number of basis functions. Thus, even if a relatively fast microprocessor could be dedicated to a single calculation, storage considerations might make large basis set calculations on molecules with eight to ten first-row atoms impractical. It is likely, however, that new storage technologies, and continued decreases in the price of computer memory, will change this situation within the next few years.

1.2.1.2. Electron Correlation. Three principal methods have been used in organic chemistry to go beyond the Hartree-Fock approximation. The first, configuration interaction (CI), constructs an explicit superposition of configurations (i.e., of antisymmetrized products with varying orbital occupancies) and solves for the optimum mixing coefficients by diagonalizing a Hamiltonian matrix. Since complete expansions of this sort are not feasible for large molecules, restrictions on the nature of the excitations are common; for example, many calculations on closed-shell species use single and double excitations from a Hartree-Fock wavefunction (this is called CISD), and many also include empirical corrections for the effects of higher excitations.

A second approach is related to this one but begins from a *multiconfiguration* reference state, that is, a wavefunction that consists of more than one antisymmetrized product, and for which both the orbital shapes and the mixing coefficients have been simultaneously optimized. This reference state is called "multiconfiguration self-consistent field," or MCSCF. Two popular special cases have their own initials: the generalized valence-bond (GVB) approach corresponds to a particular set of configurations in the MCSCF procedure, which are chosen by analogy to classical valence-bond ideas of molecular structure. The complete active-space (CAS) SCF approach performs an MCSCF calculation on all configurations that can be built from a limited subspace of the total set of orbitals. It is common for each of these methods to include some additional configuration interaction as well.

A third method uses many-body perturbation theory (MBPT) to estimate the effects of electron correlation to various orders of perturbation theory. These expansions often converge quite well when a single reference configuration dominates the wavefunction, as is often the case for closed-shell organic molecules. Perturbation theory has the advantage that it can be "automated" to follow a Hartree-Fock calculation, so that a large number of molecules have now been computed at a consistent level of theory and can be compared.

In the past, the cost of carrying out correlated calculations on medium-sized rings has been prohibitive, but this may change soon. A recent discussion of relative timings illustrates some of the difficulties.[7] Large basis sets are required, since it makes little sense to study correlation effects in systems that still have large basis set deficiencies. For a much smaller molecule, methylamine, using a 6-31G* basis set requires about 10 minutes of CPU time on a VAX 11/780 to carry out a single point calculation at the Hartree-Fock level. A CISD calculation takes about 17 times as long and perturbation theory to second and fourth order 1.5 and 5.8 times as long as Hartree-Fock. Most of these relative ratios will get worse as the size of the molecule increases. Geometry optimization at this level of theory would require many such point calculations, and would be correspondingly more expensive. It is reasonable to assume that (at least for the next few years) direct calculations of correlation effects on medium-sized rings will be limited to a few cases of special interest.

1.2.1.3. Semiempirical Methods Based on Neglect of Differential Overlap. Until quite recently, the computer expense involved in computing *ab initio* wavefunctions for all but the smallest molecules was prohibitive. For this reason, a great deal of effort has been expended in developing more approximate methods. As a general rule, these are built upon the framework of LCAO molecular orbital theory, but treat only the valence electrons explicitly and ignore many of the two-electron integrals required in *ab initio* calculations. Many or all of the remaining integrals are taken to be empirical parameters, to be fit to some selection of experimental data.

There have been an enormous number of semiempirical calculations over the past 20 years, and it is not feasible to give a comprehensive review here. Many comparisons of these methods with each other and with experiment may now be found in textbooks[10] and in review collections.[11] Unfortunately, most of these methods perform somewhat poorly in estimating nonbonded and hydrogen-bond interactions that are important in understanding conformations of cyclic systems. Hence these methods have achieved much wider application in studies of chemical reactions in small acyclic systems than they have in problems in conformational analyses in rings.

The earliest semiempirical approaches of this sort were the *complete* and *intermediate neglect of differential overlap* methods (CNDO[12] and INDO[13] respectively). Each uses a minimal basis set of valence atomic orbitals. When molecular orbitals are expressed as linear combinations of these basis orbitals, the molecular Coulomb and exchange integrals become linear combinations of two electron integrals over atomic orbitals. In a fairly standard notation.

$$\iint f^*_r(1) f_s(1)(r_{12}^{-1}) f^*_t(2) f_u(2)\, dv_1\, dv_2 = (rs|tu) \tag{1-3}$$

where the f's are valence atomic orbitals. The zero differential overlap approximation assumes that

$$(rs|tu) = \delta_{rs}\delta_{tu}(rr|tt) \tag{1-4}$$

The CNDO method makes this approximation for all pairs of atomic orbitals, so that the only nonvanishing integrals are of the form $(rr|tt)$. In the INDO method, differential overlap between atomic orbitals on the same center is not neglected, but is still neglected for two-center integrals, i.e., where the orbitals are centered on different atoms.

Although the CNDO and INDO methods have achieved considerable success in interpreting electronic spectra (in parametrizations known as CNDO/S and INDO/S), these methods present severe problems in treating interactions involving nonbonded lone pairs, both for rotations about bonds[14] and for interactions between lone pairs that are more widely separated.[15] Since these effects are particularly important for understanding the conformations of heterocyclic ring systems, it is not surprising that these original NDO methods are now used only sparingly for such systems. There is still some application of the INDO method to determine torsional preferences in acyclic compounds containing heteroatoms and the resulting ideas can be transferred to cyclic compounds as well.[16]

Related to these are the modified INDO (MINDO[17]) and modified neglect of differential overlap (MNDO[18]) methods developed in Dewar's group. These neglect different categories of integrals, and (more important) were parametrized to reproduce experimental heats of formation for a large number of small organic compounds. Although these methods give results for reaction energies that are significantly better than those of other semiempirical methods, they are still not entirely satisfactory for conformational analyses of rings. For example, both methods predict cyclopentane to be planar and give poor estimates of the amount of puckering in cyclohexane (*cf*. Table 1-1).[19] As with the CNDO and INDO methods, rotational barriers are predicted only moderately well, especially for bonds involving atoms with lone pairs: for example, the barriers

TABLE 1-1. GEOMETRIES FOR CYCLOHEXANE

Parameter	Observed	4–31G[a]	AM1[b]	MNDO[c]	MINDO/3[d]
r (C—C)	1.536 ± 0.002	1.534	1.515	1.539	1.517
r (C—H)	1.116 ± 0.004	1.087	1.121	1.114	1.123
θ (CCC)	111.4 ± 0.2	111.4	111.3	114.1	—
θ (HCH)	107.5 ± 1.5	106.9	107.4	105.7	—
τ (CCCC)	54.9 ± 0.4	55.0	55.1	46.3	62.7

[a] Ref. 40.
[b] Ref. 20.
[c] Ref. 18.
[d] Ref. 17.

to *trans* orientation in hydrazine and hydrogen peroxide are small or nonexistent in the MNDO and MINDO methods. These methods also generally give hydrogen bonds that are somewhat too weak and overestimate the energies of "crowded" molecules.

Recently, a "third generation" program was developed in Dewar's group, called AM1 (Austin Model 1).[20] This attempts to correct the main errors in MNDO, which appears to overestimate the repulsions between nonbonded atoms when they are separated by approximately their van der Waals distance. As with earlier parametrizations, barriers to rotation about saturated single bonds are generally underestimated, as are hydrogen-bond strengths. AM1 appears to give much better results for cyclohexane than one obtains for MNDO or MINDO, as shown in Table 1-1. It is too early to make secure judgments about the potential usefulness of the AM1 method to conformational analysis in heterocyclic rings, but it is likely that a number of applications will be reported in the near future that will clarify its prospects.

An alternative semiempirical method is PRDDO, partial retention of diatomic differential overlap, developed by Halgren and Lipscomb.[21] This method neglects a different set of integrals than do other semiempirical techniques, and was parametrized to reproduce *ab initio* STO-3G calculations rather than experimental values. In most cases, the comparisons to STO-3G are quite good,[9] although only a few calculations have been carried out that are sensitive to nonbonded repulsions or torsional barriers.

A method quite different from those discussed so far is PCILO (perturbative configuration interaction using localized orbitals).[22] Here one constructs localized bonding and lone-pair and antibonding orbitals to use in a configuration interaction treatment. Perturbation theory is used to compute the energy. The CNDO or INDO integral approximations are often used to simplify the calculations. This method has been extensively applied to conformations of biological molecules (including peptides) and will be discussed in conjunction with empirical calculations on peptides.

Another class of calculations uses an effective local potential to replace the exchange interactions in Hartree-Fock theory. The best known of these methods is the $X\alpha$ method introduced by Slater, primarily for use in solid-state problems.[23,24] Extensive applications to molecules began in the 1970s, using both LCAO approximations and a "scattered wave" approach derived from solid-state band theory.[25,26] More recent approaches use "density functional theory" to approximate both exchange and correlation effects into a theory that still has the computational and interpretative advantages of a molecular orbital-based wavefunction.[27] These methods have been applied mostly to problems in inorganic chemistry and in computations of organic molecules that are devoted to photoelectron and optical spectroscopy. Numerical difficulties in both the scattered-wave and LCAO versions of the theory have hindered applications to conformational energy problems, but continuing developments in methodology may make such calculations feasible soon.

1.2.1.4. Comparisons of Quantum Mechanical Methods. The literature of quantum chemistry is immense, and it is hardly possible to give even brief mention to all that appears in there. A great achievement of Pople's group has been to carry out (and to encourage others to carry out) systematic calculations on many molecules with consistent choices of basis set and level of theory. Much of this information is contained in a recent book.[7] Schleyer recently reviewed a number of bibliographies and data bases of *ab initio* calculations, including some in machine-readable form.[28] Levine[10] has prepared a very useful brief survey of comparative results. In the following paragraphs, a few comparisons are outlined that seem especially important for conformational analysis of cyclic systems.

Equilibrium Geometries. Ab initio Hartree-Fock theory generally does a good job of predicting bond lengths between heavy atoms, for both single and multiple bonds. Hehre et al.[7] report extensive comparisons to experiment that show mean absolute errors of 0.02–0.03 Å for a wide variety of molecules. The errors are generally smaller for basis sets with polarization, such as 3-21G$^{(*)}$ or 6-31G*. Furthermore, many of the errors are systematic, so that empirical corrections can be applied to give even more accurate predictions. Skeletal bond angles (for acyclic systems) are also in good agreement with experiment, with mean absolute errors of about 1°. The nonplanarity of cyclobutane and cyclopentane is substantially underestimated at the STO-3G level, but 6-31G* is much better.[29]

As noted above, Dewar and co-workers have reported extensive comparisons with experiment for MINDO/3,[17] MNDO,[18] and AM1.[20] Barriers to rotation about single bonds (and about the N—C bond in formamide) are systematically underestimated, which may limit the quantitative application of these models to conformational analyses of heterocyclic rings and peptides. Only AM1 yields a good equilibrium geometry for cyclohexane. As with *ab initio* calculations, further systematic studies on larger rings would be most useful.

Barriers to Internal Rotation. One of the most useful applications of quantum chemistry to conformational analysis has been the creation of orbital models that allow one to understand conformational preferences for rotations about single bonds. Early calculations by Radom, Hehre, and Pople[30] spanned a wide range of first-row atoms, and encouraged analysis in terms of Fourier components of the barriers to rotation. Particular attention has been paid to the "anomeric" effect, in which substituents separated by a saturated carbon interact strongly. Computational approaches to modeling these interactions have been reviewed,[31] and extended to second-row atoms.[32] These calculations, along with a localized bond analysis of such rotations by Brunck and Weinhold,[33] have had a significant influence on the development of empirical potential energy functions. A recent book by Hehre, Radom, Schleyer, and Pople[7] gives a good overview of such calculations.

Hydrogen Bonds. Semiempirical methods generally give very poor descriptions of hydrogen bonds, yielding bonds that are too weak and too long.[34,35] AM1 has smaller errors, but in the same direction. On the other hand, *ab initio* Hartree-Fock calculations generally give good qualitative accounts of hydrogen bonds, although large basis sets (like 6-31G*) are required to obtain a good geometry for the water dimer.[34,35] Correlation effects probably contribute 1–2 kcal/mol to the stabilization of the water dimer.

The analysis of hydrogen bonding interactions is greatly facilitated by analyses that (approximately) divide the interaction into component parts, such as electrostatics, exchange repulsion, and polarization.[36,37] These show that the *directionality* of the electrostatic component of the hydrogen bond, and its changes for different geometries and different complexes, are good predictors of the strength and directionality of the complete interaction. This is in accord with the way many empirical force fields are constructed (see below), and is confirmed by more recent studies of gas-phase hydrogen-bonded dimers.[38]

Vibrational Frequencies. The development of analytical techniques to compute the second derivatives of the energy (with respect to the nuclear coordinates) from *ab initio* wavefunctions has led to a number of studies aimed at predicting vibrational frequencies.[39] Force constants at the Hartree-Fock level of theory are generally too large, but the errors appear to the systematic, so that empirical corrections can be made to give useful results. A recent 4-31G calculation on cyclohexane illustrates the type of result one may expect for larger rings as well.[40] When the computed C—H frequencies were scaled by a factor of 0.92 and the remainder by a factor of 0.88, the root-mean-square (rms) error for cyclohexane and cyclohexane-d_{12} was only 22 cm^{-1}. A more extensive fitting procedure, with six scaling factors, yielded an rms error of 10 cm^{-1}, and also gave a good account of observed intensities. This level of agreement should be adequate in many cases to aid in spectral assignments. Calculations of this sort can also help in more conventional force-field analyses by indicating approximate relationships between various force constants that otherwise might be independently varied.

Relative Energies of Isomers. Computing reliable relative energies of stereoisomers turns out to be a challenging task, and most such calculations to date are not directly relevant to heterocyclic rings. One important test case has been *n*-butane, for which the energy difference between the *syn* and *anti* forms provides an important test for parameterization of empirical potentials. At the Hartree-Fock level, basis set effects are small but not negligible: STO-3G gives an energy difference of 6.0 kcal/mol[41] and 6-31G gives 6.4 kcal/mol.[42] In this instance, however, dispersion effects are important, since the *syn* form is more compact than *anti.* Estimates at the 6-31G CISD level predict a barrier of 4.5 kcal/mol,[42] or a lowering of $3k_BT$ from the Hartree-Fock result. The experimental result is not accurately known, although Raman studies[43] also yield a barrier of 4.5 kcal/mol.

It is likely that differences in dispersion energy will be important in other cases where the "compactness" of the molecule differs from one conformer to another; the "alanine dipeptide" case described below is a second example. For many other stereoisomerization processes, however, one can expect Hartree-Fock results to give answers of useful accuracy, at least if adequate basis sets are used. The almost routine application of such calculations to medium-sized rings should be only a few years away.

1.2.1.5. Conclusions. Computers are just now becoming fast enough and cheap enough for quantum chemistry to make substantial contributions to conformational analyses in organic chemistry. Most of the comparisons outlined bear only indirectly on medium-sized heterocyclic rings, since even with modest basis sets such calculations would require hours to days on a machine such as a VAX 11/780. Semiempirical calculations are much cheaper and could easily be applied to such molecules, but most of these methods have severe drawbacks when it comes to conformational analysis: they represent noncovalent and H-bond interactions poorly, and generally underestimate barriers to rotation about single and partial double bonds. Most available evidence suggests that *ab initio* calculations should fare much better, and that Hartree-Fock calculations at the 4-31G or 6-31G* level will provide a great deal of useful information that is difficult or impossible to gather experimentally. Truly reliable calculations, which systematically include electron correlation effects and have errors less than k_BT, are probably much further away.

1.2.2. Molecular Mechanics Potentials

The most useful theoretical approach to the conformational analysis of medium-sized rings has been the use of empirical functions to represent the molecular potential energy surface. This approach is called "molecular mechanics," and attempts to represent the Born-Oppenheimer surface by an expression of the general form,

$$V = \sum V_{\text{stretch}} + \sum V_{\text{bend}} + \sum V_{\text{torsion}} + \sum V_{\text{nonbon}} \qquad (1\text{-}5)$$

where the sums are over all bonds, angles, torsion angles, and remaining pairs of atoms (the "nonbonded interactions") in the molecule. There are two characteristic features of such fits to potential energy surfaces: first, the functional forms of the various terms are simple enough that they can be evaluated (along with their first and second derivatives) very rapidly, even for large molecules; and second, the constants (or parameters) are determined empirically by fits to experimental data. The first feature allows sophisticated methods to be used to find local minima and transition states, so that straightforward calculations of geometries, heats of formation, vibrational frequencies, and relative energies of different isomers can be carried out for medium-sized rings, using computers as small as the IBM PC. The empirical nature of the potential implies, however, that applications must generally be limited to those resembling molecules included in the set from which the "fit" was originally obtained.

A great deal of information about molecular mechanics is contained in the book by Burkert and Allinger[4]; Niketic and Rasmussen have given may details of the implementation of the consistent force field (CFF),[44] and other general reviews of applications to ring systems are available.[45,46] The first two chapters of Riddell's book[5] give a nice overview of molecular mechanics approaches to conformations in heterocycles, and, of course, the other chapters in the present volume provide a great deal of information about specific systems.

1.2.2.1. Force Fields for Alkanes. There are at least five force fields that are used fairly frequently to represent saturated hydrocarbons, including the cyclic species of interest here. These are connected with the names of Allinger (MM2),[47] Boyd,[48] Schleyer,[49] Ermer and Lifson (CFF-3),[50] and Bartell (MUB-2).[51] Of these, only the CFF parameters were designed to be able to reproduce vibrational properties as well as geometries and relative energies.[52] The remaining force fields were fit primarily to experimental geometries and heats of formation, and differ principally in the nonbonded interactions associated with hydrogens. The MM2 force field has been the most widely used, and gives excellent results for geometries and heats of formation for a wide range of cyclic and acyclic saturated hydrocarbons; Boyd's force field (and variants of it) has also been widely applied to cyclic alkanes. Burkert and Allinger[4] give many comparisons of results from different force fields, and their book should be consulted by anyone with a serious interest in relative results. A few comparisons of particular relevance to cyclic systems are given in the following.

Recently, a modification of the MM2 force field was proposed, called MM2', which is designed to look more specifically at torsional interactions in cyclic systems.[53] Some of the specific differences between this force field and MM2 (which it closely resembles) are outlined below. Another offshoot of MM2 is the STRFIT function, which represents all interactions in terms of two-body central forces.[54] This has some computational advantages, for both energy minimization and the application of statistical mechanical theories.

A new general force field, called MM3, is under development in Allinger's group, and should be available soon. It is likely that this will have a major effect on a wide variety of conformational calculations in organic chemistry.

Relative Energies of Cyclooctane Conformers. Eight-membered rings are large enough to have considerable conformational flexibility, and cyclooctane has been a popular test case in the development of empirical force fields. As discussed below, many of the conventional "named" conformers probably do not correspond to local minima, and there appear to be only four low-energy local minima: the boat–chair (BC) and twist–boat–chair (TBC) forms, which alternate on a pseudorotation pathway with relatively low activation energy, and a pair of conformers, the twist–chair–chair (TCC) and twist–boat (TB), which are separated by larger barriers from the BC conformer (see Fig. 1-1).

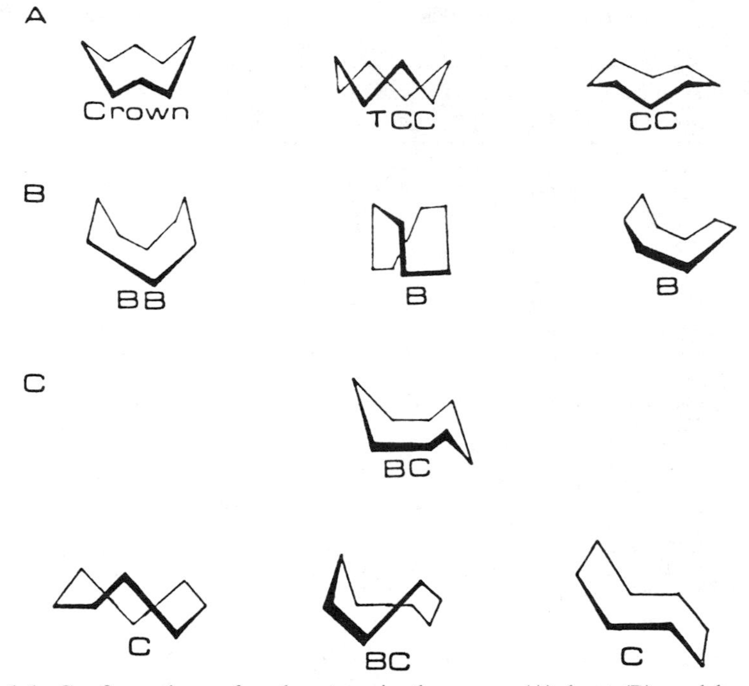

Figure 1-1. Configurations of cyclooctane in the crown (A), boat (B), and boat–chair (C) families. In the MM2 and AMBER force fields, only the configurations in the center column correspond to local minima on the potential energy surface. Adapted from Ref. 58.

Table 1-2 gives relative energies computed from the Schleyer, Allinger, and Boyd[55] force fields, as well as from four other recent force fields. That by Pakes, Rounds, and Strauss[56] is characteristic of potentials developed by the Strauss group to study ring deformations of a variety of ring systems.[57] It assumes constant C—C and C—H bond lengths, and computes the positions of "idealized" hydrogens based on the heavy atom positions. The AMBER results[58] use the hydrocarbon portion of a more general force field (discussed below) designed to study proteins and nucleic acids. The MM2'[59] and STRFIT potentials were mentioned above.

Aside from the Schleyer force field, which shows little energy distinction among the various conformers, the results of the various calculations are in fairly good agreement. The PRS potential suggests that the TCC conformer might be populated at room temperature, but the authors noted that its energy could be modified significantly by changing torsional terms that are relatively poorly defined in their model. (It seems most likely experimentally that the TCC form is not seen in equilibrium mixtures, and hence has a relative energy greater than $k_B T$.[60,61]) Where comparisons can be made, the structural parameters are nearly identical as well: for example, for 10 of the 11 arrangements reported by both Nguyen and Case and by Anet and Krane, the maximum difference in computed torsion angles is 3° and most values are within 1°. (This set of

TABLE 1-2. RELATIVE ENERGIES OF CYCLOOCTANE CONFORMERS (KCAL/MOLE)

Conformer Symmetry	BC C_s	TBC C_2	TCC D_2	TB S_4	(BC ⇆ TBC) C_1
Schleyer[b]	0.2	—	0.1	—	—
MM2[a]	0.0	1.7	1.0	3.1	2.8
Boyd[c]	0.0	1.7	0.8	2.8	3.3
PRS[d]	0.0	1.7	0.2	2.9	—
AMBER[e]	0.0	1.8	0.7	3.8	2.9
MM2′[f]	0.0	2.4	1.2[g]	1.7[h]	3.4
STRFIT2[i]	0.0	2.6	1.4	4.4	—

[a] Ref. 4.
[b] Ref. 49; values are relative to the crown conformer, computed to be lowest.
[c] Reported by Anet and Krane.[55]
[d] Pakes, Rounds and Strauss.[56]
[e] Reported by Nguyen and Case.[58]
[f] Reported by Ivanov and Osawa.[59]
[g] This refers to the crown (D_{4d}) conformation, which is the only minimum found in this region in the MM2′ force field.
[h] This refers to the boat-boat (D_{2d}) conformation, which is the only minimum found in this region in the MM2′ force field.
[i] Reported by Saunders and Jarret.[54]

arrangements includes a variety of transition states not listed in Table 1-2.) The one exception is for the TB state, for which Anet and Krane report unique torsion angles of 57° and 49° whereas Nguyen and Case obtain 68° and 32°. The origins of this difference are not clear, although it is likely that the potential energy surface is quite flat in this region, allowing similar force fields to give different locations for the exact position of the minimum.

Vibrational Spectra of Cyclooctane. Table 1-3 gives results for several calculations of the low-frequency vibrations of the BC form of cyclooctane. The low-frequency modes are probably the most important for conformational analysis, since they represent distortions that can be carried out with small energy penalties; in favorable cases, comparisons with experiment can be used to help to refine potential energy functions. The classic alkane vibrational force field of Snyder and Schactschneider[62] gives a fairly good account of higher-frequency vibrations but fares poorly for the low-frequency ring deformation modes, probably because no explicit account is made of nonbonded interactions. Results from two molecular mechanics potentials are also given in Table 1-3, one from the Strauss group,[63] which includes bend–bend and bend–torsion interaction terms in addition to the usual diagonal terms, and a modification of the AMBER protein force field, which includes Urey-Bradley 1–3 interaction terms.[64] Although further improvements are certainly possible, both molecular mechanics potentials give a good qualitative account of the ring-deformation energetics. The modified AMBER calculations perform much more poorly for modes that involve significant CH motion. It can be hoped that future parameterizations may be able to combine the relatively good low-frequency behavior already available with terms that will also allow higher-frequency infrared (IR) and Raman bands to be interpreted.

TABLE 1-3. Observed and Calculated Vibrational Frequencies (cm^{-1}) for Cyclooctane

	Observed[a]		Calculated		
Character	IR	Raman	SS[a,b]	PRS[a,c]	NC[d]
τ		125	110	135	129
θ, τ	220	212	138	220	225
θ, τ, CH rock	255	245	234	261	263
θ, τ	295	292	256	311	289
θ, τ, CH rock	317	326	352	300	310
θ, τ, CH rock	367	367	352	391	355
θ, τ, CH rock	475	482	390	419	417
θ, CH rock	514	517	450	564	455
θ, CH rock	—	540	467	516	531
θ, CH rock	—	664	530	641	549

[a] Ref. 63.
[b] Computed from Snyder and Schactschneider alkane force field.[62]
[c] Computed from empirical potential energy function of Pakes, Rounds, and Strauss.[63]
[d] Computed from modified AMBER potential.[64]

1.2.2.2. Inclusion of Heteroatoms. A number of additional considerations come into play when heteroatoms are included in a molecular mechanics force field. First is the straightforward problem that one needs many more parameters to characterize torsional preferences, nonbonded interactions, and so on, so that a self-consistent parameterization must include a wider variety of test molecules. More troublesome in a fundamental sense is the fact that electrostatic interactions can no longer be ignored, as they often are for hydrocarbons. This means that some account must be taken of the interactions of bond dipoles or partial charges that characterize molecular charge distributions. It is, of course, not at all difficult to add terms based on Coulomb's law to a molecular mechanics potential, but two special features of electrostatic interactions complicate efforts to make such representations realistic: (1) the constituent atoms of a molecule are *polarizable*, so that appropriate charges or dipoles depend upon the environment; and (2) solvent effects on electrostatic interactions (through specific hydrogen bonds and more general dielectric effects) are often much greater than for other types of interactions. It is fair to say that no general solution to these problems has appeared, but considerable progress has been made for specific bonding situations; a description of some of this work is given below.

Silicon. This is probably the simplest heteroatom to add to a hydrocarbon force field, since silicon is tetrahedral, like carbon, and electrostatic interactions are generally not very important. There is, however, much less experimental data for calibration than for hydrocarbons. An early parameterization along the lines of the MM1 force field was that of Tribble and Allinger.[65] More recently, attempts have been made to extend the range of molecules that are studied, and to develop parameters appropriate to the MM2 force field.[66,67]

A slightly different parameterization[68] was recently used to study the conformations of tetradecamethylcycloheptasilane $(SiMe_2)_7$.[69] Burkert and Allinger[4] have reviewed the silicon results from the earlier force field.

Ether Oxygen. In many ways, this is the next simplest heteroatom to add to cycloalkanes, since the resulting conformations are generally easily related to those of the parent hydrocarbon, and many of the differences can be traced to H—H steric repulsions that are removed when a methylene group is replaced with oxygen. Important test cases for parameterization of ether force fields include methyl ethyl ether, 2-methyl-1,3-dioxane, and tetrahydrofuran. Popular force fields include that in MM2[70] and one developed by the Strauss group specifically for cyclic seven- and eight-membered rings.[71,72] As with hydrocarbons, the former force field is influenced primarily by calculations of geometries and heats of formation whereas the latter gives special emphasis to vibrational spectra. Kollman, Wipff, and co-workers have developed similar force fields to model crown ethers and spherands, along with their complexes to metal ions.[73,74]

A number of force fields for carbohydrates have been proposed, which contain parameters that in principle could be applied to larger rings as well. Starting from the pioneering work of the Ramachandran group,[75] representative potentials have been derived from groups headed by Rao,[76] Rees,[77] and Rasmussen.[78] Similar comments apply to potentials developed primarily to look at ribose rings in nucleic acids.[79] It should be interesting to see how potentials derived primarily for five- and six-membered rings will fare when applied to larger systems.

The principal theoretical questions facing those developing force fields for ethers revolve around the best way to represent torsional preferences for CCOC, OCCO, and COCOC fragments. These arise from a combination of van der Waals repulsions (whch may include explicit lone-pair contributions), electrostatic interactions, and explicit torsional potentials. The mix among these varies quite a bit among force fields, but most find that explicit torsional potentials with low periodicity are required to obtain good results. Explicit lone pairs on oxygen are often not used, and an early Allinger force field that had important lone-pair interactions[80] was later replaced with one in which the lone pairs were much less significant.[4] Hydrogen bonding interactions generally seem to be well characterized by electrostatic and van der Waals interactions, without the need to add special terms; this is consistent with pioneering studies on hydrogen bonding in amides.[81,82]

Other Oxygen Environments. Cyclic ketones with seven- to 10-membered rings have received some attention from molecular mechanics calculations.[83,84] Generally, fairly simple force fields appear to give adequate results, with the 1,4 nonbonded interactions represented by van der Waals repulsions plus explicit torsional terms.[4] The torsional interaction terms can reproduce the preference (by 1 kcal/mol or more) of C—C bonds to eclipse adjacent C=O bonds.[85] Careful consideration of electrostatic effects is important when oxygen can interact with other polar groups, as in halogenated ethers and ketones.[86]

Sulfur. Substitution of sulfur for methylene groups in hydrocarbon rings leads to two principal effects that often work in opposite directions. First, the C—S bond is significantly longer than a C—C bond (*ca.* 1.81 Å versus 1.53 Å); this tends to reduce steric interactions between substituents on either side of the sulfur atom. Second, however, the C—S—C bond angle of around 100° is generally significantly less than the tetrahedral angle appropriate to the parent hydrocarbon; this has the general tendency of increasing such steric interactions. A fairly straightforward force field fit to monothiols and monosulfides by Allinger and Hickey[87] gave good results without requiring explicit consideration of electrostatic interactions or sulfur lone pairs. This force field was applied to the cyclic systems thiacyclopentane and thiacyclohexane, and is expected to work well for single sulfur substitution in larger rings as well.

Since hydrogen bonds to sulfides and thiols may often be important, particularly in aqueous solution, Weiner et al. considered methods of deriving empirical force fields that could represent these interactions reliably.[109] They began by considering the complexes HSH–OH_2 and H_2S–HOH, looking at the angle above the plane (AAP) between the H_2S plane and the O—H bond in H_2S–HOH. Molecular orbital calculations suggest that dimers between second-row hydrides and water have larger values for AAP than the corresponding first-row hydrides.[88] Thus, for example, the quantum mechanical structure for H_2S–HOH has an AAP of 78° whereas empirical potentials with atom-based charges gave much smaller values, near 15°. Furthermore, the small-AAP structures showed computed quantum-mechanical energies about 1–1.5 kcal/mol above the large-AAP structures, an error that was deemed too large to be acceptable. (By contrast, water dimers show a much smaller energetic dependence on the AAP angle.[89]) One remedy for this behavior is to include explicit lone pairs on sulfur, which was found to allow good qualitative agreement with the quantum mechanical structures for hydrogen bonds to sulfur.

Molecules containing disulfide and polysulfide moieties require further consideration, since the torsional potential about an S—S bond has a minimum near 90°. Even with force fields that include explicit lone pairs, additional twofold torsional potentials are required to reproduce this behavior. Cyclic molecules then adopt conformations that strike a balance between favorable disulfide torsions and the conformational requirements of the hydrocarbon portion of the molecule. The MM2 parameterization for disulfides[90] works well for acyclic molecules as well as for di- and tetrathianes; a related parameterization gives good results for seven- to 10-membered polysulfur rings.[91]

The vibrational spectra of disulfides is of some interest since the S—S stretching frequency can often be observed in large systems such as proteins, and its frequency can be used to monitor local conformations. Recent analyses of the IR and Raman spectra of a series of eight- to 12-membered rings containing disulfide bonds have clarified this relationship.[92,93] A vibrational force field of the Urey-Bradley type gives remarkably good agreement for 12 S—S bonds with considerable variation in the local geometry. The force constants are fixed for this set of molecules, so that only the molecular geometry affects the coupling

TABLE 1-4. VIBRATIONAL FREQUENCIES (CM^{-1}) OF 1,2,5,6-TETRATHIACYCLOOCTANE[a]

Observed	Calculated	Motion
521	520	82% SS #1, 6% SCC
504	505	88% SS #2, 10% SS #1
428	425	SCC, CS
421	412	SCC, 10% SS #1
288	297	ring mode
272	295	ring mode
265	258	SSC
248	241	SSC
156	154	SSC, SCC, SS torsion
138	139	SSC, SCC, SS torsion
—	103	SS and CC torsions
—	77	CS and CC torsions

[a] From Ref. 93.

between internal coordinates in particular modes, and hence the observed frequencies. This extent of transferability is pleasing, although it should be noted that a somewhat different set of force constants is obtained from fits to acyclic dialkyl disulfides.[94]

Table 1-4 shows some of the computed and observed low-frequency vibrations for 1,2,5,6-tetrathiacyclooctane (Fig. 1-2). This molecule contains two disulfide bonds in rather different local environments, one with the two adjacent C—S torsions near −50° and the other with the same angles near 105°. The Urey-Bradley force field gives a good account of the low-frequency vibrations and, in particular, the difference in the two S—S stretch vibrations. Shifts of the two S—S stretches in the perdeuterio isotopomer are also well reproduced. These calculations suggest rules by which the vibrational spectra of other disulfide groups can be inferred from IR and Raman spectra.

Use of vibrational force fields of this type, however, assumes that the equilibrium geometry of each molecule is known. It should be possible in principle to develop a molecular mechanics force field that can *predict* both molecular geometries and S—S stretching frequencies in agreement with experiment. Preliminary calculations with the AMBER protein force field (discussed below) gave quite reasonable geometries for molecules like 1,2,5,6-tetrathiacyclooctane, but failed to reproduce the observed pattern of S—S vibrational frequencies.[95] This is not surprising in light of the Urey-Bradley results, which show that the conformational dependence of the disulfide stretching frequency originates from a coupling between the SS stretch and SCC bend that changes as a function of the CCSS torsion angle: most current molecular mechanics potentials contain either no stretch–bend interactions, or only rudimentary ones. Attempts to remedy this situation are currently underway,[96] but (as the cyclooctane results also indicated) we are still some way from achieving a "consistent force field" that could be used simultaneously for geometries, energies, and vibrational calculations.

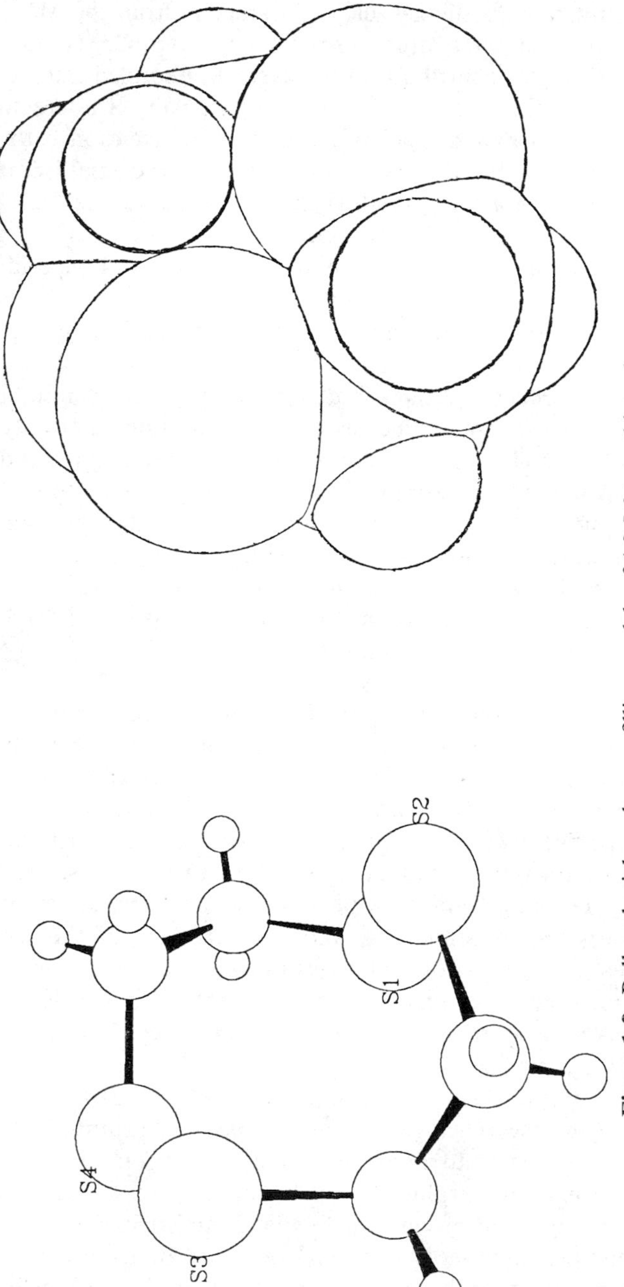

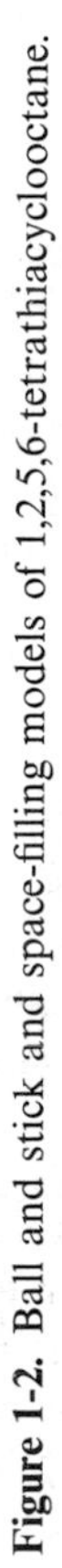

Figure 1-2. Ball and stick and space-filling models of 1,2,5,6-tetrathiacyclooctane.

Nitrogen. The most extensive molecular mechanics study of aliphatic amines is that of Profeta and Allinger, using parameters from the MM2 force field.[97] Compared with parent hydrocarbon rings, shorter C—N bond lengths and smaller CNC angles contribute to increased nonbonded interactions in many cases. For piperidine, the axial H—H and lone-pair–H interactions appear to be attractive, rather than repulsive, and the conformation with an equatorial amine proton is predicted to be more stable than the axial isomer by 0.3 kcal/mol. This bears on a long-standing discussion about the "size" and orientation of amine lone pairs,[98] and appears to be in qualitative agreement with experimental data. Calculations on methyl-substituted piperidines also give good results.

Force fields representing amides will be discussed in conjunction with peptides.

Phosphorus. Generally, phosphorus compounds have not been as widely studied by molecular mechanics as other heteroatoms, partially because five-coordination (which can sample both trigonal-bipyramidal and square-pyramidal conformations) is somewhat hard to model. Berry pseudorotation has received some attention, however.[99] Tetracoordinate phosphate force fields have been developed for use in modeling nucleic acids,[100,109] and three-coordinate phosphaalkanes (including phosporinanes) have been modeled.[101] As more secure experimental data become available, it is likely that greater attention will be paid to problems in this area.

1.2.2.3. Potentials for Peptides and Proteins. The importance of proteins and peptides in biology has spurred the development of a wide number of empirical potentials for these systems. As might be expected from the large number of side chains one needs to describe, the potentials are not so uniform as for the "organic chemistry" force fields described above. Furthermore, there are fewer calibration systems available that can be used to test the resulting functions—for one thing, almost all peptide and protein experiments refer to solution or crystalline conditions, and the influence of the condensed-phase environment is often very difficult to estimate. Nevertheless, a great deal of work has gone into the development of these potentials, and it is likely that many of the ideas developed can be used not only for cyclic peptide but for other heterocyclic systems as well.

The earliest systematic attempts to create potentials for peptides and proteins came from the Scheraga[102,103] and Lifson[104] groups. In the mid- to late 1970s, potentials based to a greater or lesser extent on these were proposed by Levitt,[105] Gelin and Karplus,[106] and Hermans et al.[107] Since that time, there has been an explosion of work in this field, and it is hardly possible even to list all of the proposed molecular mechanics potentials. For peptides, the most popular ones have been those mentioned, based on the work of Scheraga, Levitt, Lifson, and Karplus (and their co-workers), plus several more recent proposals: a "valence force field" developed by Hagler and co-workers[108]; "AMBER" potentials, from Kollman's group, at both the united-atom[109] and all-atom levels[110]; a peptide potential of White and Guy[111]; and the "CHARMM" potentials from the Karplus group.[112] Very recent, but quite promising, is the

"OPLS" potential from the Jorgensen group,[113] which is primarily based on extensive liquid-state Monte Carlo simulations.[114]

Comparisons Among Protein Force Fields. Hall and Pavitt recently reported a comparison of many of these force fields for their ability to reproduce the crystal structures of three cyclic hexapeptides, *cyclo*-(-Ala-Ala-Gly-Ala-Gly-), *cyclo*-(-Ala-Ala-Gly-Ala-Gly-Gly-), and *cyclo*-(-D-Ala-D-Ala-Gly-Gly-Gly-Gly-).[115] This test suite is also sensitive to the nature of the water–protein interaction potential that is assumed (since the crystals contain waters of hydration), but clearly tests only a small part of the side-chain potentials required for a protein force field. The AMBER potential function gave the best results in this comparison, but the differences among several potentials were minor. The OPLS potentials have also been tested against this set of conformers, with good results.[113]

Whitlow and Teeter[116] have used the small protein crambin (which has a very-high-resolution crystal structure) as a test case to examine the performance of a variety of protein potentials. These comparisons are particularly sensitive to the form of the hydrogen-bond interaction and to the nature of the van der Waals and electrostatic parameters. Some combinations work better than others, but a complete protein force field contains so many parts that it is often difficult to sort out which parameters are influencing particular results.

The "Alanine Dipeptide." The calculations described above compared *geometries* predicted by several potential functions, concluding (as have many other studies) that these are only mildly sensitive to details of the potential function. Relative *energies* of different conformations are expected to be much more variable. As a simple example, Table 1-5 reports relative conformational energies from a number of calculations on the "analyl dipeptide," *N*-acetylal-anyl-*N*-methylamide. The conformations of this molecule can be specified approximately by the values of the conventional torsion angles ϕ and ψ about the central carbon atom. Several "allowed" regions have local minima on the potential energy surface: C_7^{eq} near $(\phi, \psi) = (-75°, 65°)$, C_7^{ax} near $(70°, -65°)$, C_5 near $(-160°, 170°)$, and α_R near $(-60°, 40°)$. It seems clear that the conformation adopted in solution depends upon solvent polarity,[117] and the relative energies in the gas phase are not known. Theoretical work dates back to early calculations of Brandt and Flory,[118] and a variety of more recent calculations is given in Table 1-5. These include potentials based on the Scheraga,[119] Karplus,[120,121] Hagler,[122] and AMBER[109] empirical force fields, as well as quantum mechanical studies using the PCILO[123] and molecular orbital[124,125] methods. All of the results except the PCILO calculations predict the C_7^{eq} conformer to be lowest in energy, but there is considerable uncertainty about the penalty one must pay to populate alternate conformers. The large value (8.8 kcal/mol) for C_7^{ax} in the ECEPP calculation is almost certainly a result of assuming fixed bond lengths and angles, since small variations in these (which are allowed for in the other empirical calculations) can relieve the steric strain in this rather crowded conformer. Results for other sets of calculations are qualitatively consistent with each other, but the disagreements are larger than k_BT at room temprature (0.6 kcal/mol), so that predictions of equilibrium constants do not yet appear to be feasible for conformational changes of this sort.

TABLE 1-5. Intramolecular Energies (kcal/mol) of Some Minimum Energy Structures for Analyl Dipeptide

Authors	Reference	Method	C_7^{eq}	C_7^{ax}	C_5	α_R
Empirical Potential Functions						
Weiner et al.	109	AMBER ($\epsilon = 1$)	0.0	1.1	4.0	4.7
Zimmerman et al.	119	ECEPP	0.0	8.8	0.4	1.1
Rossky and Karplus	120	EPF[a]	0.0	0.0	4.5	~8
Stern et al.	122	EPF[a]	0.0	1.3	—	4.2
Pettitt and Karplus	121	"model 4"	0.0	0.3	4.8	
Quantum Mechanical Studies						
Pullman and Pullman	123	PCILO	0.3	0.0	1.6	2.4
Scarsdale et al.	124	4-21G	0.0	2.6	1.4	4.9
Weiner et al.	125	4-31G	0.0	2.4	0.2	—
Other						
Weiner et al.	125	AMBER ($\epsilon = 1$) ΔG (RRHO)[b]	0.0	1.2	2.9	3.6
Weiner et al.	125	4-31G +dispersion correction[c]	0.0	2.2	1.9	—

[a] Empirical potential function.
[b] AMBER ΔE values corrected for zero-point-energy and entropy differences in the rigid-rotor harmonic-oscillator approximation.
[c] Quantum mechanical results plus an empirical dispersion correction.

Several additional points about Table 1-5 are worthy of comment. First, there is a surprising difference in the C_5 energy between the 4-21G and 4-31G basis sets, suggesting that neither is yet sufficiently close to the Hartree-Fock limit to neglect basis set errors. (At least 0.8 kcal/mol of this difference is intrinsic to the calculations, i.e., is not related to the slight differences in predicted geometries between the two calculations. Similar results have been obtained for the glycine dipeptide.[125]) Larger basis set calculations are only now becoming feasible for molecules of this size, so that uncertainties concerning the effects of basis set on computed results may be expected to continue for some time.

Second, it is important to note that molecular orbital calculations ignore electron correlation effects, in particular, the dispersion energies that favor compact structures over extended ones. Empirical estimates of the magnitude of these dispersion corrections (Table 1-5) show that they favor the C_7 conformers over C_5 by a significant amount: in the Hartree-Fock 4-31G results, C_7^{eq} and C_5 are about isoenergetic, whereas with dispersion corrections, C_5 is about 1.9 kcal/mol (or $3k_BT$) above the C_7 structures Unfortunately, these empirical dispersion estimates themselves have considerable uncertainty, so that the "correct" gas phase values, even at this level of basis set, are not yet established.

In addition to intermolecular energies, two additional pieces of information are necessary to estimate observed equilibrium populations. The first are additional contributions to free energies, arising from differences in zero-point energy and entropy between conformers. An estimate of these (using the harmonic oscillator–rigid rotor approximation) is included in Table 1-5. It may be seen that these corrections favor the C_5 and α_R conformations (relative to C_7) by about 1.1 kcal/mol. This arises primarily becausee the C_7 conformers have the strongest internal hydrogen bond, and hence the least internal flexibility. These estimated corrections are based on an empirical potential energy function, but it is likely that the (gas-phase) corrections would be similar for other energy surfaces as well.

Solvation is also expected appreciably to influence the equilibrium behavior of peptides in solution, and the calculations discussed here do not include such effects, except to the extent that empirical fitting of parameters in the molecular mechanics force fields represents in some average way experimental data derived from solution measurements. It is only recently that theoretical methods with secure foundations have become available to estimate these effects. Such calculations will be discussed in section 1.3.3.

1.2.3. Conclusions

Computations of molecular potential energy surfaces have come a long way in the past two decades, first for alkanes and later for heterocyclic systems. The fairly detailed analyses given for cyclooctane and the alanine dipeptide serve as reminders that universally agreed-upon results are not yet available even for these prototype systems. The problems are especially acute for peptides, where many different heteroatoms are present in a variety of bonding situations, and where electrostatic interactions among groups significantly affect conformational preferences. For "simpler" heterocycles, computed results are generally more reliable, although surprises continue. Progress in the next few years will probably involve more direct quantum mechanical calculations on cyclic systems, and molecular mechanics potentials that give a better account of electrostatic interactions between heteroatoms.

1.3. Using Potential Energy Surfaces for Conformational Analysis

The previous sections discussed a method of calculating, essentially on a point-by-point basis, the internal energy of heterocyclic systems as a function of nuclear positions. A second major task of theory is to relate such numbers to observable properties of the system. For conformational analysis, such relations are based primarily upon statistical thermodynamics and upon transition state theories of isomerization rates. For qualitative purposes, the relation between internal energies and experiment is often very simple; for example, we expect the lowest-energy conformer to be the one observed and that barrier heights between conformers can be identified with Arrhenius activation energies for isomerization rates. As we seek more exact and quantitative relations, however, we need to be as careful in elucidating the connections between potential energy surfaces and experiment as we are in computing the energy surfaces themselves.

This is no easy matter, particularly for condensed phases. In the following sections, recent work aimed at developing this aspect of conformational analysis is outlined.

1.3.1. Finding Stationary Points

The simplest and generally most useful way of characterizing a molecular potential energy surface is to describe its *stationary points*, i.e., its local minima and the transition states between them. These are points on the surface where the gradient, $\partial V/\partial x$, vanishes. For medium-sized rings, such characterization can present formidable problems, even when molecular mechanics potentials are being used so that a large number of points on the surface can be scanned quickly. For relatively simple heterocycles, the "allowed" conformations can be related to those of the parent cycloalkane, whose conformational properties are fairly well understood. If, however, there are many heteroatoms and/or unsaturated bonds (as, for example, in cyclic peptides), exploring the potential energy surface in any systematic fashion can be a difficult task. Some recent work along these lines is described in the following.

Finding Local Minima. Procedures for finding nearby minima from initial guesses are by now quite well known.[126] Conjugate gradient methods, which require only first derivatives, usually work very well, even for systems with a large number of degrees of freedom. Near a minimum, the classic Newton-Raphson procedure generally converges very quickly. In this approach, the local potential energy V is expanded about a point whose Cartesian coordinates are x_0:

$$V(x + x_0) = V(x_0) + g \cdot x + (\tfrac{1}{2})x^T \cdot H \cdot x \tag{1-6}$$

Here g and H are the first-derivative (gradient) and second-derivative (Hessian) arrays respectively. Setting $\partial V/\partial x = 0$ yields the Newton-Raphson update

$$x = -H^{-1} \cdot g \tag{1-7}$$

provided that H has an inverse. This method runs into difficulties when the length of the step predicted is so large that the truncation of the Taylor series is insufficiently accurate. Two major methods exist to alleviate this problem: the first adds a line search to the algorithm, i.e., the *magnitude* of x is varied to obtain the best energy lowering. In the second method, the Hessian matrix is altered; this approach is discussed in the section on finding transition states.

No matter which minimization method is used, it is important to analyze the second derivative matrix at stationary points to assure that a particular point is indeed an energy minimum. (A minimum energy structure will have a Hessian with no negative eigenvalues whereas transition states will have at least one negative eigenvalue.) A number of common programs (such as MM2) use (approximate) Newton-Raphson minimization methods that can converge to transition states rather than to true minima. Other methods of analysis look only for conformations with some element of symmetry. However, configurations of high point-group symmetry (such as the D_{4d} "crown" conformation of cyclooctane) often correspond to transition states on the potential energy surface. (Indeed, for cyclooctane MM2 and AMBER, calculations indicate that only four of the 10 commonly "named" configurations correspond to local minima.[58,59]) Fortunately, analysis of the eigenvalue spectrum of the Hessian matrix is now becoming a much more common procedure, eliminating uncertainties of this sort.

Finding Transition States. A variety of procedures are available to locate transition states if some initial guess at their structure is available.[127] For molecular mechanics calculations, the classic procedure described by Wiberg and Boyd[128] "drives" one or more torsion angles by the addition of a penalty function that has a sharp minimum at various values of the driven angle; the remaining degrees of freedom are minimized. This procedure (and its variants[129,130]) continues to be widely applied, although it can occasionally encounter difficulties when correlated motions of several torsions are required to obtain a low-energy pathway. Furthermore, this procedure generally will not find the exact position of the transition state, i.e., the stationary point on the potential energy surface in the absence of penalty functions.

Recently, algorithms that "walk" along a potential energy surface from one minimum through a transition state to another minimum have been described. These use the eigenvectors of the Hessian matrix to provide directional information, and have the advantage that the location of the transition state does not need to be specified in advance, even in an approximate fashion. Early work along these lines was described by Crippen and Scheraga[131] and by Hilderbrandt,[132] and most recent work follows a reanalysis of this problem by Cerjan and Miller.[133] The last method constrains the step length in the Newton-Raphson method to a fixed value Δ by means of a Lagrange multiplier λ. Consider the Lagrangian L,

$$L(x, \lambda) = V(x + x_0) + (\tfrac{1}{2})\lambda(\Delta^2 - x \cdot x) \tag{1-8}$$

Setting $\partial L/\partial x = \partial L/\partial \lambda = 0$ yields[134]

$$x = (\lambda I - H)^{-1} \cdot g \tag{1-9}$$

$$\Delta^2 = g^T \cdot (\lambda I - H)^{-2} \cdot g \tag{1-10}$$

where I represents the unit matrix. For a given value of Δ, eq. (1-10) can be solved for λ, then eq. (1-9) yields the update step x. There may be several solutions for the multiplier λ, and, remarkably, these lead to moves (i.e., coordinate updates x) with quite different properties. One choice of λ will walk *up* "river beds" on potential energy surfaces, leading from local minima to transition states; other choices are appropriate for downhill walks that generally work much better than the unmodified Newton-Raphson method. Nguyen and Case[135] applied these ideas to the alanine dipeptide and to cyclooctane, and found that all of the low-energy minima and transition states on these surfaces could be found in an "automatic" fashion. For cyclooctane, for example, walking along the lowest Hessian eigenvector direction (starting from the *BC* conformer) led to the usual *BC*, *TBC* pseudorotation pathway, whereas walks along higher eigenvector directions led to minima in the boat and chair families. This method provides no guarantee of finding all relevant minima and transition states but it does appear to be a useful way of characterizing local portions of a potential energy surface when little is known about its behavior. Other transition state-finding routines are probably more efficient when a good guess can be made at their location.

Global Searches of Conformation Space. For ring systems that cannot simply be mapped to a parent cycloalkane, the task of carrying out a global search—e.g., to find the lowest of all the local minima—is a difficult one. Most attention has been paid to cyclic peptides, but the techniques outlined here could be applied to other cyclic systems as well.

Most "global" search strategies use a systematic procedure to sample all feasible ranges of torsion angle space, coupled with ordinary minimization techniques to find the exact positions of local minima. The computational difficulties lie in deciding how to "prune" the enormous number of structures generated by the systematic search into a reasonable number for further study. White and Morrow pointed out in 1979 that "global optimization is in a very primitive state of development,"[136] and it is fair to say that the situation has not improved a great deal since then. The following refers to a few recent computational approaches to indicate the types of calculations being attempted.

The generation of initial structures for cyclic systems has special problems since not all dihedral angles can be accommodated, that is, the chain-closure constraints restrict torsion angle possibilities in nontrivial ways. The procedure described by Go and Scheraga[137] has been widely used; this procedure computes torsion angles that allow fixed end points to be connected by the chain. A recent update by Bruccoleri and Karplus[138] allows limited alterations in bond angles as well. A very different procedure uses distance geometry methods[139] to generate a "random" sampling of initial coordinates. Weiner et al.[140] have tested this method on cyclooctane and some crown ethers. Although special constraints were required to generate all the allowed conformations, distance geometry was shown to be a very useful technique for generating initial configurations for further study.

The algorithms devised to process these initial configurations further are quite complicated, and still the focus of much study. Three reports that illustrate recent work are those of Paine and Scheraga,[141] Bruccoleri and Karplus,[142] and White.[143] It is likely that within a few years generally accepted methods for carrying out global searches on medium-sized ring systems will be available, although they may still require large amounts of computer time. Even expensive algorithms might be well worth the effort if they allowed truly systematic surveys of conformational preferences to be carried out.

1.3.2. Computing Equilibrium Constants

In the canonical ensemble, the relative populations of two isomers at temperature T is given by the ratio of their partition functions:

$$K_{\text{eq}} = \frac{[A]}{[B]} = \frac{Z_A}{Z_B} \tag{1-11}$$

where

$$Z_A = \int_A e^{-H/k_BT}\, d\Gamma \tag{1-12}$$

and the integral is over all phase space points "associated" with conformer "A." If the barrier between A and B is large compared to k_BT the equilibrium constant will be insensitive to the exact position of the dividing surface chosen to separate the conformers (see Fig. 1-3). On the other hand, if the barrier is low, then the observed "equilibrium constant" will probably depend upon the nature of the experimental apparatus, and its ability to discriminate one conformer from another.

For relatively rigid molecules in the gas phase, the configuration integrals can be evaluated in the rigid-rotor, harmonic-oscillator approximation, yielding "textbook" formulas for computation of equilibrium constants.[144] A classic paper by Herschbach, Johnston, and Rapp[145] showed how these partition functions could be expressed in terms of local properties, in which each atom contributes an effective volume based on average vibrational amplitudes and the geometric configurations of the neighboring atoms. This approach is particularly appropriate for larger molecules, where many local parts of the conformation may be unchanged in going from one conformer to another; in this case those portions of the partition functions cancel in forming equilibrium constant ratios. Both of these approaches also allow quantum corrections to be easily evaluated, something that can be quite difficult to carry out for more complicated models of molecular motion.

For "floppy" molecules with relatively large-amplitude anharmonic motions, or for systems in condensed phases, more complex models must be used to estimate partition functions. The simplest extension is probably the "quasi-harmonic" approach.[146,147] Here, as in the harmonic approximation, the potential energy is approximated as a quadratic function of the coordinates q:

$$V = (\tfrac{1}{2})q \cdot F \cdot q \tag{1-13}$$

In the harmonic approximation F is given by

$$F_{ij} = \frac{\partial^2 V}{\partial q_i \partial q_j} \tag{1-14}$$

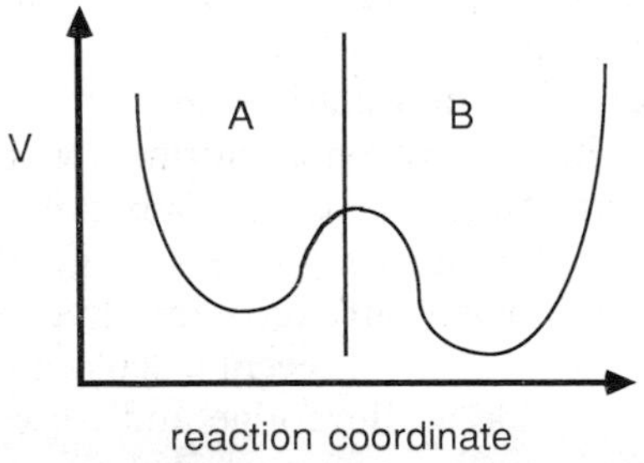

Figure 1-3. Schematic energy profile for an isomerization reaction. The center dividing line separates regions considered to be "reactants" from those labeled "products." The vertical axis might represent either potential (internal) energy or free energy.

evaluated at the minimum energy conformation of the system. In the quasi-harmonic approximation, on the other hand, the elements of F are

$$F_{ij} = k_B T \left[\sigma^{-1}\right]_{ij} \tag{1-15}$$

where

$$\sigma_{ij} = \langle (q_i - \langle q_i \rangle)(q_j - \langle q_j \rangle) \rangle \tag{1-16}$$

is the variance-covariance matrix of the internal coordinate fluctuations. These matrix elements can be determined from Monte Carlo or molecular dynamics simulations that may include anharmonic potentials or solvent contributions; in general they will be temperature dependent (unlike the second derivatives of V), so that the resulting free energies may be quite different from those obtained in the pure harmonic approximation. Brady and Karplus have applied this procedure to estimate configurational entropy differences among various conformers of the alanine dipeptide in aqueous solution.[148] They find that entropy differences among conformers are 1–2 cal/(mol-K), corresponding to free energy differences of about 0.5 kcal/mol at room temperature. These are not negligible, but are much smaller than expected values for solvation enthalpies. These initial results still have sizable statistical errors, but longer simulation runs are expected to provide data with less noise. (Of course, as with all simulations, longer sampling times will not remove *systematic* biases arising from possible errors in the potential functions.)

It has been realized fairly recently that more direct approaches to computing free energy differences are possible by using molecular dynamics or Monte Carlo simulations to sample thermal distributions of molecular configurations in solution. For conformers separated by low barriers (as might be common for medium-sized rings), the simulations then directly give relative population ratios, and hence free energy differences, for various conformers. For example, in simulations on *n*-butane, Jorgensen and co-workers found about 360 effective transitions from *gauche* to *trans* during their Monte Carlo simulation, with about an equal number in each direction.[149] They were able to establish a population ratio that was independent of the starting configuration assumed, and hence directly to compute (thermally averaged) solvation free energies for the two conformers. A number of such calculations on small organic molecules have been carried out over the past few years,[150,151] and applications to cyclic systems should now be feasible.

When the barriers between conformations are larger, direct simulations will generally fail, since transitions between conformers will be very infrequent. In this case, however, biasing ("umbrella") potentials can be added to the simulations to force sampling in regions of high free energy; the effect of the biasing potential can later be "subtracted out" to obtain free energy profiles even for large barriers.[152,153] Representative of recent calculations of this sort are simulations of the alanine dipeptide by Beveridge and co-workers.[154,155] Here the lowest energy conformer in the gas phase (probably) has a seven-member ring closed by a hydrogen bond (see Table 1-5). In aqueous solution, however, the α_R conformer is computed to be more favorably solvated than C_7 by 3.6 kcal/

mol. Continued calculations of this general sort are likely to be crucial in developing a secure understanding of molecular conformational preferences in solution. More approximate integral equation approaches to solvation (which are much cheaper than explicit simulations) may help to make these ideas more accessible to "routine" calculations on organic molecules.[156,157]

1.3.3. Computing Isomerization Rate Constants

The same "umbrella" techniques that are used to obtain free energy differences between stable conformers can also be used to determine *activation* free energies in solution. These barriers then allow simple transition-state theory estimates of rate constants to be made:

$$k_{TST} = (\tfrac{1}{2})\langle |d\delta/dt| \rangle_{\text{eq}}\, \rho(\delta^{+}) \left[\int \rho(\delta)\, d\delta \right]^{-1} \qquad (1\text{-}17)$$

Here δ is a reaction coordinate measuring the progress of the reaction and ρ is the equilibrium probability distribution. The time derivative is evaluated at the top of the barrier (where $\delta = \delta^{+}$) and the integral is over the initial state. The factor $\frac{1}{2}$ simply reflects the fact that, at equilibrium, crossings in the forward and reverse directions are equally probable. Hence the first two factors in eq. (1-17) give the mean rate of crossing the barrier in the forward direction, and the final two factors represent the probability of finding the system at the top of the barrier. All of the quantities in eq. (1-17) may be derived from equilibrium molecular dynamics or Monte Carlo simulations.

Jorgensen and co-workers used these ideas to carry out an influential set of simulations on a model S_N2 reaction, where the activation parameters are very different in the gas phase and in solution,[158,159] and similar calculations continue to be reported. For quantitative interpretations of rate constants, however, one wants to go beyond transition state theory, introducing dynamical corrections, which (it can be shown) always lower the transition state estimate of the rate. The nature of these corrections has been understood for some time,[160,161] and a number of calculations of solution reaction and isomerization rates have been reported.[162]

The basic ideas of these corrections are illustrated in Fig. 1-4. Transition state theory assumes that all trajectories crossing the barrier have "straight-through" behavior, i.e., no recrossings. Thus schematic motions such as that shown in Fig. 1-4*a* and *b* are included in the theory. Other types of motion are possible, most notably nonreactive trajectories (shown in Fig. 1-4*c* and *d*, where the system reaches the transition state, but then returns to potential energy well from which it began), and reactive trajectories that have multiple crossings (such as shown in Fig. 1-4*e* and *f*). In this approach, "correcting" transition state theory involves computing dynamical trajectories starting at points on the transition state surface, looking for recrossings (which almost always will occur, if at all, within the first picosecond), and determining the required correction to k_{TST}. For most simple solution reactions, it appears that these corrections

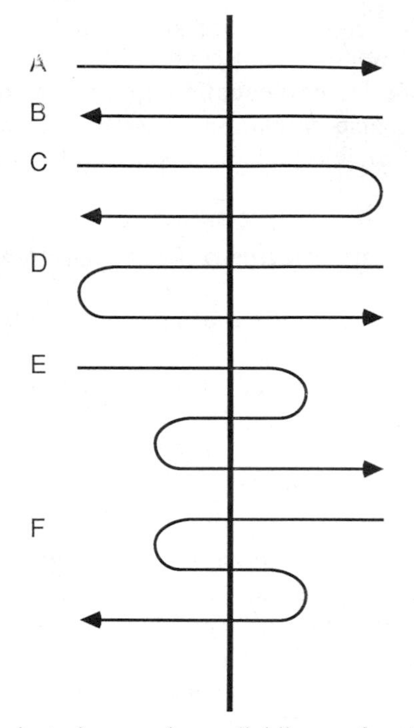

Figure 1-4. Schematic trajectories crossing a dividing surface (center line). Trajectories A and B satisfy the requirements of the transition state theory approximation and the rest do not.

will change k_{TST} by at most a factor of 5–10, but it is too early to make secure generalizations.

Although the theory outlined here does not include quantum mechanical effects, such as tunneling and zero-point vibrational motion, it does (in principle) provide exact classical rate constants, and should begin to make possible truly quantitative comparisons of theory to experiment in condensed phases.

1.4. Conclusions

In a sense, this is an inauspicious time to be writing a "review" of computational methods, since the impact of the scientific computer revolution is just beginning to be felt in the area of medium-sized heterocyclic rings. It is fair to expect several crucial developments in the next few years: (1) Reliable quantum mechanical calculations (incorporating some description of electron correlation) should become available to solve some long-standing problems related to *gas-phase* potential energy surfaces. It is likely that such calculations will be carried out first on cycloalkanes, with heterocyclic rings likely not to be too far behind. (2) New force-field parameterizations (such as MM3) will build upon the two decades of experience we now have with such calculations, allowing (in many cases) predictions of strain energy with expected errors less than $k_B T$ at room temperature, and extending the range of bonding situations to which the theory can be applied. (3) Monte Carlo and molecular dynamics simulations will continue to provide microscopic descriptions of solvent influence on equilibrium populations and rate constants. This parallels the increased ability of experimental techniques such as IR, Raman, and nuclear magnetic resonance spectroscopy to study moderately large molecules in the gas phase, so that both experiment and computer simulation can be carried out in both phases. The next decade should be an exciting one.

Acknowledgments

The author is indebted to Dzung Nguyen, Barbara Rudolph, Boris Weiss-Lopez, William Fink, Andy McCammon, Charles Nash, Tim Havel, and Peter Kollman for many helpful comments. During the time that this chapter was written, the author's research was supported by grants from the National Science Foundation.

References

1. Hassel, O. *Quart. Rev.* **1953**, *7*, 221.
2. Barton, D. H. R. *Experientia* **1950**, *6*, 313.
3. Hendrickson, J. B. *J. Am. Chem. Soc.* **1961**, *83*, 4537.
4. Burkert, U.; Allinger, N. L. "Molecular Mechanics." American Chemical Society: Washington, D.C.: 1982.
5. Riddell, F. G. "The Conformational Analysis of Heterocyclic Compounds." Academic Press, Inc.: New York, 1980.
6. Strauss, H. L. *J. Chem. Educ.* **1971**, *48*, 221.
7. Hehre, W. J.; Radom, L.; Schleyer, P.v.R; Pople, J. A. "*Ab initio* Molecular Orbital Theory." John Wiley: New York, 1986.
8. Hehre, W. J. *J. Am. Chem. Soc.* **1975**, *97*, 5308.
9. Halgren, T. A.; Kleier, D. A.; Hall, J. H., Jr.; Brown, L. D.; Lipscomb, W. N. *J. Am. Chem. Soc.* **1978**, *100*, 6595.
10. Levine, I. N. "Quantum Chemistry" 3rd ed. Allyn and Bacon, Inc.: Boston, 1983.
11. Segal, G. A., ed. "Semiempirical Methods of Electronic Structure Calculation, Parts A and B." Plenum: New York, 1977.
12. Pople, J. A.; Santry, D. P.; Segal, G. A. *J. Chem. Phys.* **1966**, *43*, S129.
13. Pople, J. A.; Beveridge, D. L.; Dobosh, P. A. *J. Chem. Phys.* **1967**, *47*, 2026.
14. Veillard, A. *Chem. Phys. Lett.* **1975**, *33*, 15.
15. Gregory, A. R.; Paddon-Row, M. N. *J. Am. Chem. Soc.* **1976**, *98*, 7521.
16. See, for example, Brunck, T. K.; Weinhold, F. *J. Am. Chem. Soc.* **1979**, *101*, 1700; Cosse-Barbi, A.; Dubois, J-E. *J. Am. Chem. Soc.* **1987**, *109*, 1503.
17. Bingham, R. C.; Dewar, M. J. S.; Lo, D. H. *J. Am. Chem. Soc.* **1975**, *97*, 1285.
18. Dewar, M. J. S.; Thiel, W. *J. Am. Chem. Soc.* **1977**, *99*, 4899.
19. Dewar, M. J. S.; Thiel, W. *J. Am. Chem. Soc.* **1977**, *99*, 4907.
20. Dewar, M. J. S.; Zoebisch, E. G.; Healy, E. F.; Stewart, J. J. P. *J. Am. Chem. Soc.* **1985**, *107*, 3902.
21. Halgren, T. A.; Lipscomb, W. N. *J. Chem. Phys.* **1973**, *58*, 1569.
22. Malrieu, J. P. in "Semiempirical Methods of Electronic Structure Calculation, Part A," Segal, G. A., ed. Plenum: New York, 1977, chap. 3.
23. Slater, J. C. *Phys. Rev.* **1951**, *81*, 385.

24. Slater, J. C. "Quantum Theory of Molecules and Solids, Vol. 4." New York: McGraw-Hill, 1974.
25. Johnson, K. H. *Annu. Rev. Phys. Chem.* **1975**, *26*, 39.
26. Case, D. A. *Annu. Rev. Phys. Chem.* **1982**, *33*, 151.
27. Dahl, J. P.; Avery, J., eds. "Local Density Approximations in Quantum Chemistry and Solid State Physics." Plenum: New York, 1984.
28. Schleyer, P.v.R. *J. Computat. Chem.* **1986**, *7*, 380.
29. Pople, J. A. in "Applications of Electronic Structure Theory." Schaefer, H. F., ed. Plenum: New York, 1977.
30. Radom, L.; Hehre, W. J.; Pople, J. A. *J. Am. Chem. Soc.* **1972**, *94*, 2371.
31. Schleyer, P.v.R.; Kos, A. J. *Tetrahedron* **1983**, *39*, 1141.
32. Schleyer, P.v.R.; Jemmis, E. D.; Spitznagel, G. W. *J. Am. Chem. Soc.* **1985**, *107*, 6393.
33. Brunck, T. K.; Weinhold, F. *J. Am. Chem. Soc.* **1979**, *101*, 1700.
34. Dewar, M. J. S.; Ford, G. P. *J. Am. Chem. Soc.* **1979**, *101*, 5558.
35. Scheiner, S. *Theor. Chim. Acta* **1980**, *57*, 71.
36. Morokuma, K. *Accts. Chem. Res.* **1977**, *10*, 294.
37. Kollman, P. A. *Accts. Chem. Res.* **1977**, *10*, 365.
38. See, for example, Legon, A. C.; Millen, D. J. *Accts. Chem. Res.* **1987**, *20*, 39.
39. Gogarasi, G.; Pulay, P. *Annu. Rev. Phys. Chem.* **1984**, *35*, 191.
40. Wiberg, K. B.; Walters, V. A.; Dailey, W. P. *J. Am. Chem. Soc.* **1985**, *107*, 4860.
41. Allinger, N. L.; Profeta, S., Jr. *J. Computat. Chem.* **1980**, *1*, 181.
42. Van-Catlege, F. A.; Allinger, N. L. *J. Am. Chem. Soc.* **1982**, *104*, 6272.
43. Compton, D. A. C.; Montero, S.; Murphy, W. F. *J. Phys. Chem.* **1980**, *84*, 3587.
44. Niketic, S. R.; Rasmussen, K. "The Consistent Force Field: A Documentation." Springer-Verlag: Berlin, 1977.
45. Ermer, O. *Structure and Bonding (Berlin)* **1976**, *27*, 161.
46. Dale, J. *Top. Stereochemistry* **1976**, *9*, 199.
47. Allinger, N. L. *J. Am. Chem. Soc.* **1977**, *99*, 8127.
48. Chang, S.; McNally, D.; Shary-Tehrany, S.; Hickey, M. J.; Boyd, R. H. *J. Am. Chem. Soc.* **1970**, *92*, 3109.
49. Engler, E. M.; Andose, J. D.; Schleyer, P.v.R. *J. Am. Chem. Soc.* **1973**, *95*, 8005.
50. Ermer, O.; Lifson, S. *J. Am. Chem. Soc.* **1973**, *95*, 4121.
51. Fitzwater, S.; Bartell, L. S. *J. Am. Chem. Soc.* **1976**, *98*, 5107.
52. Lifson, S.; Warshel, A. *J. Chem. Phys.* **1970**, *53*, 582.
53. Jaime, C.; Osawa, E. *Tetrahedron* **1983**, *39*, 2769.
54. Saunders, M.; Jarret, R. M. *J. Comput. Chem.* **1986**, *7*, 578.
55. Anet, F. A. L.; Krane, J. *Tetrahedron Lett.* **1973**, *50*, 5029.
56. Pakes, P. W.; Rounds, T. C.; Strauss, H. L. *J. Phys. Chem.* **1981**, *85*, 2469.
57. Strauss, H. L. *Annu. Rev. Phys. Chem.* **1983**, *34*, 301.
58. Nguyen, D. T.; Case, D. A. *J. Phys. Chem.* **1985**, *89*, 4020.
59. Ivanov, P. H.; Osawa, E. *J. Comput. Chem.* **1984**, *5*, 307.
60. Anet, F. A. L.; Basus, V. J. *J. Am. Chem. Soc.* **1973**, *95*, 4424.
61. Meiboom, S.; Hewitt, R. C.; Luz, Z. *J. Chem. Phys.* **1977**, *66*, 4041.
62. Snyder, R. G.; Schactschneider, J. H. *Spectrochim. Acta* **1965**, *21*, 169.
63. Pakes, P. W.; Rounds, T. C.; Strauss, H. L. *J. Phys. Chem.* **1981**, *85*, 2476.
64. Nguyen, D. T. Ph.D. thesis, University of California, Davis, 1986.
65. Tribble, M. T.; Allinger, N. L. *Tetrahedron* **1972**, *28*, 2147.
66. Profeta, S., Jr.; Unwalla, R. J.; Nguyen, B. T.; Cartledge, F. R. *J. Comput. Chem.* **1986**, *7*, 528.
67. Frierson, M. R.; Allinger, N. L. Unpublished data cited in Ref. 4.
68. Damewood, J. R., Jr.; West, R. *Macromolecules* **1985**, *18*, 159.
69. Shafiee, F.; Damewood, J. R.; Haller, K. J.; West, R. *J. Am. Chem. Soc.* **1985**, *107*, 6950.
70. Allinger, N. L.; Chang, S. H-M.; Glaser, D. H.; Honig, H. *Isr. J. Chem.* **1980**, *20*, 51.
71. Bocian, D. F.; Strauss, H. L. *J. Am. Chem. Soc.* **1977**, *99*, 2876.
72. Pakes, P. W.; Rounds, T. C.; Strauss, H. L. *J. Phys. Chem.* **1981**, *85*, 2469.
73. Wipff, G.; Weiner, P.; Kollman, P. *J. Am. Chem. Soc.* **1982**, *104*, 3249.
74. Kollman, P.; Wipff, G.; Singh, U. C. *J. Am. Chem. Soc.* **1985**, *107*, 2212.
75. Rao, V. S. R.; Sundararajan, P. R.; Ramakrishnan, C.; Ramachandran, G. N., in "Conformation of Biopolymers," Ramachandran, G. N., ed. Academic: London, 1967, p. 721.
76. Rao, V. S. R.; Vijayalakshmin K. S.; Sundararajan, P. R. *Carbohyd. Res.* **1971**, *17*, 341.
77. Rees, D. A.; Smith, P. J. C. *J. Chem. Soc., Perkin Trans.* **1975**, *2*, 830.
78. Rasmussen, K. *Acta Chem. Scand. Ser. A* **1982**, *36*, 323.

79. See, for example: Olson, W. *J. Am. Chem. Soc.* **1982**, *104*, 278; Levitt, M.; Warshel, A. *J. Am. Chem. Soc.* **1979**, *100*, 2607; Weiner, S. J.; Kollman, P. A.; Case, D. A.; Singh, U. C.; Ghio, C.; Alagona, G.; Profeta, S., Jr.; Weiner, P. *J. Am. Chem. Soc.* **1984**, *106*, 765.
80. Allinger, N. L.; Chung, D. Y. *J. Am. Chem. Soc.* **1976**, *98*, 6798.
81. Hagler, A. T.; Huler, E.; Lifson, S. *J. Am. Chem. Soc.* **1974**, *96*, 5319.
82. McGuire, R. F.; Momany, F. A.; Scheraga, H. A. *J. Phys. Chem.* **1972**, *76*, 375.
83. Allinger, N. L.; Tribble, M. T.; Miller, M. A. *Tetrahedron* **1972**, *28*, 1173.
84. Anet, F. A. L.; St. Jacques, M.; Henrichs, P. M.; Cheng, A. K.; Krane, J.; Wong, L. *Tetrahedron* **1974**, *30*, 1629.
85. Profeta, S. Jr.; Kollman, P. A.; Wolff, M. *J. Am. Chem. Soc.* **1982**, *104*, 3745.
86. Dosen-Micovic, L.; Jeremic, D.; Allinger, N. L. *J. Am. Chem. Soc.* **1983**, *105*, 1723.
87. Allinger, N. L.; Hickey, M. J. *J. Am. Chem. Soc.* **1975**, *97*, 5167.
88. Kollman, P.; McKelvey, J.; Johansson, A.; Rothenberg, S. *J. Am. Chem. Soc.* **1975**, *97*, 955.
89. Umeyama, H.; Morokuma, K. *J. Am. Chem. Soc.* **1977**, *99*, 1316.
90. Allinger, N. L.; Hickey, M. J.; Kao, J. *J. Am. Chem. Soc.* **1976**, *98*, 2741.
91. Kao, J.; Allinger, N. L. *Inorg. Chem.* **1977**, *16*, 35.
92. Nash, C. P.; Olmstead, M. M.; Weiss-Lopez, B. E.; Musker, W. K.; Ramasubbu, N.; Parthasarathy, R. *J. Am. Chem. Soc.* **1985**, *107*, 7194.
93. Weiss-Lopez, B. E.; Goodrow, M. H.; Musker, W. K.; Nash, C. P. *J. Am. Chem. Soc.* **1986**, *108*, 1271.
94. Sugeta, H. *Spectrochim. Acta* **1975**, *A31*, 1729.
95. Nguyen, D. T.; Case, D. A. Unpublished calculations.
96. Nguyen, D. T., Ph.D. thesis, University of California, Davis, 1986; Giammona, D. A.; Karplus, M. Personal communication.
97. Profeta, S., Jr.; Allinger, N. L. *J. Am. Chem. Soc.* **1985**, *107*, 1907.
98. Blackburne, I. D.; Katritzky, A. R. *Accts. Chem. Res.* **1975**, *8*, 300.
99. Deiters, J. A.; Gallucci, J. C.; Clark, T. E.; Holmes, R. R. *J. Am. Chem. Soc.* **1977**, *99*, 5461.
100. Nilsson, L.; Karplus, M. *J. Comput. Chem.* **1986**, *7*, 591.
101. Allinger, N. L.; von Voithenberg, H. *Tetrahedron*, **1978**, *34*, 627.
102. Momany, F.; McGuire, R.; Burgess, A.; Scheraga, H. *J. Phys. Chem.* **1975**, *79*, 2361.
103. Nemethy, G.; Pottle, M. S.; Scheraga, H. A. *J. Phys. Chem.* **1983**, *87*, 1883.
104. Warshel, A.; Levitt, M.; Lifson, S. *J. Mol. Spectrosc.* **1970**, *33*, 84.
105. Levitt, M. *J. Mol. Biol.* **1974**, *82*, 93.
106. Gelin, B. R.; Karplus, M. *Biochemistry* **1979**, *18*, 1256.
107. Hermans, J.; Ferro, D.; McQueen, J.; Wei, S. in "Environmental Effects on Molecular Structure and Properties," Pullman, B., ed. Reidel: Dordrecht, Holland, 1976.
108. Dauber, P.; Goodman, M.; Hagler, A. T.; Osguthorpe, D.; Stern, P. *ACS Symp. Ser.* **1981**, *173*, 161.
109. Weiner, S. J.; Kollman, P. A.; Case, D. A.; Singh, U. C.; Ghio, C.; Alagona, G.; Profeta, S.; Weiner, P. *J. Am. Chem. Soc.* **1984**, *106*, 765.
110. Weiner, S. J.; Kollman, P. A.; Nguyen, D. T.; Case, D. A. *J. Computat. Chem.* **1986**, *7*, 230.
111. White, D. N.; Guy, M. H. P. *J. Chem. Soc. Perkin Trans.* **1975**, *II*, 43.
112. Brooks, B. R.; Bruccoleri, R. E.; Olafson, B. D.; States, D. J.; Swaminathan, S.; Karplus, M. *J. Computat. Chem.* **1983**, *4*, 187.
113. Jorgensen, W. L.; Tirado-Rives, J. *J. Am. Chem. Soc.* (submitted for publication).
114. Jorgensen, W. L.; Swenson, C. J. *J. Am. Chem. Soc.* **1985**, *107*, 569, and references therein.
115. Hall, D.; Pavitt, N. *J. Computat. Chem.* **1984**, *5*, 441.
116. Whitlow, M.; Teeter, M. M. *J. Am. Chem. Soc.* **1986**, *108*, 7163.
117. Madison, V.; Kopple, K. D. *J. Am. Chem. Soc.* **1980**, *102*, 4855.
118. Brandt, D. A.; Flory, P. J. *J. Mol. Biol.* **1967**, *23*, 47.
119. Zimmerman, S.; Pottle, M.; Nemethy, G.; Scheraga, H. *Macromolecules* **1977**, *10*, 1.
120. Rossky, P. J.; Karplus, M. *J. Am. Chem. Soc.* **1979**, *101*, 1931.
121. Pettitt, B. M.; Karplus, M. *J. Am. Chem. Soc.* **1985**, *107*, 1166.
122. Stern, P.; Chorev, M.; Goodman, M.; Hagler, A. T. *Biopolymers* **1983**, *22*, 1885.
123. Pullman, B.; Pullman, A. *Adv. Protein Chem.* **1974**, *28*, 348.
124. Scarsdale, J.; Van Alsenoy, C.; Klimkowski, V.; Schafer, L.; Momany, F. *J. Am. Chem. Soc.* **1983**, *105*, 3438.
125. Weiner, S. J.; Singh, U. C.; O'Donnell, T. J.; Kollman, P. A. *J. Am. Chem. Soc.* **1984**, *106*, 6243.
126. Scales, L. E. "Introduction to Non-Linear Optimization." Springer-Verlag: New York, 1985.
127. For a review, see Bell, S.; Crighton, J. S. *J. Chem. Phys.* **1984**, *80*, 2464.
128. Wiberg, K. B.; Boyd, R. H. *J. Am. Chem. Soc.* **1972**, *94*, 8426.

129. Thomas, M. W.; Emerson, D. *J. Mol. Struct.* **1973**, *16*, 473.
130. van de Graaf, B.; Baas, J. M. A. *J. Computat. Chem.* **1984**, *5*, 314.
131. Crippen, G. M.; Scheraga, H. A. *Arch. Biochem. Biophys.* **1971**, *144*, 462.
132. Hilderbrandt, R. L. *Comput. Chem.* **1977**, *1*, 179.
133. Cerjan, C. J.; Miller, W. H. *J. Chem. Phys.* **1981**, *75*, 2800.
134. Simons, J.; Jorgenson, P.; Taylor, H.; Ozment, J. *J. Phys. Chem.* **1983**, *87*, 2745.
135. Nguyen, D. T.; Case, D. A. *J. Phys. Chem.* **1985**, *89*, 4020.
136. White, D. N. J.; Morrow, *C. Comput. Chem.* **1979**, *3*, 33.
137. Go, N.; Scheraga, H. A. *Macromolecules* **1970**, *3*, 178.
138. Bruccoleri, R. E.; Karplus, M. *Macromolecules* **1985**, *18*, 2767.
139. Crippen, G. M. "Distance Geometry and Conformational Calculations," Research Studies Press—Wiley: New York, 1981.
140. Weiner, P. K.; Profeta, S., Jr.; Wipff, G.; Havel, T.; Kuntz, I. D.; Langridge, R.; Kollman, P. A. *Tetrahedron* **1983**, *39*, 1113.
141. Paine, G. H.; Scheraga, H. A. *Biopolymers* **1985**, *24*, 1391.
142. Bruccoleri, R. E.; Karplus, M. *Biopolymers* **1987**, *26*, 137.
143. White, D. J. N. *J. Mol. Graphics* **1986**, *4*, 112.
144. McQuarrie, D. A. "Statistical Mechanics." Harper & Row: New York, 1976,
145. Herschbach, D. R.; Johnston, H. S.; Rapp, D. *J. Chem. Phys.* **1959**, *31*, 1652.
146. Karplus, M.; Kushick, J. N. *Macromolecules* **1981**, *14*, 325.
147. Levy, R. M.; Karplus, M.; Kushick, J.; Perahia, D. *Macromolecules* **1984**, *17*, 1370.
148. Brady, J.; Karplus, M. *J. Am. Chem. Soc.* **1985**, *107*, 6103.
149. Jorgensen, W. L.; Binning, R. C., Jr.; Bigot, B. *J. Am. Chem. Soc.* **1981**, *103*, 4393.
150. For references to recent work, see: Zichi, D. A.; Rossky, P. J. *J. Chem. Phys.* **1986**, *84*, 1712.
151. Jorgensen, W. L. *J. Phys. Chem.* **1983**, *87*, 5304.
152. Patey, G. N.; Valleau, J. P. *Chem. Phys. Lett.* **1973**, *21*, 297.
153. Torrie, G. M.; Valleau, J. P. *J. Computat. Phys.* **1977**, *28*, 187.
154. Mezei, M.; Mehrotra, P.K.; Beveridge, D. L. *J. Am. Chem. Soc.* **1985**, *107*, 2239.
155. Ravishankar, G.; Mezei, M.; Beveridge, D. L. *J. Comput. Chem.* **1986**, *7*, 345.
156. Hirata, F.; Rossky, P. J.; Pettitt, B. M. *J. Chem. Phys.* **1983**, *78*, 4133.
157. Pettitt, B. M.; Karplus, M.; Rossky, P. J. *J. Phys. Chem.* **1986**, *90*, 6335.
158. Chandrasehkar, J.; Smith, S. F.; Jorgensen, W. L. *J. Am. Chem. Soc.* **1984**, *106*, 3049.
159. Chandrasekhar, J.; Smith, S. F.; Jorgensen, W. L. *J. Am. Chem. Soc.* **1985**, *107*, 154.
160. Keck, J. C. *Discuss. Faraday Soc.* **1962**, *33*, 173.
161. Chandler, D. *J. Chem. Phys.* **1978**, *68*, 2959.
162. Bergsma, J. P.; Gertner, B. J.; Wilson, K. R.; Hynes, J. T. *J. Chem. Phys.* **1987**, *86*, 1356, and references therein.

2

Medium-sized Oxygen Heterocycles

Frank A. L. Anet
Department of Chemistry and Biochemistry
University of California, Los Angeles, CA

2.1. Introduction

Molecules containing rings with more than seven members made a rather late appearance in organic chemistry, for it was not until the 1920s that such compounds were either isolated from natural sources or synthesized.[1,2] Before that time, it was thought that both small and large rings were very strained, in accordance with the Baeyer angle strain theory. The application of this theory to large rings was not justified, since such rings do not have to be planar. A real understanding of the strain in medium-sized cycloalkanes was obtained only after it was realized that three factors are involved: (1) torsional effects in single bonds (Pitzer strain), (2) nonbonded repulsions (van der Waals terms), and (3) bond angle strain.[3] The presence of heteroatoms, either as substituents or in the ring, introduces additional effects that are discussed in detail in this chapter.

Some synthetic methods—e.g., the cyclization of dicarboxylic acids to ketones—produce molecules with eight- to 10-membered rings in extremely small yields. The cause of this medium ring effect has been extensively studied, and both enthalpic and entropic causes have been identified. The competition between intermolecular condensations and intramolecular cyclization can be made to favor the latter by high-dilution conditions. The presence of structural features, such as *gem*-dimethyl groups, heteroatoms, or aromatic rings, profoundly affects the ease of these reactions. Some cyclizations, such as the acyloin reaction, are found to work equally well for any ring size whereas other cyclizations are clearly favored by complexation to cations. Medium rings can show a great selectivity when new stereocenters on the rings are formed, and can be valuable synthetic intermediates, even when the target molecules do not contain such rings.[4–6] Regiochemically and stereochemically specific transannular reactions occur occasionally in medium rings and are also synthetically useful.[7] An excellent book by Riddell on the fundamentals of conformational analysis in heterocyclic compounds (especially five- and six-membered rings) was published a few years ago; the present review extends the presentation given therein on medium rings containing oxygen and brings it up to date.[8]

2.2. Definitions of Terms

A variety of terms, such as "medium ring," "conformation," and "pseudorotation," needs to be defined, even if not precisely, because these terms often have different meanings as used in the chemical literature.

2.2.1. Medium Rings

The medium rings have generally been considered to be those with eight to 11 or 12 members,[3] but for the present purpose, rings with seven to 12 members have been included in this category. The justification for including seven- and 12-membered rings is not very strong, but molecules with these two ring sizes are useful for comparison purposes; also, they do have some features in common with eight- to 11-membered rings, which are indisputably "medium rings," at least in organic chemistry.[9] Some mention of the properties of five- and six-membered rings, as well as of large rings, also needs to be made to provide a proper perspective. Although this review is restricted to heterocycles containing oxygen in the ring, continual comparisons with the corresponding carbocycles are made. Discussions of the analogous nitrogen- and sulfur-containing heterocycles are very limited, and molecules containing both oxygen and other elements, such as silicon or phosphorus, are almost completely ignored. Molecules that have an oxygen-containing ring extensively fused and bridged to other rings, as in strychnine, are in general not covered. However, unsaturated and benzo-fused heterocycles are treated in some detail. On the whole, emphasis is placed on the simpler monocyclic, bicyclic, and tricyclic molecules and on general conformational principles.

2.2.2. Isomer, Configuration, Conformation, and Related Terms

A precise and correct definition (that is, fully in accordance with quantum mechanics) of a molecule is not a simple task, especially when isomeric molecules are considered.[10–12] Fortunately, nearly all molecules have fixed bonding schemes, so that the "constitution" is a well-defined property, at least on the time scale of human life and at room temperature (fluxional molecules, such as semibullvalene[13] and oxepin,[14] are exceptions). Because of the large energy barrier to configurational change at sp^3 and sp^2 hybridized carbons, well-defined stereoisomers derived from any given constitutional isomer exist, although there are exceptions, mostly involving sp^2 hybridized atoms—in amides, for example—where isolation of stereoisomers is sometimes just possible at room temperature. Actually, the placing of constitutional isomerism above stereoisomerism is not logically required, and, indeed, Mislow has suggested that the reverse scheme has advantages[15]; in any case, the arbitrariness of the choice should be kept in mind.

The accepted chemical nomenclature (*Chemical Abstracts*, IUPAC) has been strongly influenced by practical considerations, such as the ability to separate mixtures into "pure compounds" with reasonable ease at room temperature. Since conformations in general interconvert rapidly under these conditions, they do not represent "chemically" different compounds, although to spectroscopists and to those who work with compounds at very low temperatures, different conformations can be as different in their properties as are stereoisomers and constitutional isomers, or indeed nonisomeric molecules. Eliel[16] has pointed out that conformational and configurational isomers do not differ in principle, but only in terms of lifetimes, which depend on temperature. Furthermore, each method used to investigate a compound has its own time scale as to what is fast and what is slow, and this is particularly evident in nuclear magnetic resonance (NMR) spectroscopy. Actually, the definition of an unambiguous time scale is not trivial, and in NMR there are possible time scales based on chemical shift differences, on coupling constants, and on relaxation phenomena.[17] Most commonly, the NMR time scale, as used in the literature, refers to times of the order of 1 millisecond to 1 second, and it is clearly shorter than the normal preparative time scale, but very much longer than time scales relevant to vibrational or electronic spectroscopy. Molecules with very low barriers to conformational change pose difficult problems, but even otherwise, the general unification of configuration and conformation under one concept is beset with almost insuperable practical difficulties, as Eliel has pointed out.[16]

The terms "conformation" and "form" are used interchangeably in the present review to represent any molecular geometry, although often the lowest energy minimum is clearly meant, as, for example, in "the conformation of cyclohexane is a chair"; "conformer" is used to refer to a distinct local energy minimum when two or more conformers of a molecule are being discussed whereas "conformational isomer" is used especially when isolation of a conformer is feasible.

2.2.3. Flexible and Rigid Conformations

Each of the N atoms in a molecule has three degrees of freedom—for example, motion along Cartesian X, Y, and Z coordinates. Six degrees of freedom are associated with the translation and rotation of the molecule as a whole and thus there are $3N - 6$ independent internal coordinates and the same number of vibrational normal modes. It is instructive to begin with a nonlinear acyclic molecule consisting of a chain of N atoms.[18] The number of bonds is $N - 1$, the number of bond angles is $N - 2$, and the number of torsional angles is $N - 3$. Therefore, the total number of these internal coordinates is $3N - 6$, and is precisely the allowed degrees of internal freedom in the acyclic molecule. All the bond lengths, bond angles, and torsional angles can be changed independently of one another, and the conformation of the molecule is easy to describe in terms of these internal coordinates.

An extra bond is created when the noncyclic molecule is converted into a cyclic one by joining the two end atoms, and this results in N bond lengths, N bond angles, and N torsional angles. Only $3N - 6$ of these $3N$ internal coordinates are independent of one another, and thus cyclization introduces six constraints, but the choice of $3N - 6$ appropriate internal coordinates is not obvious. If the N bond lengths and the N bond angles are fixed, then only $N - 6$ torsional angles can be changed in an arbitrary way. When this number ($N - 6$) is negative, as in cyclopentane, even the bond angles are not independent of one another. In a six-membered ring, $N - 6$ is zero, and this leaves no torsional degrees of freedom, so that such a molecule must in general be "rigid"; however, special six-membered rings can be flexible, and this is discussed below.[18] For seven-membered rings, $N - 6$ is 1, and thus such molecules are flexible, but this does not mean that all conformations of a given ring can be interconverted without any change in bond angles. Eight-membered and larger rings not only are flexible, but all conformations of a given molecule can be interconverted without any angle distortion. If there are torsional constraints, then the situation is, of course, different; this is discussed in detail below.

It is important to realize that the rigidity that is being discussed here is a mathematical one, and assumes that bond lengths and angles are absolutely unchangeable and that torsional and nonbonded constraints are absent. Dreiding molecular models have fixed bond lengths, but small bond-angle distortions are possible and are not always apparent; on the whole, however, these models satisfy the requirements of the foregoing mathematical treatment.

Actually, rings in organic molecules are more complex than is described here, because of the presence of substituents. In an N-membered cycloalkane, there are $3N$ atoms, and thus $9N - 6$ parameters have to be specified, but if the CH_2 group is treated as a rigid unit (i.e., as a pseudoatom), then only $3N - 6$ parameters are required. This assumption is reasonable and has often been made, but it leads to significant errors in the calculated energies,[19] even though the shape of the energy surface may be approximately correct. Such a treatment should be considered only as a first step to fuller calculations, especially in medium rings, where nonbonded interactions can be large.

The question of flexibility and rigidity in cyclohexane was first investigated by Sachse near the end of the last century.[20,21] He showed that a six-membered ring could exist in both rigid and flexible forms. At the time, unfortunately, knowledge about molecular geometry was almost nonexistent, although the tetrahedral disposition of the bonds at a saturated carbon was well established.

Nothing quantitative was known about the energy required to stretch bonds or to distort bond angles. Free rotation about single bonds was accepted, and the concept of torsional strain did not arise until very much later. Thus it is hardly surprising that conformational analysis failed to make progress at that time, and that Sachse's theory was ignored, especially as attempts to use it to determine the number of stereoisomers in substituted cyclohexanes failed.

The acceptance of ring planarity, even for bicyclic bridged terpenes, dominated the interpretation of cyclic molecules up to the early 1920s. Although Mohr[22] had predicted in 1922 that boat-to-chair interconversions should occur easily, and that isomeric decalins should exist, it came as a shock when *cis* and *trans* isomers of decalin were isolated in 1925.[23] Nevertheless, this opportunity for conformational analysis was largely lost, and *cis*-decalin was written in an incorrect boat–boat form that persisted in textbooks for many years. The possibility of isolating chair and boat forms of cyclohexane at room temperature still had some proponents as late as 1950, although the work of Hassel, Pitzer, Barton, and others had by that time firmly established that cyclohexane had a chair conformation and that this had some rigidity, but could undergo a ring inversion easily at room temperature.[24,25]

With the application of dynamic NMR to conformational analysis of cyclic molecules starting in the early 1960s,[26] the mathematical rigidity and flexibility of rings received new attention. In 1972, Dunitz and Waser reviewed this subject,[18] and referred to previous work by mathematicians and by structural engineers interested in the mechanical rigidity of polygonal structures. Dunitz and Waser pointed out that the strict mathematical requirement for flexibility in a ring such as cyclohexane is that there be a "nonintersecting" C_2 symmetry axis (such an axis does not pass through atoms or bonds). Thus the boat, twist–boat, and all in-between conformations have such an axis, and (geometrically speaking) are perfectly flexible, whereas the chair does not have this axis and is rigid. However, the presence of nonbonded and even small torsional constraints in real molecules can change the picture completely. For example, the twist–boat form of *cis,cis*-1,5,-cyclooctadiene, which should be perfectly flexible, actually has pseudorotation barriers in the range of 4–5 kcal/mol.[27] In this molecule, the two carbons in a double bond can be treated as a pseudoatom, as described in section 2.3. Conversely, the comparative ease of small bond angle distortions in real molecules can lead to low barriers for mathematically rigid geometries when there is an approximate nonintersecting axis, as is expected, for example, in the boat forms of tetrahydropyran. Nevertheless, the concept of geometric rigidity and flexibility under the special conditions of fixed bond lengths and bond angles is often a useful starting point for conformational discussions.

2.2.4. Ring Puckering, Pseudorotation, and Inversion

Cyclopentane has a nonplanar conformation and has given rise to the concept of (ring) pseudorotation.[28] The puckering in this cycloalkane is dynamic and can be considered to arise from two degenerate normal modes, where the CH_2

groups act as units and move more or less perpendicular to the general plane of the molecule. The energy barrier to this motion is nearly zero, so that the pseudorotation is virtually free; the amplitude of the motion and the atomic masses determine the pseudorotation frequency. In derivatives of cyclopentane, however, the pseudorotation is more or less hindered, and certain geometries are favored, although the pseudorotation barrier is at most only a few kilocalories per mole. The definition of the conformations of such five-membered rings, especially in the case of furanose sugar derivatives in the crystalline state, where the molecules appear to have fixed static geometries, has attracted much attention.[29] Actually, the problem is more general and concerns the description of puckering in rings of any size.

The simplest way to introduce ring puckering is to start with a planar ring and then to introduce infinitesimal out-of-plane displacements. Cartesian coordinates with the X- and Y-axes in the ring plane can be defined; the Z-axis is then normal to the plane. The N bond lengths can be kept fixed, but the bond and the torsional angles can be allowed to vary. Although there are N out-of-plane (Z) displacements, only $N - 3$ linear combinations of these displacements are normal modes and leave the molecule unshifted along the Z-axis and unrotated along the X- and Y-axes. Similarly, there are $N - 3$ degrees of freedom for infinitesimal in-plane displacements (normal modes) that leave the molecule unshifted along the X and Y directions and unrotated along the Z-axis. The $N - 3$ out-of-plane parameters are sufficient completely to describe the conformation under these special conditions. Alternatively, $N - 3$ combinations of torsional angles can be used, and there are mathematical relationships between infinitesimal puckering amplitudes and the torsional angles.[30] For a finite ring puckering, however, both the out-of-plane and in-plane displacements must be taken into account, but an approximate description of the conformation, especially for rings that are not large or very puckered, can be obtained with $N - 3$ parameters.

Three main approaches have been used to describe ring puckering in terms of the previously discussed $N - 3$ parameters: the first depends on the out-of-plane distances of ring atoms, the second depends on both in-plane and out-of-plane parameters, and the third approach uses ring torsional angles. The first approach has already been mentioned for the special case of cyclopentane, and an extension of this procedure to larger rings has been made by Cremer and Pople,[31,32] who devised a simple and mathematically precise way of defining a "mean plane" that is applicable even when the bond lengths and bond angles are not equal. The out-of-plane amplitudes of N atoms arranged around a circle projected from a regular polygon onto the reference plane in the Cremer–Pople treatment can be considered to sample a puckering wave, and this motion can be dissected by Fourier analysis or Fourier transformation in term of waves as a function of an angular parameter, ϕ (0 to 2π) measured from the center of the molecule. The lowest or zero "frequency" wave (as a function of ϕ, not time) consists of all the ring atoms moving together in the same direction, either above or below the plane, but since this is a translation of the molecule as a whole rather than a conformational change, the wave amplitude can be set equal to zero. The next two frequencies are degenerate and have single nodes perpendicular to one another, with atoms on one side of the ring moving in one direction and those on the other side moving in the opposite direction; they correspond to rotations of the molecule about the X- or Y-axis, and thus their amplitudes can also be set to zero. These three constraints define the mean plane of the ring. The remaining $N-3$ waves are independent of one another and can have arbitrary (small) amplitudes. The possible conformations of the molecule are then described by linear combinations of these $N-3$ extreme or basis forms, and, of course, $N-3$ is exactly the number of the out-of-plane normal modes in an N-atom ring.

The conformations described by the $N-3$ Cremer–Pople parameters represent unique geometries when the molecule is a regular polygon, otherwise a given conformation represents an infinite number of (closely related) geometries. The highest frequency wave is unique (nondegenerate) when the ring is even membered and corresponds to an alternate up–down motion of adjacent ring atoms; this is called an *inversion mode* by Pople and Cremer and there is no phase associated with this motion. All the other waves are pairwise degenerate and are associated with two degrees of freedom, and thus can be described by either two amplitudes with no arbitrary phase factor or by one amplitude and one phase factor. It is convenient to define a phase relationship so that the sum of the two degenerate waves can be treated as a single wave having a definite phase with respect to the labeled ring atoms. A change in the phase factor then leads to a pseudorotation mode. There is a close connection, mathematically speaking, between the puckering waves mentioned above and the well-known Hückel molecular orbitals of annulenes consisting of N sp^2 carbons, in particular; in both cases, the puckering wave or MO with the largest number of nodes is unique only when N is even.

The Cremer–Pople treatment has been applied to rings with N up to 8. In cyclopentane there are $N-3$ (i.e., 2) independent conformations (C_s and C_2) and all other geometries can be expressed as linear combinations of these two. In cyclohexane there are three independent conformations: chair, boat, and twist–boat. All other conformations, including those with four and five carbons in a plane, can be expressed as linear combinations of these, and all the possible conformations having a given total puckering amplitude can be represented by unique points on the surface of a sphere, with the chairs at the poles, and the boat and twist–boat distributed along the equator. In cycloheptane there are $N-3=4$ independent conformations, and, as will be discussed below, these are the chair, the boat, and their respective twist partners; all possible conformations can be represented by points on a four-dimensional surface. With cyclooctane the surface becomes five-dimensional, and with cyclododecane it is nine-dimensional! In contrast to five-, six-, and seven-membered rings, where the important local energy minima are always basis geometries, the lowest energy conformation (boat–chair) in cyclooctane is described as a linear combination of three different basis geometries,[33,34] none of which is a local energy minimum. The simplicity seen in the lower cycloalkanes is absent here, and it is usually easier and more practical to describe the conformations and conformational processes in other ways.[35]

The Cremer–Pople treatment has received some criticism,[34,36,37] but it has been defended.[38,39] Although it is well defined mathematically, this does not mean that it is necessarily useful chemically. For example, it is not well suited for describing changes in torsional angles,[34] which are of such great importance in determining NMR coupling constants. Also, it ignores differences in bond lengths in dealing with ring puckering. The ring atoms can be considered to sample a continuous puckering around a circle, and if the bond lengths are not the same, then the sampling interval is no longer constant, and this fact should really be taken into account. However, since bond lengths in practice are similar, the neglect of such differences is not serious. Application of the Cremer–Pople treatment to large rings can be ambiguous, it should be noted, if there is a possibility that the polygon that is projected on the mean plane can be concave as well as convex. Stated in another way, the puckering in such rings can be so large that it cannot be described satisfactorily in terms of out-of-plane displacements only, and in-plane displacements have to be considered, as mentioned previously. It should also be kept in mind that any treatment that is given in terms of $N - 3$ (or even N) parameters cannot possibly describe all the $3N - 6$ degrees of freedom in the molecule. The reduced number of parameters are both a strength and a weakness, and one cannot have one without the other.

Strauss and co-workers[19,40–45] have made extensive studies of the vibrational spectra and conformations of six- to eight-membered rings, with and without ring oxygens. They have used out-of-plane displacements for qualitative descriptions of ring shapes, but for more quantitative analysis, they have included some in-plane variables. Since they start from the normal modes of planar rings, the definition of a plane does not arise. Somewhat complicated and arbitrary internal coordinates are used, and these are reduced in a nonlinear fashion to $N - 3$ parameters that resemble those of Cremer and Pople, but have only approximately the correct molecular symmetry. All bond lengths are kept fixed and CH_2 groups are treated as pseudoatoms, and this, together with the use of a force field developed to reproduce low-frequency ring vibrations, allows the energy surface for conformational change to be described as for the Cremer–Pople treatment. The main virtue of the Strauss approach is the demonstration of pseudorotation in some conformational transition states, and the development of a proper theoretical approach to calculate the entropy for pseudorotating molecules in general.[40] Unfortunately, the force-field structure used is unusual from the point of view of molecular mechanics, and the geometric constraints imposed lead to rather poor conformational potential energies. Also, the parametrization of the force field is not based on a unified treatment of many molecules, but has been arbitrarily modified to fit individual molecules. Furthermore, dipole–dipole and anomeric effects in molecules containing two oxygens are neglected. Thus the generality of the force field is not well established. A combination of the entropy treatment used by Strauss and co-workers with fuller molecular mechanics calculations (section 2.4) would be desirable. Finally, attempts to use more symmetrical coordinates and to simplify further the Strauss method have been suggested,[46] but are only applicable in special cases.[47]

The third approach to ring puckering makes use of torsional angles, and this has been applied extensively in five-membered rings. The reduction of the five torsional angles to two parameters has been done by approximate methods that depend slightly on the labeling of the ring.[29] The reduction in other ring sizes has been done using Cremer and Pople's Fourier analysis method applied either to the torsional angles themselves,[37] or to the sines of half the torsional angles.[34] These methods use $N-3$ parameters that are combinations of torsional angles, and just like the Cremer–Pople treatment, they are not well suited to the medium rings.

The term *inversion* ("reversal" is preferred by some workers) is used here to describe conformational processes where the torsional angles maintain their absolute values but change their signs.[26]

2.2.5. Conformational Nomenclature for Medium Rings

Conformational nomenclature is not needed in chemical commerce or law, and ignoring conformational isomers greatly simplifies the indexing of molecules. Actually, the naming of stereoisomers is not always given proper recognition even by *Chemical Abstracts*, and a generally acceptable and logical chemical nomenclature of molecules is elusive at best, one might say.[48] With conformations, special problems arise when the barrier separating two conformations becomes very low. Occasionally, the barrier can become so high that "conformational isomers" (often called atropisomers) can be separated at room temperature, as in sterically hindered biphenyls. Such isomers can be quite stable and have different melting points and other characteristic chemical properties, and are clearly different chemical compounds that require separate names. The nomenclature for atropisomers can be used for conformations in analogous molecules with lower barriers.[49]

Hendrickson has introduced a medium-ring conformational terminology that is based on the combination of the words "chair" and "boat" for plane-symmetrical conformations, as in boat–chair cyclooctane or boat–chair–boat cyclodecane.[50] Conformations without a plane of symmetry often have a C_2-axis and are generally closely connected by a pseudorotation step to a conformation that does have such a plane. The prefix "twist" can then be used, as in twist–boat–chair cyclooctane.

Difficulties with Hendrickson's scheme arise in a few cases, but Dale's nomenclature for local energy minima and conformational transition states,[35] which is best suited to large rings, or at least to the larger medium rings, can then be used. For example, the cyclononane D_3 conformation (**1**) is called [333], where the 3's are the number of bonds (not atoms) along a "side" of a conformation. Sides are joined together at corners, which are isoclinal carbons, that is, where the two adjacent torsional angles are both gauche and have the same

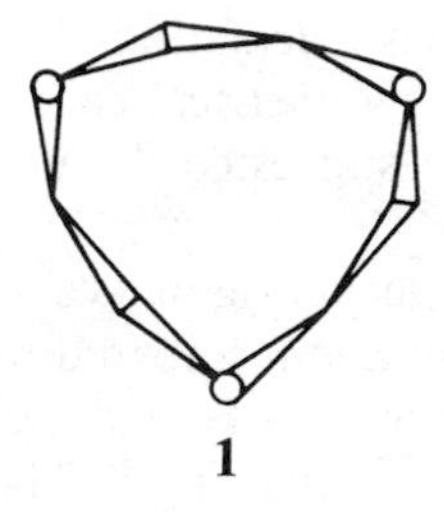

sign; torsional angles between two adjacent side atoms are theoretically 180°, but in medium rings these angles can be 100°, or even lower, and Dale occasionally has made use of one-bond sides that do not fit the above definition, but look like polygon sides on models.

Although the coordinates of the atoms, or the torsional angles, or the Pople–Cremer puckering parameters, can all be used to define the conformation of a ring, convenient and more or less self-evident names, such as those of Hendrickson and Dale, are useful for discussion and qualitative purpose. Special nomenclatures have been proposed for distorted six-membered rings[51] and for five- to seven-membered rings in carbohydrate chemistry.[52] Details of the method for seven-membered rings are given in section 2.6.

The last point in conformational nomenclature concerns the naming of the exocyclic bonds and substituents on a ring. In the chair conformation of six-membered rings, "axial" and "equatorial" are universally used, and if the six-membered ring is unsaturated, as in cyclohexene, the substituents on the 3 and 6 carbons are called pseudoaxial and pseudoequatorial. In other cases, and for other ring sizes, these conformational labels, as well as terms such as quasiaxial, have also been employed, even though they do not necessarily have the same meanings as in six-membered rings. The two substituents on a carbon atom that lies on an actual (or local) C_2-axis are equivalent (or nearly equivalent, respectively), and are called isoclinal.

Cremer[32] has proposed a scheme based on the Pople–Cremer mean plane for naming conformational substituent positions. Although the method is well defined and is consistent with previous usage in six-membered rings, it is not well suited for medium and large rings, and it is both complicated and is more geometrical than chemical in spirit. The important feature of an axial bond in cyclohexane is not that this bond is parallel to the principal (threefold) axis in the chair, but that the substituent on this bond is chemically different from an equatorial substituent, because of steric hindrance at the axial site. This steric hindrance occurs even when a single substituent is present on the ring, and is caused by a 1:3 interaction with the β atoms and their substituents. It is a local interaction and should not be related to a mean plane, which is affected even by distant atoms in the ring. A simpler and unambiguous naming method that is also consistent with previous usage is the following.[53] To determine whether a substituent is axial or equatorial, consider the system where the two substituent bond lengths are taken to be the same and measure the distances from the midpoint of the line joining the two β ring atoms to the two substituents,

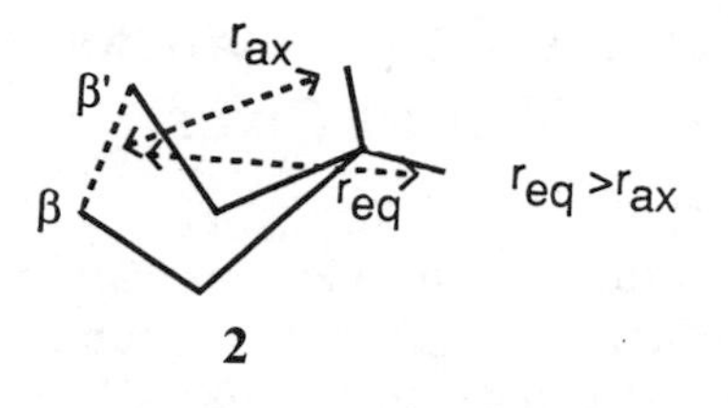

as shown in **2**. The substituent with the shorter distance is axial and the one with the longer distance is equatorial. If the distances are the same, or nearly the same, the substituents are in isoclinal positions. Pseudoaxial and pseudo-equatorial positions have smaller differences in the two distances than have regular axial and equatorial positions. All ring positions, independent of ring size, can be labeled in a chemically meaningful manner.

2.3. Conformational Effects of Strong Torsional Constraints

Rings can be classified on the basis of the number of strong torsional constraints, such as an endocyclic *cis* double bond.[18,35,54] Similar constraints occur upon fusion of a medium ring to an aromatic ring or to a three-membered ring. In the present context, a torsional constraint is considered strong if the barrier to rotation is about 10 kcal/mol or more. The O—CO bond in a lactone is treated here as a strong constraint, even though the barrier to rotation about this bond is probably less than the limit given above. Formally single bonds having a low double-bond character, as the single bond between two sp^2 hybridized carbons in a 1,3-diene or between an oxygen and a sp^2 carbon, have low barriers (<6 kcal/mol), and are not considered strong constraints—nor are, of course, any bonds to a saturated carbon.

The conformational similarity of a CH_2 to a *cis*-ethylenic group in a ring gives rise to an important qualitative principle that analogous conformations can be anticipated from the replacement of one of these groups by the other, as in the series cyclohexane, *cis*-cycloheptene, and the three *cis,cis*-cyclooctadienes. Similarly, a *trans* double bond when present in a ring is like an anti-$(CH_2)_4$ chain (*anti* butane conformation) in its conformational effect on the rest of the ring. For a monocyclic ring with N members, $N - 6$ strong torsional constraints of the *cis*-alkene type are required for a "rigid" form, as defined in section 2.2.3, to exist. The subsequent discussion of the conformational properties of medium rings is organized according to the number of such strong constraints present. *Trans* double bonds are quite common in 10- and higher-membered rings, but are strained in eight-membered rings and do not occur at all in seven-membered rings that are stable at room temperature. Twelve-membered rings can be fused in a meta fashion to aromatic rings, and it should be noted that this arrangement effectively contains a *trans* double bond in the large ring, as does also 1:4 bridged aromatic molecules ([n]-paracyclophanes).

2.4. Overview of the Conformations of Carbocyclic Medium Rings

The strain possessed by the medium-ring hydrocarbons is most clearly shown by heats of formation (Table 2-1).[55] The major contributions to the strain in medium rings arise from bond angle, torsional, and nonbonded effects. Dipole–dipole interactions, and, for unsaturated molecules, out-of-plane distortions, are also of importance. The presence of more than one heteroatom, whether in the ring or not, can lead to interactions, such as the anomeric effect, which are attributable in part to dipole–dipole interactions and in part to electronic effects that can be described in terms of either no-bond double-bond resonance structures or as n to σ^* delocalization.

The number of electrons in medium rings are sufficiently large that *ab initio* quantum mechanical calculations[56] of conformational properties require a great deal of computer time, but such calculations, done with extended basis sets, and even with the inclusion of some configuration interaction, are currently feasible, especially for the smaller medium rings, although only a few results have yet emerged.[14] Semiempirical quantum mechanical calculations (e.g., MINDO/3 or MNDO) require relatively little computer time but in general have insufficient accuracy for conformational purposes, because differences in energies between conformers are often small. Furthermore, such semiempirical calculations do not reproduce barriers to rotation in conjugated systems.[57]

Molecular mechanics calculations (either of static structures or of the dynamics of molecular motions) are based on empirical force fields, and are quite useful,[58,59] but have a confidence level that is sometimes difficult to judge. A variety of force-field parameters as well as force-field functions have been employed, although in recent years the MM2 parameters, which are available for a wide variety of groups, have become popular, especially in synthetic organic chemistry. The accuracy and area of applicability of such calculations should continue to increase in the future.

TABLE 2-1. Heats of Formation and Relative Strain Energies in Some Cycloalkanes

Ring Size (n)	ΔH_f° (Gas Phase),[a] kcal/mol	Relative Strain Energy,[b] kcal/mol
6	−29.5	0.0
7	−28.2	6.2
8	−29.7	9.6
9	−31.7	12.5
10	−36.3	12.9
11	−42.9	11.2
12	−55.0	4.0
	(−50.5)	(8.5)

[a] Data from Ref. 55.
[b] Relative to cyclohexane, i.e., $\Delta H_f^\circ(n) - \Delta H_f^\circ$ (cyclohexane)$(n/6)$.

Early quantum mechanical calculations tended to make use of a fixed geometry, but more recently the importance of optimizing the geometry has come to be appreciated, and, fortunately, efficient techniques for this purpose have appeared.[56] It is also important in any kind of calculations to know whether a given converged geometry corresponds to a (local) energy minimum, a conformational transition state (energy saddle point), or a higher-order energy extremum, with the last having no thermodynamic or kinetic significance. The best way to do this is to calculate the vibrational frequencies of the normal modes in the harmonic approximation.[60] Nonlinear molecules have $3N - 6$ real (fundamental) frequencies, whereas transition states have $3N - 7$ real frequencies and one imaginary frequency, and higher-order extrema have more than one imaginary frequency. Finally, the number of degrees of freedom in medium rings is not so high that there is any real difficulty in finding all the low-energy local minima, and hence the global minimum, or in finding all the significant conformational transition states. However, such problems can be difficult to solve in a general, efficient, and correct way.[61,62] Conformational calculations were discussed in detail in Chapter 1.

In the absence of reliable theoretical data on medium rings, the main evidence concerning structure comes from X-ray diffraction results on crystalline compounds; some data on molecules in the gas phase are available from electron diffraction, but for molecules as large as the medium rings, its inherently low resolution makes the analysis difficult to carry out and results are sometimes ambiguous, especially when more than one conformation is present. Infrared (IR) spectra, particularly far-IR spectra, can give evidence of the presence or absence of low-energy pseudorotation modes, and are most useful for comparisons of solution- and crystal-phase spectra. Solution- or liquid-phase IR spectra occasionally show extra lines compared with those present in the crystal spectra, and this is a good indication of the presence in solution or the liquid of conformation(s) not found in the crystal.

The NMR spectra, except for those of molecules dissolved in a liquid-crystal phase,[63] do not provide accurate structural data, although valuable information on symmetry (actually, the lack of certain symmetry elements), as well as the presence of more than one kind of conformation, can often be obtained. Coupling constants in some molecules can provide much information on torsional angles.[52,64,65] Some information on the presence of gauche torsional angles in C—C—C—C or related heteroatom moieties can be obtained from ^{13}C chemical shifts (shielding γ effect).[66] Modern two-dimensional NMR (especially COSY and NOESY and their numerous variants) has proved useful in determining the conformations of carbohydrates, polypeptides, polynucleotides, and other natural products,[65] but it has not been applied to any extent in medium-ring conformational problems, except for some cyclic peptides. COSY is useful for assigning resonances in complex but nearly first-order spectra. Measurements of NOE effects, which can be done in either one- or two-dimensional NMR, give approximate relative distances, but the interpretation of the results is not necessarily simple or very reliable, especially if more than one conformation is present, as can be the case in medium rings.

NOESY spectra provide not only NOE data but also site-exchange data, and these two effects have to be carefully separated when both occur together. Applications of ^{17}O NMR to heterocycles have been limited, but offer considerable promise, and very high magnetic fields are clearly advantageous.[67]

Dynamic NMR is a very powerful tool for obtaining conformational information on molecules where barriers are in the range of about 5 to 25 kcal/mol; extensive reviews on this subject exist,[8,26,49,68] and so only a few brief comments will be made here. For small barriers, NMR spectra need to be measured at temperatures as low as $-170°C$, whereas for large barriers, temperatures greater than 150°C are necessary. Intermediate barriers (11–16 kcal/mol) can result in line broadening at room temperature, especially in high-field spectrometers. Conformationally degenerate systems, such as cyclohexane, are the easiest to study, since the thermodynamic parameters are determined by symmetry and are independent of temperature. In other cases, values of $\Delta G°$, $\Delta H°$, and $\Delta S°$ have to be obtained by peak intensity measurements. Line-shape analysis based on an appropriate kinetic scheme can then be used to obtain the corresponding activation parameters ($\Delta G^‡$, $\Delta H^‡$, and $\Delta S^‡$), but accurate rate constants, preferably obtained over a wide temperature range, must be available, and careful attention is needed to avoid systematic errors—which, unfortunately, do not show up in statistical error analysis. The free-energy parameters (ΔG's) are relatively easy to obtain, because an equilibrium constant or rate constant measurement at a single temperature is sufficient, and approximate methods, such as obtaining the rate constant from a coalescence temperature, suffice. The enthalpies and entropies, by contrast, are very prone to errors, since their determinations effectively require extrapolations of the data to absolute zero. Strongly one-sided equilibria can be difficult to study since the dynamic NMR effect can then give rise to only a very small broadening and to very weak peaks; in the limit where only one conformation is populated (i.e., an anancomeric system), no thermodynamic or kinetic information can be obtained by NMR.

Chemical reactions, such as transannular atom transfers,[7] depend in general on the conformational energy surfaces of both the ground and transition states, and have to be interpreted with care, especially if the chemical and the relevant conformational rates are expected to be comparable in magnitude.

2.4.1. Seven-membered Carbocycles

Dreiding models, which are so useful in picturing the conformations of cyclohexane, are equally valuable for cycloheptane. Because of the extra ring bond, cycloheptane has one more degree of torsional freedom as compared with its lower homolog, and therefore no rigid conformation of cycloheptane can exist, but, curiously, there are still two conformational families that cannot be interconverted without increasing bond angles, just as in cyclohexane. One family consists of the chair (**3**), the twist–chair (**4**), and all the intermediate forms between these; the other family can be similarly described in terms of a boat (**5**) and twist–boat (**6**). According to molecular mechanics calculations, there is

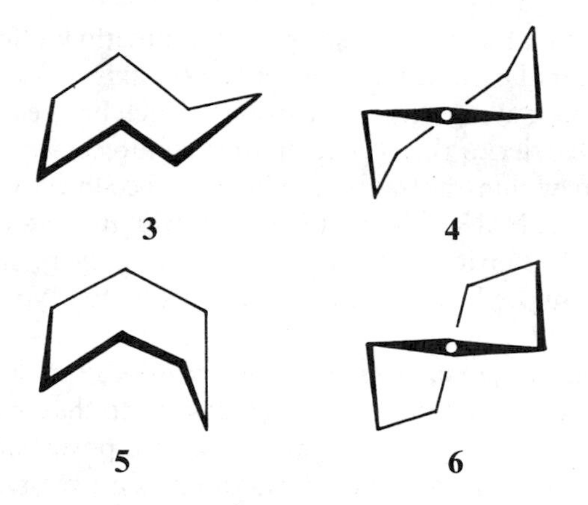

a substantial barrier (about 8 kcal/mol) preventing the interconversion of conformations belonging to the two families,[43,50,58] but very low barriers to pseudorotation within either family.

The twist forms of cycloheptane have C_2-point group symmetry and are energy minima, whereas the chair and boat have C_s symmetry and are transition states for the slightly hindered pseudorotations that occur in both families. The pseudorotation process can be described simply by starting from a plane symmetrical form. This conformation has one bond whose torsional angle is 0°, and this bond is bisected by the plane of symmetry (σ). The bonds on each side of this plane exist in pairs and have opposite signs of torsional angles. The twist conformation, on the other hand, has a C_2-axis passing through one ring atom and bisecting the opposite bond. Torsional angles related by 180° rotation about the C_2-axis must, of course, have the same sign. The essential change in a pseudorotation step from the C_s to the C_2 form involves making the 0° torsional angle of the C_s form either positive or negative; other torsional angles do change somewhat, but their signs do not change, which is the important fact. Each C_s form is directly linked to two chiral and enantiomeric C_2 forms. Conversely, each twist form is linked to two C_s forms that have different atoms at a particular conformational position. In both the chair and boat families, $14C_2$ and $14C_s$ forms exist, and the energy surface for their interconversions can be represented on a torus, which is effectively equivalent to the surface of a four-dimensional sphere.

In the chair family, the directly linked pseudorotation partners have their symmetry feature (plane or twofold axis) shifted by $\frac{1}{14}$ revolution, whereas in the boat family the shift is $\frac{1}{7}$ revolution. Another difference is that the change in torsional angles during a pseudorotation step is much less in the boat than in the chair, and it is therefore not surprising that the energy difference between the boat and twist–boat is calculated to be almost zero; i.e., the pseudorotation is nearly free. By comparison, the pseudorotation in the twist–chair is estimated to have a barrier of 1.5 kcal/mol. All the evidence is consistent with cycloheptane existing very largely in the twist–chair, and the experimental evidence for an appreciable population of boat-family conformations is inconclusive at best.

Two sizable geminal substituents, as in 1,1-dimethylcycloheptane, can be placed on the isoclinal carbon that lies on the C_2-axis in the twist–chair with little steric repulsion, but not on other carbons, which have equatorial and axial substituents. The barrier to ring inversion in such molecules, which takes place by pseudorotation within the same family, then becomes high enough to be measured by dynamic NMR.[26] Substituted cycloheptanes, especially when fused to other rings, can take up a variety of geometries, and the relationships between the ring torsional angles have been used to describe the type of conformation present.[69,70]

When one strong torsional constraint is introduced in a seven-membered ring, the conformational picture becomes analogous to that in cyclohexane, and therefore cycloheptene and analogous molecules can potentially exist in a rigid chair (**7**) or in flexible boat-family conformations, with the boat (**8**) and twist–boat (**9**) as extremes. Cycloheptene, cycloheptene oxide, and most benzocycloheptenes[71,72] exist in the chair conformation only, but 1,1,4,4-tetramethyl-6,7-benzocycloheptene exists in both chair and boat forms.[73] Twist–boat forms are found in sulfur-containing heterocyclic analogs,[74] and as will be discussed in section 2.6.2.2, are common in medium-ring cyclic acetals and ketals. Conformational barriers vary from about 6 kcal/mol in cycloheptene, to about 11 kcal/mol in benzocycloheptene; the increase in barriers caused by fusion to an aromatic ring is general and has been ascribed to differences in the torsional potential of the allyl single bond in propene (barrier = 2 kcal/mol) and the benzyl single bond in toluene (barrier = about 0.01 kcal/mol).[58] Ring inversion in the cycloheptene chair (**7**), which has much torsional strain in its allylic single bonds, involves boat forms as intermediates, and occurs by a flattening of C-6, C-7, C-1, C-2, C-3, and C-4 to give a geometry with essentially no allylic torsional strain, although bond angle strains are rather large. An alternate path that looks more favorable in molecular models consists of flattening C-3, C-4, C-5, C-6, and C-7.

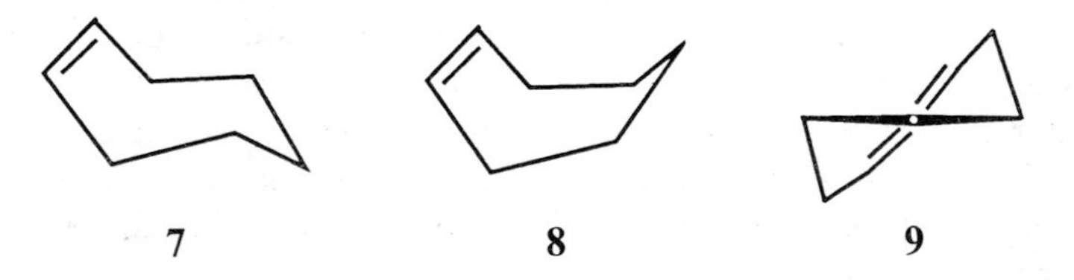

7 **8** **9**

The cycloheptadienes are rather flexible and only slightly puckered,[75,76] as expected from the conformational analogy with cyclopentane, but fusion to benzene rings gives biphenyl derivatives with relatively high barriers. In 1,3-cycloheptadiene, conjugation makes the C_s be of lower energy than the C_2 form, despite the lower angle strain of the latter. Cycloheptatriene has a nonplanar ring (**10**), and, being conformationally analogous to cyclobutane, shows a simple ring inversion process, but the barrier (about 6 kcal/mol) is much higher than in a four-membered ring.[58,75] Fusion with aromatic rings raises the barrier, and the tribenzo derivative of **10** has a $\Delta G^{\ddagger}$ of 23.8 kcal/mol because of interference between the ortho hydrogens in the transition state.[77]

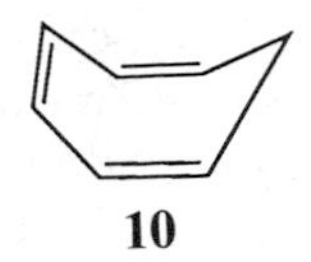

2.4.2. Eight-membered Carbocycles

Eight-membered rings without torsional constraints are geometrically flexible, as molecular models readily show. Dreiding molecular models are not as useful as with cyclohexane and cycloheptane, because the carbons on opposite sides of the model of an eight-membered ring can easily come much too close to one another. Molecular mechanics calculations show that there are three families of rapidly pseudorotating conformations, with each family separated from the other two by substantial barriers, of the order of 8–10 kcal/mol.[78] In the case of the parent hydrocarbon cyclooctane, the boat–chair (**11**, about 95%) and a crown-family conformation, most likely the twist–chair–chair (**12**, about 5%), are populated at room temperature. Substituents can change the proportions between these two types, but the third family (boat–boat, **13**) is virtually unpopulated in all cases.

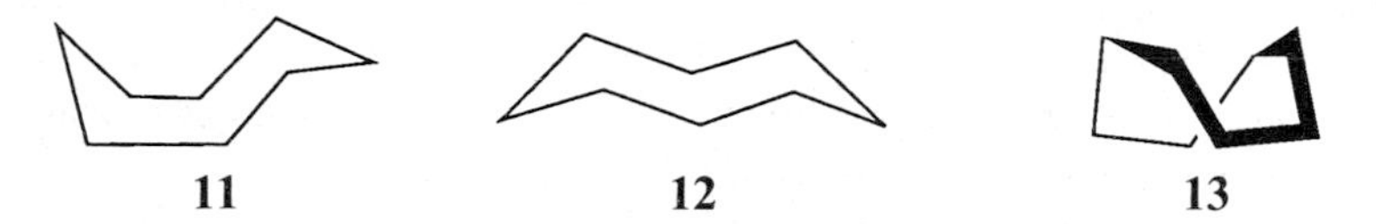

The boat–chair is characterized by a fairly easy pseudorotation, which is unique among all the cycloalkanes in that it does not cause all the protons to have one chemical shift when the pseudorotation process is fast; rather, two chemical shifts result, and a second, higher-energy process (ring inversion) is required to produce a single, averaged chemical shift. The intermediate in the pseudorotation of the boat–chair (C_s symmetry) is the twist–boat–chair, which has C_2 symmetry, and the transition state lacks symmetry. This C_s–C_2 dichotomy also occurs in the pseudorotation of cyclopentane and in both the chair and boat forms of cycloheptane. However, in the cyclooctane twist–boat–chair, the twofold axis does not pass through atoms as it does in odd-membered rings. The transition state in the ring inversion of the cyclooctane boat–chair is the chair (not to be confused with the crown-family chair–chair) or twist–chair. Large geminal substituents can increase the pseudorotation barrier substantially in the boat–chair.[26]

The introduction of one torsional constraint in an eight-membered ring, as in *cis*-cyclooctene, leaves the molecule flexible in all its conformations. *Cis*-cyclooctene has an unsymmetrical form corresponding roughly to the transition state for the boat–chair to twist–boat–chair pseudorotation in cyclooctane, and it shows two conformational processes by dynamic NMR.[79] In the case of cyclooctene oxide, the oxygen is forced to occupy the exo rather than the sterically crowded endo position, and only one process (a ring inversion) is observable.[80] The conformation of *trans*-cyclooctene is related to that of the chair–chair in cyclooctane, but there is much distortion of the double bond caused by the strain of placing a *trans* double bond in an eight-membered ring.[60] This compound is chiral and can be resolved because interconversion of the enantiomers is very slow, even at 100°C.

When two torsional constraints are present in an eight-membered ring, the situation becomes analogous to that in cyclohexane, and both geometrically rigid and flexible forms occur. In the case of *cis,cis*-1,5-cyclooctadienes, the rigid conformation is the chair, and the flexible one consists of the twist–boat (C_2 symmetry), which can undergo two different kinds of pseudorotations, with transition states of C_{2v} and D_2 symmetries. Only the twist–boat is populated in *cis,cis*-1,5-cyclooctadiene itself, but in related systems, such as the diepoxide or the dibenzo derivative, both the chair and the twist–boat are found. Two conformational processes have been measured in 1,5-cyclooctadiene, but molecular mechanics calculations have given conflicting results concerning the identity of the transition states.[26,81,82] With other isomers, namely, the *cis*, *cis*-1,3 and *cis,cis*-1,4 dienes[83] and their derivatives, two conformers generally occur, and Dreiding molecular models show that one form is rigid whereas the other is flexible. In the actual molecules, however, the distinction between rigid and flexible is not sharp, as shown by both dynamic NMR and molecular mechanics calculations. Even so, the barriers in the flexible forms are invariably lower than those in the rigid forms. The 1,3 diene, when fused to benzene rings, becomes a biphenyl derivative and exhibits high conformational barriers and atropisomers can be isolated at room temperature.

The fusion of an eight-membered ring to the peri (1,8) position of naphthalene results in two torsional constraints, but of an unusual kind, where the two constrained bonds are adjacent to one another (cyclic allenes also have two adjacent torsional constraints, but with perpendicular rather than planar geometries). In the peri naphthalene case, rigid-chair and flexible-boat forms can exist, with the boat, which has staggered bonds (cf boat–boat cyclooctane), being the more stable.

Eight-membered rings with three torsional constraints can be of the 1,3,5 or 1,3,6 triene type.[84] Molecular models indicate that (all-*cis*) 1,3,5-cyclooctatriene (**14**) should exist in a single boat conformation of C_s symmetry, but this is deceptive, for the same reasons that such models show that cyclobutane should be planar. The CH_2–CH_2 group in **14** is eclipsed in the symmetrical form, which is actually a transition state for the interconversion of two unsymmetrical twist–boats. Molecular mechanics calculations show that the torsional angle in the above group is about 60°, and the distortion in the twist–boat not only relieves the eclipsing strain, but also increases the conjugation between

14

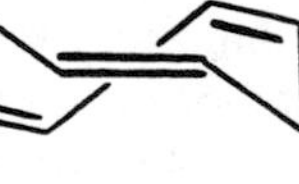

15

the double bonds. The twist–boat can also undergo a ring inversion. In the 1,3,6 triene, only a single conformation (**15**) exists.

With four *cis* double bonds, only a single symmetrical (D_{2d}) form occurs, and a ring inversion, as well as a bond shift, is possible. Since there are no known heterocyclic analogues of cyclooctatetraene with a ring oxygen (the nitrogen systems, however, are known), the conformational properties of this system will not be further discussed. However, there is an oxygen analogue of the 10-π electron aromatic cyclooctatetraene dianion, and both of these structures are planar.[85]

2.4.3. Nine-membered Carbocycles

Cyclononane is known to exist in three different conformations whose relative populations are temperature dependent.[86] At −170°C, the highly symmetrical (D_3) [333] form (**16**) is present to the extent of over 90%, whereas the less symmetrical and higher entropy (C_2) [225] (also called [22221]) form (**17**) is probably the major one at room temperature. A percent or two of the twist–chair–chair ([144] or [9]) form (**18**) is also present. Cyclononylamine hydrobromide and the mercuric chloride complex of cyclononanone in the crystalline states have the [225] form, but the conformational equilibria for these molecules in solution are not known. Still another form, [234], (**19**) is found in 4,4,7,7-tetramethylcyclononanone.[87,88]

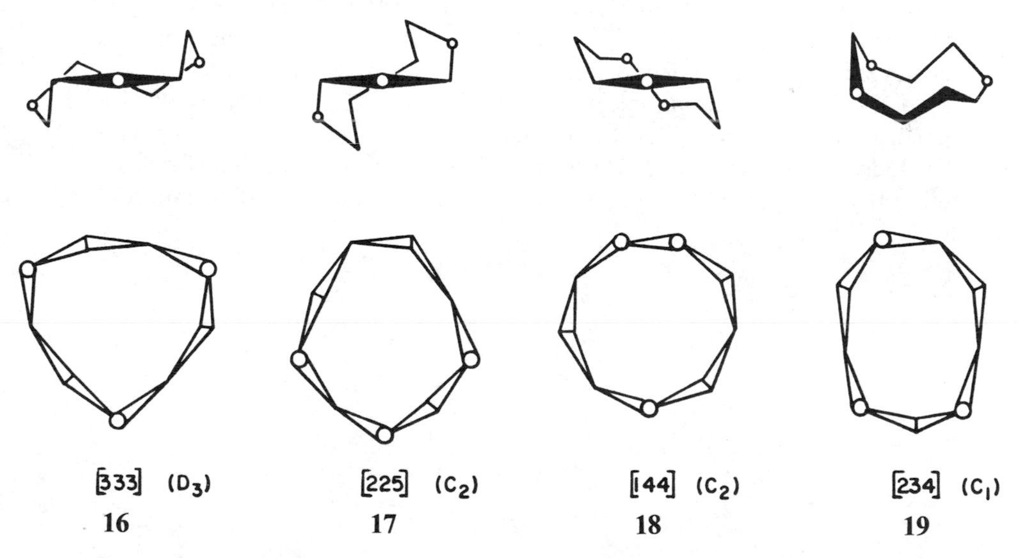

16 **17** **18** **19**

Cis- and *trans*-cyclononene have not been extensively investigated, except by molecular mechanics calculations, although it is known that the *trans* isomer can be resolved but racemizes at a measurable rate at 0°C.[89] Nine-membered rings with two torsional constraints include the allene, 1,2-cyclononadiene,[90] and the unconjugated diene, 1,4-cyclononadiene.[91] The conformations of these molecules are known from dynamic NMR and molecular mechanics calculations, and an X-ray structure of a derivative of the allene is consistent with these results.[92]

The symmetrical *cis,cis,cis*-1,4,7-cyclononatriene and its benzo derivatives have received much attention, in part because these molecules can be considered trishomobenzenes. Because of the presence of three torsional constraints, these molecules have rigid and flexible conformations, and are indeed conformationally analogous to cyclohexane. The parent triene and its substituted tribenzo derivative have the rigid "crown" form (C_{3v} point group) but the ketone in the tribenzo series has the flexible "saddle" form. The barrier to ring inversion in the parent triene is about 15 kcal/mol and is not well reproduced by molecular mechanics calculations, at least with the particular force field used.[93] The tri(dimethoxybenzo) derivative (cyclotriveratrylene) has such a high barrier to ring inversion that the replacement of a CH_2 by a CHOH group leads to crown and flexible forms that do not interconvert at room temperature and so are atropisomers.[94] Cyclotriveratrylene-d_9 (C_3 symmetry) is chiral and racemizes with a half-life of about one month at 20°C ($\Delta G^{\ddagger} = 26.5$ kcal/mol).[95]

2.4.4. Ten-membered Carbocycles

Cyclodecane can have a number of conformations, but the parent molecule seems to have the [2323] (or boat–chair–boat) conformation (**20**) predominantly at low temperatures,[96] although molecular mechanics calculations indicate only about 50% of this form at room temperature,[97] with about 35% of a twist–boat–chair and less than 10% of the twist–boat–chair–chair and boat–chair–chair. Substituted cyclodecanes have been shown by X-ray diffraction to have the [2323] conformation, unless this forces a substituent into a crowded axial position, in which case the [2233] conformation becomes of lower energy. Cyclodecanone has the [2323] form,[98] but 1,6-cyclodecanedione exists as a mixture of the [2323] and an "open *trans*-decalin" kind of conformation.[99,100]

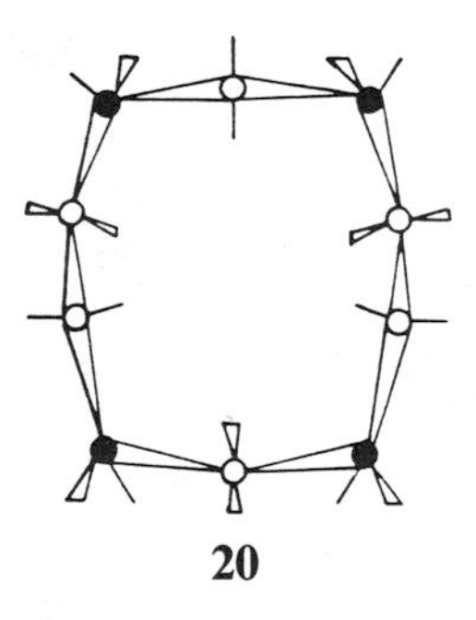

20

Both *cis*- and *trans*-cyclodecene have been investigated by dynamic NMR, using either ^{1}H or ^{19}F (on fluorinated derivatives). More highly unsaturated 10-membered carbocyclic rings have also been studied, the most important being the symmetrical *cis,cis*-1,6-cyclodecadiene derivatives[101] and the bis allene, 1,2,6,7-cyclodecatetraene,[90] and numerous unsaturated 10-membered ring sesquiterpenes such as the germacranolides.[102,103]

2.4.5. Eleven- and 12-membered Carbocycles

Eleven-membered rings are relatively rare, but there are some data on the lowest-energy form in cycloundecane from molecular mechanics calculations and limited NMR evidence.[104] Cycloundecanone has the same type of conformation as postulated for the hydrocarbon, as shown by an X-ray structure.[105]

Twelve-membered rings are much less strained than the other medium rings. Cyclododecane has a slightly distorted "square" [3333] form (nearly D_{4d}, **21**), as shown by X-ray and NMR data.[106,107] This conformation does not fit on the diamond lattice, although it can be considered to be derived from a diamond–lattice helix by flattening it to allow ring formation. Cyclododecanone in the crystalline state has the carbonyl group in a noncorner position of the [3333] form.[108] Interestingly, tetrathia-12-crown-4 also has a square conformation, but the sulfur atoms occupy all four corner positions.[109]

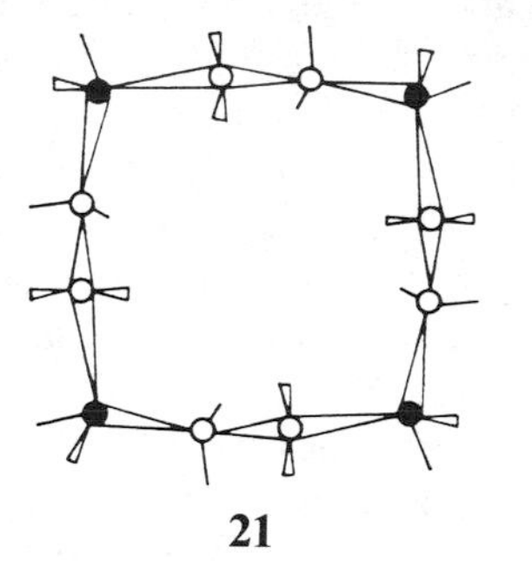

21

2.5. Conformational Consequences of Oxygens in Heterocycles

Heterocyclic molecules containing ring oxygens can be regarded as being derived from their parent carbocycles by the interchange of methylene groups with oxygen atoms and this can cause a variety of conformational effects.[110] The structural and conformational changes produced by the introduction of a single oxygen atom are illustrated by data on propane (**22**) and dimethyl ether (**23**).[111–113]

The C—O bond in **23** is shorter than the C—C bond in **22** and the bond angle, $\theta_{C—O—C}$, is slightly smaller than $\theta_{C—C—C}$. The difference of only 0.7° between these θ values hides an important fact about the ease of distortion through angle opening. The bond angle in H_2O is only 104.5° in contrast to the tetrahedral angle (109.5°) in CH_4. Molecular mechanics calculations—MM2,[58,114] for example—use a reference "unstrained" angle that is smaller for oxygen than for carbon (106° and 109.5°, respectively). In addition, the force constant for bending is chosen to be about 70% larger for oxygen than for carbon. Nonbonded repulsions are therefore less easily relieved by angle opening in heterocyclic than in carbocyclic molecules.

CH_2 / CH_3 CH_3

22

r (C-C) = 1.53 Å

θ (C-C-C) = 112°

r (C-H) = 1.096 Å

Torsional barrier = 3.4 kcal/mol

Dipole moment = 0.08 D

O / CH_3 CH_3

23

r (C-O) = 1.41 Å

θ (C-O-C) = 111.7°

r (C-H) = 1.095 (av.) Å

Torsional barrier = 2.7 kcal/mol

Dipole moment = 1.31 D

The torsional barrier in **23** is slightly smaller than in **22**, but a more significant comparison involves the anti and gauche conformers of methyl ethyl ether and butane. Although the more stable forms in both molecules have anti geometries, the anti-gauche energy difference is much larger in the ether than in the hydrocarbon (1.5 versus 0.7 kcal/mol), and this has important repercussions in both acyclic and cyclic molecules. To reproduce this difference, the torsional potential for C—C—O—C in MM2 is chosen to have a large twofold component.[114] This repulsive gauche effect does not occur in the corresponding sulfur analogue and should not be confused with the attractive gauche effect that occurs in the closely related O—C—C—O structure (see below).

The CH_3 of dimethyl ether does not have local C_3 symmetry because of the perturbation of the C—H bonds that are anti to lone pairs, which for the present purpose can be considered to reside in approximately sp^3 (n) orbitals. (This kind of dissection tends to be arbitrary, and the overall electron density hardly reflects the presence of tetrahedrally localized lone pairs; an alternative picture is to treat the system as consisting of a lower-energy σ and a higher-energy π orbital, where the latter is the more important orbital.[115]) The interaction can be described qualitatively in terms of a contributing no-bond double-bond resonance structure (**24**), or in molecular orbital parlance, as an $n - \sigma^*$ or $n(\pi) - \sigma^*$ overlap interaction (**25**). As a result, the force constant for stretching is lowered when the C—H bond is antiparallel to a lone pair,[116] and the affected C—H bond is lengthened by 0.009 Å in **23**.[112]

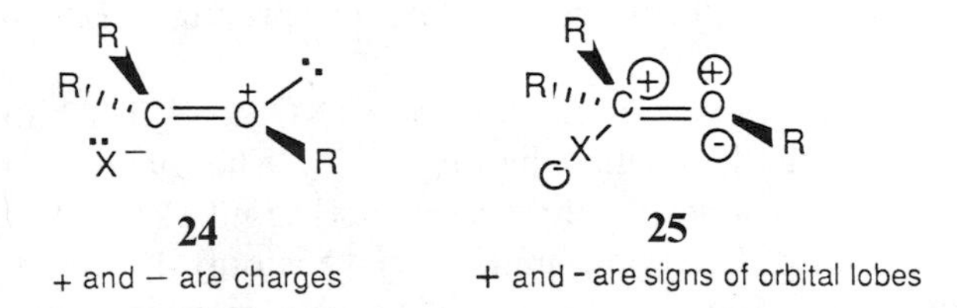

24 + and − are charges

25 + and - are signs of orbital lobes

The interaction described above is in part responsible for the well-known anomeric effect in carbohydrate chemistry, where, instead of being a hydrogen atom, it is an electronegative group (e.g., oxygen or halogen) that is antiparallel to the lone pair.[117,118] In the normal anomeric effect involving electronegative substituents, such as oxygen, there is also a more or less strong dipole–dipole interaction, so that both contributions (dipole–dipole and electron delocalization) can be important. The anomeric effect is responsible for the axial preference of oxygen and halogen substituents at the 1 position of sugars, and also for the ± gauche ± gauche conformer of dimethoxymethane (**26**) being highly favored over the anti-anti (**27**) or the anti-gauche (**28**) forms.[119,120] The same effect also leads to an unsymmetrical conformation for chloromethyl methyl ether, and the diastereotopic nature of the methylene protons in the NMR spectrum of that molecule is evident below about −170°C. The barrier ($\Delta G^{\#}$) to internal rotation, which is 4.2 kcal/mol, is higher than would be expected in the absence of an anomeric effect.[121]

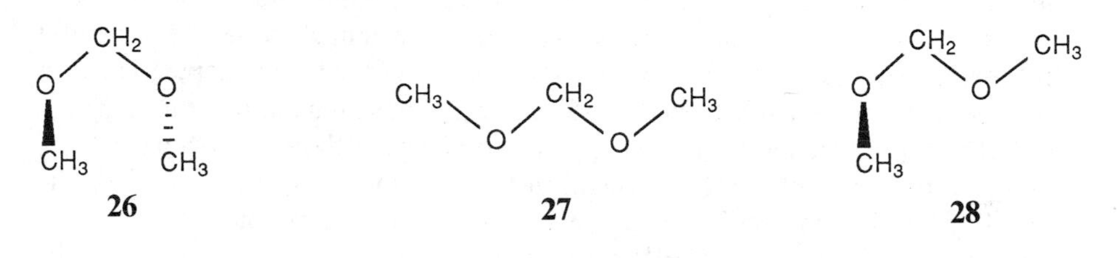

Tetrahydropyran, with only a single oxygen atom, closely resembles cyclohexane in its conformational behavior, except that electronegative substituents in the α position of tetrahydropyran have a more or less strong preference for the axial orientation because of the anomeric effect, in contrast to cyclohexane, where such substituents prefer the equatorial position. It is significant that the C—O—C angles of the ring oxygens in pyranose sugars have normal values when the electronegative 1-substituents are equatorial, but are several degrees larger when the substituents are axial, where electron delocalization is operative, causing the ring oxygens to acquire some sp^2 character.[122] The C—O bond lengths in acetals and ketals are variable because of the dependence of electron delocalization on the relative orientation of these groups. Kirby and co-workers have analyzed a large number of molecules in the Cambridge Crystallographic Data Base, and have rationalized much of the geometric data and correlated the data with chemical reactivity.[122–124] One general conclusion is that if one C—O bond in an acetal is short, then the other is long, in agreement with theoretical expectations of resonance and electron delocalization. Molecular mechanics calculations of acetals and ketals have to incorporate parameters to deal with this situation, if experimental bond lengths are to be well reproduced.[125] The presence of substituents on the central carbon of a C—O—C—O—C fragment can reduce the preference for the ± gauche ± gauche conformation.[120,126,127]

Cyclic acetals in six-membered rings, such as 1,3-dioxane (**29**), cannot adopt the preferred ± gauche ± gauche conformation. They exist as chairs and the twist–boats have very high energies (about 8 kcal/mol above the chairs).[128] The

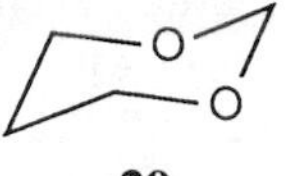

29

geometry of the ether groups in the chair results in a large dipole moment for 1,3-dioxane, in contrast to dimethoxymethane and some, but not all, of the larger ring cyclic acetals. Although 1,3-dioxane has an unfavorable alignment of the C—O—C dipoles, it does have an ideal geometry for $n - \pi^*$ electron delocalization.

Dramatic conformational differences can exist between a carbocyclic molecule and its heterocyclic analogue and a comparison of the conformational equilibria in certain substituted cyclohexanes and their corresponding 1,3-dioxane analogues offers an instructive illustration of general validity. The substituted cyclohexane (**30**) has the expected conformation with the larger *tert*-butyl group equatorial. (The free-energy difference between the axial and equatorial conformer in a monosubstituted cyclohexane is the Winstein *A* value; the *tert*-butyl and methyl groups have *A* values of about 5 and 1.7 kcal/mol, respectively.) In contrast, the similarly substituted 1,3-dioxane (**31**) has an axial *tert*-butyl group, so that in this molecule the methyl appears to be larger than a *tert*-butyl group. Two factors contribute to this reversal in the position of equilibrium: first, the axial *tert*-butyl group in **31** has a 1:3 interaction with lone pairs of electrons rather than with the larger C—H groups in **30**; second, the axial methyl in the alternate ring inverted form of **31** has unusually large nonbonded 1,3 repulsions with the CH groups across the oxygen atoms, because of the reduced C—O bond length and the stiffness of C—O—C bond angles in ethers. The conformational preferences for the equatorial positions in 5-*tert*-butyl-1,3-dioxane and in 2-methyl-1,3-dioxane are about 1.6 and 4.0 kcal/mol, respectively, and are almost the reverse of the corresponding *A* values.[8]

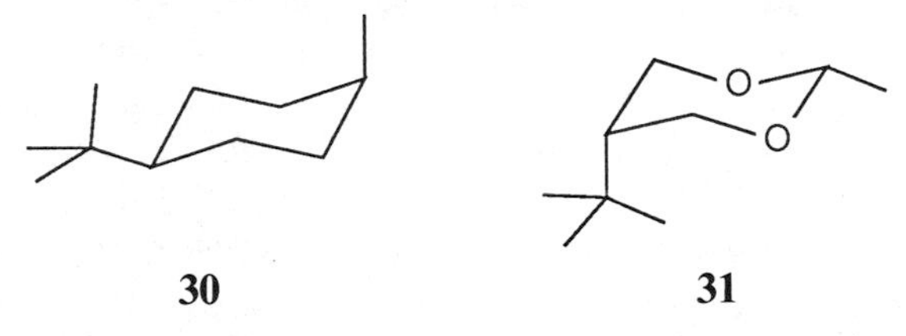

30 **31**

This example brings up the question of the parameters to be used in molecular mechanics calculations for the nonbonded (van der Waals) terms in molecules containing oxygen, as well as how to describe qualitatively the size of the oxygen atom. Some workers simply ignore the electron pairs on oxygen, whereas other workers treat electron pairs as pseudoatoms that have their own force-field parameters.[58] In any case, the oxygen atom and its associated electron pairs together are smaller than a CH_2 group when approached by an external atom, as can be seen from the smaller *A* value for methoxy than for methyl (0.6 versus 1.7 kcal/mol). On the other hand, electrons take up more space on a given atom when in the form of lone pairs than when involved in bonding protons, as is shown by the geometry of H_2O or NH_3. This is in agreement with the lone-pair orbital having much *s* character and therefore being wide but not extending far from the atom on which it resides, in contrast to the common and convenient long-lobe and "rabbit ear" depiction of the electron pairs on oxygens.

The fragment O—C—C—O, with two oxygens in 1,4 positions, is important in both acyclic molecules, such as ethylene glycol and its oligomers, and in cyclic systems such as 1,4-dioxanes and crown ethers, but it has not been much studied in medium rings, except when the two carbons are part of an aromatic moiety. The gas-phase enthalpy of 1,4-dioxane shows that it is less stable than the 1,3 isomer by about 5 kcal/mol.[129] When the group O—C—C—O is in an acyclic unit, the gauche is more stable than the anti conformer (the gauche effect[130]), although the difference in dipole moments between these geometries means that the anti-gauche free-energy difference is strongly affected by the solvent polarity. The gauche effect also comes into play when electronegative substituents are present on a heterocyclic ring.[110,131] This behavior is opposite to that in the C—C—O—C moiety, when the gauche form is strongly disfavored, as discussed previously. Also, the gauche sulfur system, S—C—C—S, is destabilized rather than stabilized.[109]

Molecules containing directly attached oxygens are peroxides and have unusual torsional barriers. In hydrogen peroxide, the energy minimum corresponds to a torsional angle of about 110°, but the barrier at 180° is very low (about 1 kcal/mol), whereas the 0° barrier is quite large (about 8 kcal/mol).[132,133] Since medium-ring conformations often have torsional angles in the range of 90 to 110°, it is not surprising that eight- and nine-membered-ring peroxides appear to be easily formed.[134,135]

Ring inversion barriers in saturated six-membered rings in the chair conformation are remarkably similar: Cyclohexane, tetrahydropyran, and 1,3- and 1,4-dioxanes all have barriers to inversion of about 10 kcal/mol.[8] This is probably the result of compensating effects, which may not be operative in other systems, such as the medium rings. Six-membered rings containing peroxide groups in general have inversion barriers that are a few kilocalories higher than the barrier in cyclohexane.[8]

The presence of unsaturation next to the oxygen atom in a heterocycle can greatly modify the conformational picture presented above, and the most important example is the O—C=O group in lactones, which are cyclic esters. Because of resonance stabilization, esters and lactones exist in syn and anti forms. The more stable form in esters is syn and the syn-anti energy difference in methyl acetate is at least 8 kcal/mol in the gas phase, whereas in methyl formate it is 4–5 kcal/mol in the gas phase[136–138] and only about 2 kcal/mol in solution.[139] (The syn-anti convention used is unfortunate and confusing, because the syn form of a lactone is analogous to a *trans* cycloalkene.) The barriers to rotation in formate esters are relatively small (8–10 kcal/mol) compared with those in amides (about 18–20 kcal/mol) or alkenes (about 50 kcal/mol). In rings of seven or fewer members, the lactone ring is forced to have the unfavorable anti conformation (**32**), but the larger ring lactones can take up the syn geometry (**33**). The six-membered ring δ-valerolactone is an analogue of cyclohexene, which exists in a half-chair conformation, with the boat a transition state for ring inversion. In the lactone, however, both the half-chair and boat (or twist–boat) seem to coexist, with the half-chair preferred by 0.5 kcal/mol, as shown by experimental data and molecular mechanics calculations.[140] A

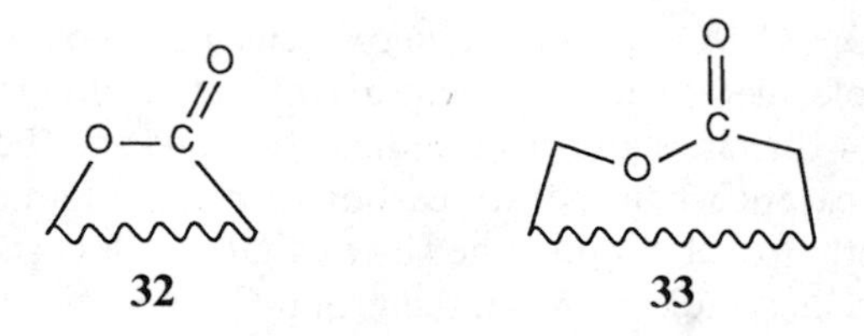

survey of X-ray data of molecules with δ-lactones shows that many have nonplanar lactone groups and that half-chairs, boats, and twist–boats occur.[141] The bond angles at the alkyl oxygen in lactones are significantly larger than in ethers, as expected from the partial double–bond character in the C—O bond of the R—O—C=O group.

Vinyl ethers are more stable thermodynamically than allyl ethers, because of the overlap of the double–bond π^* orbital with the oxygen lone pairs (n). In resonance terms, there is a contribution from the ionic structure, **34**, which places a partial negative charge on the β carbon of the molecule. The rotational barriers appear to be quite small and are not well known, but electron diffraction results on methyl vinyl ether show the presence of both a planar *s-cis* and a nonplanar (gauche) conformation, with the former preferred.[142] In dihydropyran, which exists in the half-chair form, as does cyclohexene, there is some evidence for a high-energy boat conformer as a local energy minimum, in contrast to the boat in cyclohexene, where this form seems to be a transition state for inversion of the half-chair.[143]

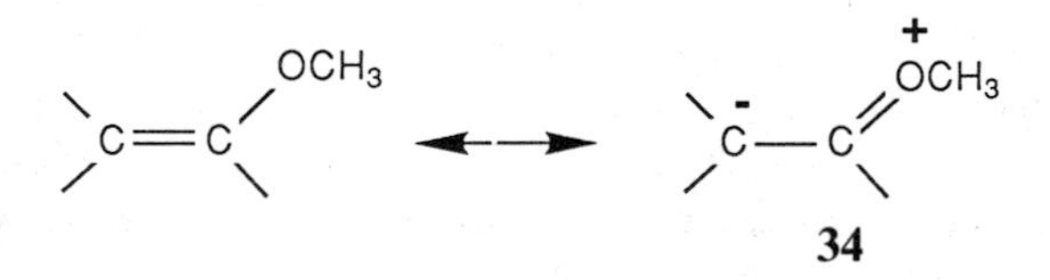

The methyl carbons in methyl ethers of phenols tend to lie in the plane of the benzene rings as long as ortho substituents are absent. The barrier to aryl-OCH_3 rotation is only about 4 kcal/mol in the gas phase, but may be as high as 6 kcal/mol in solution,[144] although *ab initio* calculations give low barriers (1 kcal/mol) and a complex potential energy surface.[145] The C—O—C bond angles in these compounds are 117–118° when the methoxy group is coplanar with the aromatic ring, but only 115° when the system is very nonplanar. The intensities of $\pi \rightarrow \pi^*$ transitions in the ultraviolet spectra of aryl ethers are sensitive to deviations of the chromophore from planarity, and this can be used to provide conformational information in heterocycles.[146]

2.6. Seven-membered Oxygen Heterocyclic Rings

2.6.1. *Rings with No Strong Torsional Constraints*

Strauss and co-workers have investigated the conformations of cycloheptane, oxepane, and 1,3-dioxepane by vibrational (Raman and IR) spectroscopy, with special emphasis on low-frequency torsional modes (10–400 cm^{-1}).[41–43,147]

They have analyzed the out-of-plane bending modes in terms of a force field specially designed for these seven-membered rings and thereby have obtained potential energy surfaces covering not only pseudorotation in the chair and boat families, but also the interconversion of these families. By treating methylene groups as rigid units, the energy surface for each molecule can be expressed in terms of two torsional parameters and forms a torus that can be projected and plotted as energy contours on a plane. Although this gives a good view of all possible low-energy processes in each molecule (there are 14 twist–chairs and equal numbers of chairs, twist–boats, and boats), the most significant results of these calculations can be expressed as appropriate cuts through the contours to show the energy changes during pseudorotation of the twist–chair (energy minimum) via the chair (transition state) and of the similar pseudorotation in the boat family. Another cut through the energy surface can show the lowest energy profile for the interconversion of the chair and boat family.

Following Bocian and Strauss,[42,43] the different sites in the cycloheptane twist–chair can be labeled as *A*, *B*, *C*, and *D*, with symmetry-related sites indicated by a prime, as in **35**. This nomenclature can also be used to describe the positions of heteroatoms in twist–chairs of analogous heterocycles. Two almost equally populated twist–chairs (**36**, **37**), with the oxygen at *B* and *C* positions, occur in oxepane. 1,3-Dioxepane exists in a single twist–chair of C_2 symmetry (**38**), with both oxygens at *C* sites next to the C_2-axis (this conformation is termed *D* by Strauss and co-workers, since they use the acetal CH_2 as a reference in the case of the dioxepane, but it would be less confusing if it were called *CC*, with oxygen as the reference atom). All the other forms (including transition states) of these two heterocycles that are within the boat or chair pseudorotation itinerary lie in the range of 2–3 kcal/mol above the twist–chair, which has the lowest energy. However, the interconversion of the two pseudorotating families requires the passage over a barrier of about 9 kcal/mol. The twist–chairs of 1,3-dioxepane have different dipole moments so that their stabilities in solution will depend on the solvent polarity, but this factor has generally been unjustifiably neglected. Also, the energies given above have been calculated without considering anomeric effects or dipole–dipole interactions, and this can introduce errors of the order of 1 kcal/mol or more. Thus the results obtained

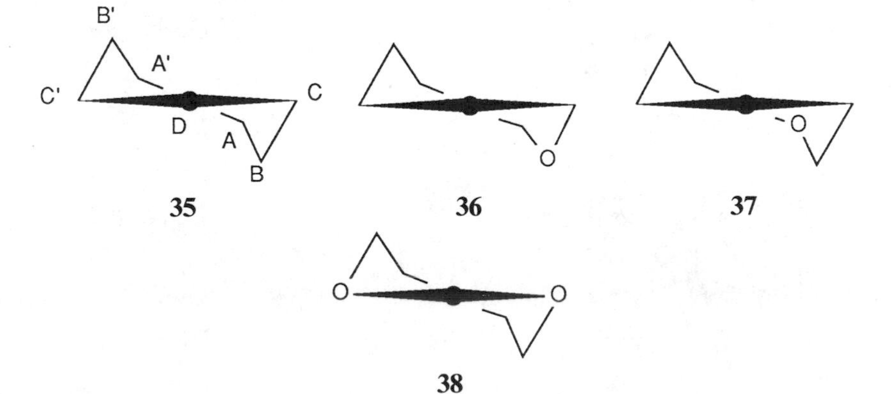

by Strauss and co-workers should be used with these considerations in mind when applied to other molecules in solution.

Saturated seven-membered heterocycles containing a single ring oxygen are mostly in the carbohydrate area, and the many oxygen substituents on the ring are important in determining the actual conformation adopted, especially if these atoms are tied together to give five-membered ketal rings, as is sometimes the case. Hexoses in general exist as equilibrium mixtures of open, furanose, and pyranose forms, with the seven-membered septanose form present in negligible amounts.[52,148] However, blocking groups allow septanose forms to be stable, and a number of X-ray structures of such compounds have been determined. Additionally, vicinal proton–proton coupling constants give some information on torsional angles for these molecules in solution, and it appears that the crystal and solution conformations are similar. Twist–chairs, sometimes distorted more or less toward the chair, are the most common conformations found, but a boat form occurs in one case.[149] A summary of the torsional angles in oxepane and the septanose derivatives (**39**,[149] **40**,[150] **41**,[151] **42**,[152] **43**,[153] **44**,[154] **45**,[155] **46**,[156] **47**,[157]) are given in Table 2-2, where Stoddart's terminology for describing the conformation of seven-membered rings, which is described below, is used.

With the usual numbering for hexoses ("1" is the anomeric carbon, which carries two oxygens, and "O" is the ring oxygen), Stoddart defines reference

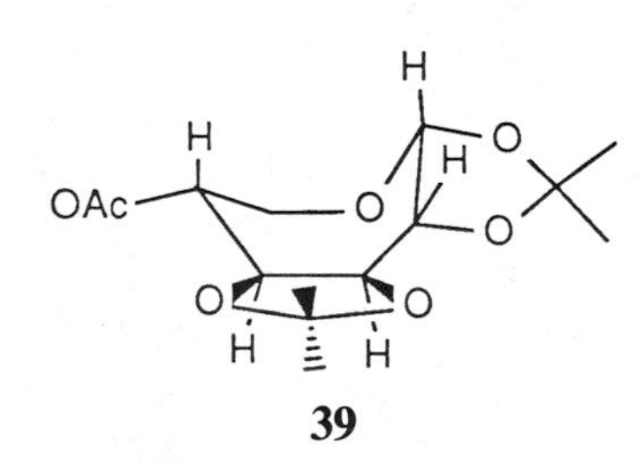

39

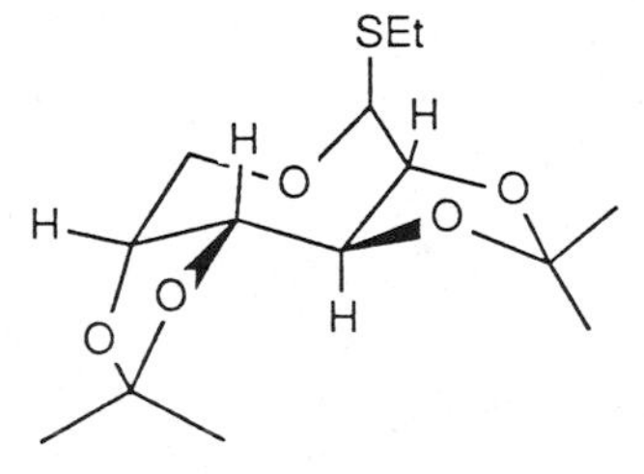

40

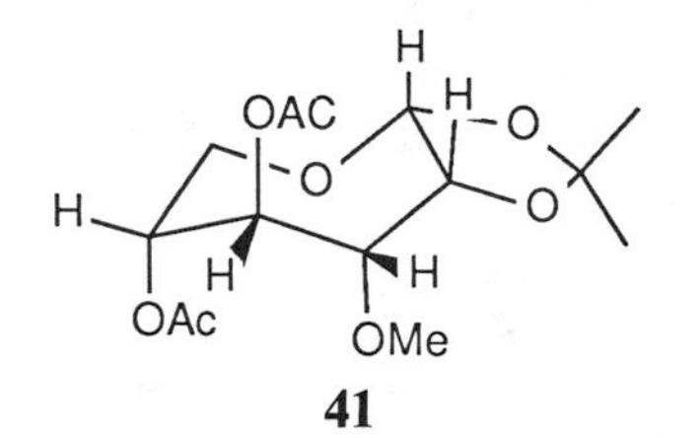

41

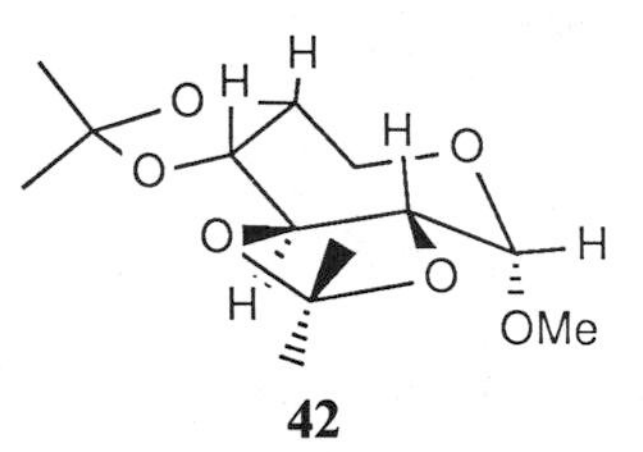

42

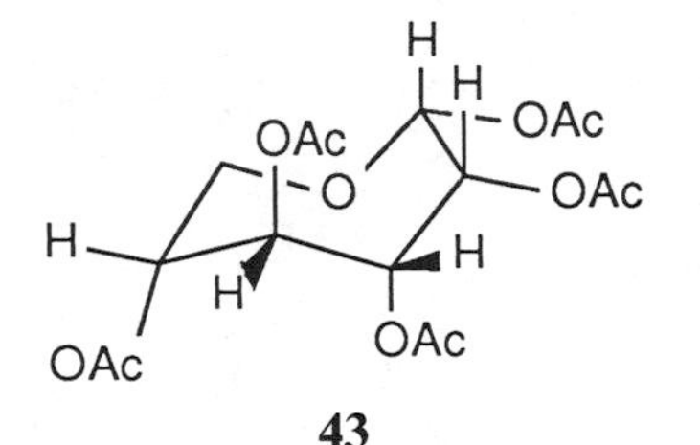

43

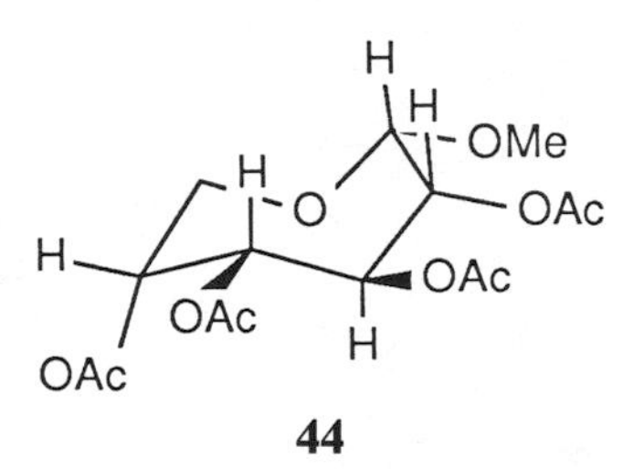

44

TABLE 2-2. Torsional Angles in Some Septanose Derivatives

Ring	Torsional Angles (degrees)									
Atoms	Oxepane (TC B)	**39**	**40**	**41**	**42**	**43**	**44**	**45**	**46**	**47**
1–2–3–4	−34	−56	−91	−74	−90	−45	−46	−42	−84	−78
2–3–4–5	84	12	60	71	51	85	84	92	68	61
3–4–5–6	−70	68	−35	−58	21	−61	−63	−65	−46	−48
4–5–6–O	54	−48	56	73	−82	50	53	50	67	73
5–6–O–1	−80	−52	−98	−96	85	−84	−85	−84	−102	−99
6–O–1–2	102	89	54	56	−63	98	95	95	55	46
O–1–2–3	−47	5	36	−22	71	−34	−31	−32	31	37
Conformation:	Twist–Chair B	$^{1,2,5}B$	$^{5,6}TC_{3,4}$	$^{5}C_{1,2}$ or $^{3,4}TC_{5,6}$	$^{1,2}TC_{6,0}$	$^{4,5}TC_{6,0}$	$^{4,5}TC_{6,0}$	$^{4,5}TC_{6,0}$	$^{5,6}TC_{3,4}$	$^{5,6}TC_{3,4}$

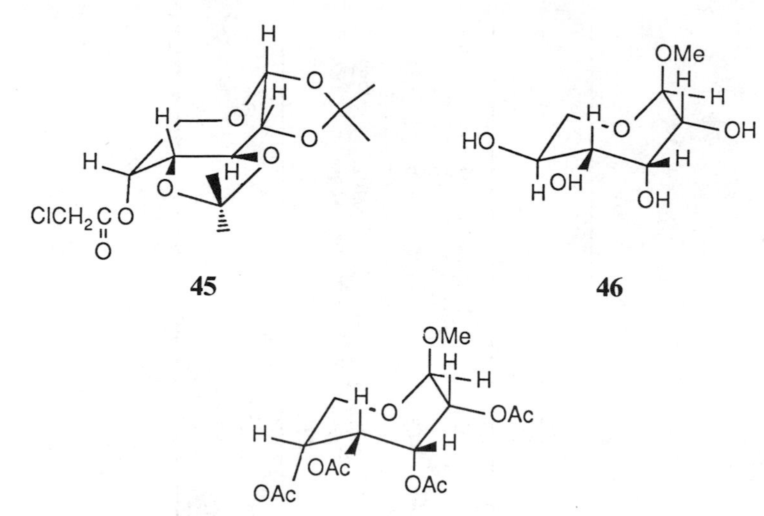

planes for the chair (plane through 2, 2′, 3, and 3′ in **48**) and the twist–chair (plane through 2–1–2 and the midpoint of the 4–4 bond in **49**). In the case of the twist–chair, which is a chiral conformation, primes are used for numbering the mirror image form of **49**. Ring atoms that lie above the reference plane are written as superscripts preceding the letter descriptor and those below the plane are written as subscripts following the descriptor, with the numbering taking place clockwise from above.

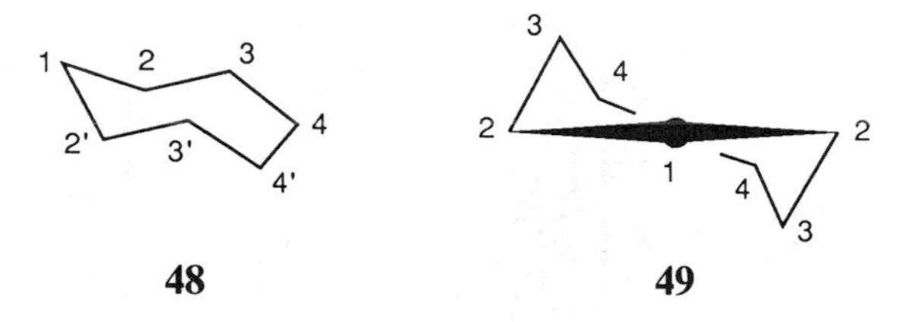

The conformational properties of 1,3-dioxepane and its 4,4,7,7-tetramethyl derivative have been investigated by NMR.[158] No dynamic NMR effect is observable for the unsubstituted molecule, as expected if only the rapidly pseudorotating twist–chair is populated, but the tetramethyl compound gives temperature-dependent ^{1}H and ^{13}C spectra from which a barrier to ring inversion in a symmetrical conformation of 10.3 kcal/mol can be calculated. The conformation is almost certainly the twist–chair, **50**, together with its mirror image, and the process observed is a conformational racemization of this chiral conformation. The increased barrier caused by *gem*-dimethyl groups has been found in other pseudorotating molecules.

The conformational properties of a series of diastereomeric 2,4 and 2,5 disubstituted 1,4-dioxepanes have been studied by a variety of methods and the results have been interpreted in terms of twist–chair conformations with the

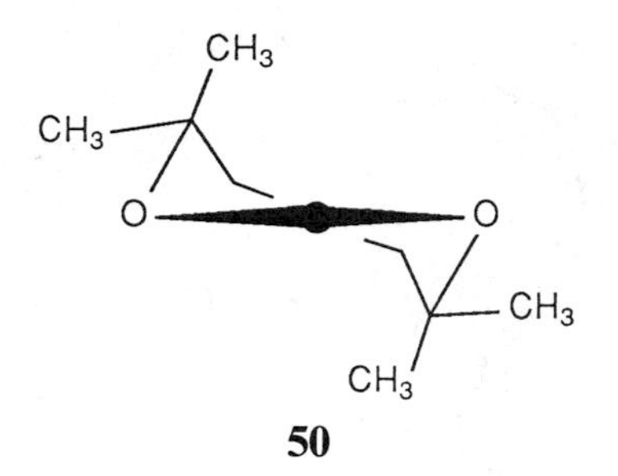

50

substituents in equatorial positions, but only rather tentative conclusions have been reached.[159]

Grindley and co-workers have used equilibration data and ^{13}C chemical shifts to investigate the conformational equilibria in a series of trisubstituted 1,3-dioxepanes (**51**, **52**, and **53**), and have made some deductions about the relative energies of the four twist–chairs of the parent dioxepane.[160] Their approach is rather indirect and makes a number of assumptions in order for a conclusion to be reached. They find that the *CC* twist–chair (or *D* in Strauss' original terminology) is the preferred conformation for **53**, but that **52** cannot exist mainly in this type of conformation, but probably is a mixture of the *AB* and the *AC* twist–chairs. Measurements of dipole moments, IR in the liquid and solid phases, and low-temperature NMR spectra are needed before the above conclusions can be considered firm.

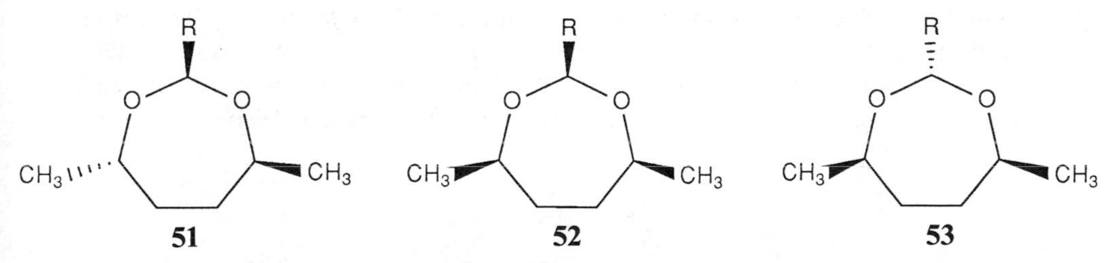

51 **52** **53**

1,3-Dioxepane units also occur in carbohydrate derivatives and can be formed in competition with 1,3-dioxanes or 1,3-dioxolanes when acetals are made from formaldehyde and polyols. Burkert has carried out molecular mechanics calculations on systems that have fused 1,3-dioxepane and 1,3-dioxolane rings (e.g., **54**), and has found that the seven-membered rings are calculated to have chair rather than twist–chair conformations, but the energy minima are quite shallow.[161] The calculations are in agreement with NMR coupling constants, and with equilibrium data involving the above [5.3.0] and the isomeric and more stable [4.4.0] system, but unfortunately there are neither dynamic NMR

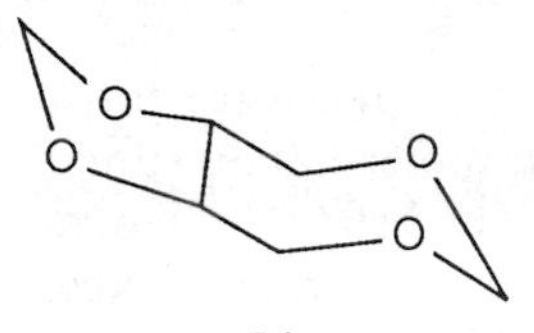

54

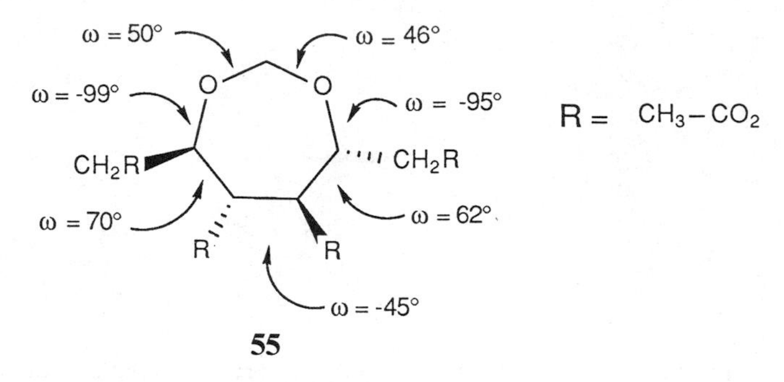

55

data nor X-ray structures on these molecules, although an X-ray structure is available for a monocyclic formal (**55**) derived from mannitol; this has the *BB′* conformation with an approximate C_2-axis passing through the O—CH_2—O carbon.[162]

2.6.2. Rings with One Strong Torsional Constraint

2.6.2.1. Rings with One Oxygen. The seven-membered-ring lactone caprolactone is an analogue of cycloheptene and caprolactam, and all exist in chair conformations. Noe and Roberts have made use of the γ,γ difluorinated lactone, **56**, to determine the barrier to ring inversion by ^{19}F dynamic NMR. The barrier ($\Delta G^\ddagger$) is surprisingly large (10.0 kcal/mol) as compared with that in the corresponding cycloheptene analogue (7.4 kcal/mol), but it is similar to the barrier in the corresponding lactam,[163,164] and this may be due to the previously mentioned differences in bond angles and bond lengths at carbon and oxygen (and also nitrogen), although differences in torsional barriers may also play a role.

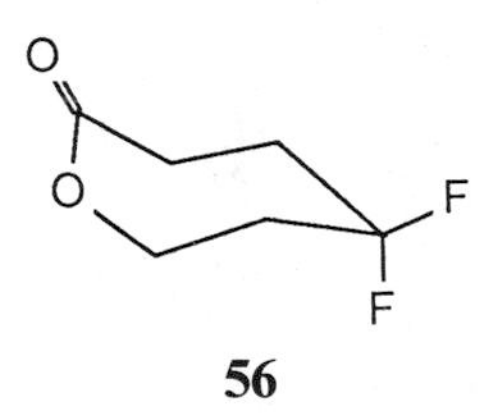

56

The mechanism of the inversion process involves a boat intermediate and a transition state lying between the chair and the boat. The bond angles are much enlarged in this transition state, with several atoms lying in, or close to, a plane. Allinger has done molecular mechanics calculations on caprolactone and finds that the chair is by far the lowest-energy conformation.[165] A caprolactone derivative (**57**) with exo unsaturation next to the ring oxygen also exists in a chair conformation.[166]

The conformations of none of the three possible isomeric oxacycloheptenes have been investigated; these molecules presumably have low inversion barriers, as does cycloheptene itself. The benzo derivative of 5-oxacycloheptene (**58**) has a free-energy barrier to ring inversion of 9.5 kcal/mol, and has been studied by dynamic ^{1}H NMR.[167] The relatively high value of this barrier is not consistent

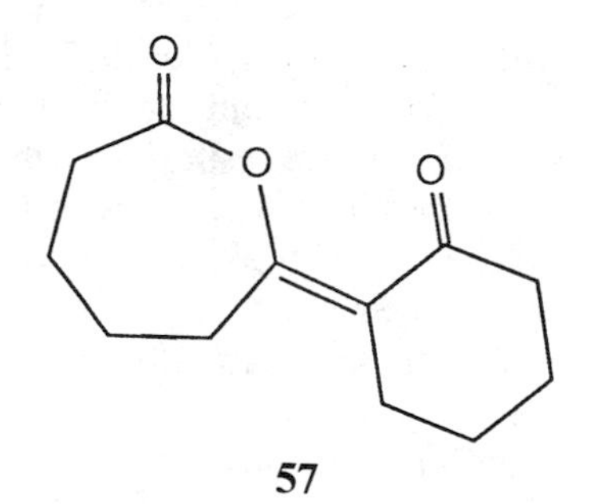

57

with a boat or twist–boat and its similarity to the ring inversion barrier in benzocycloheptene indicates a chair. The seven-membered-ring vicinal coupling constants in the above two compounds are similar, but this evidence does not distinguish between a chair and a twist–boat for **58**. An unusual deuterium chemical shift isotope effect of 0.005 ppm observed in diastereomeric tetradeuterio derivatives of **58** probably originates, in part at least, from a perturbation of the ring inversion equilibrium.[168]

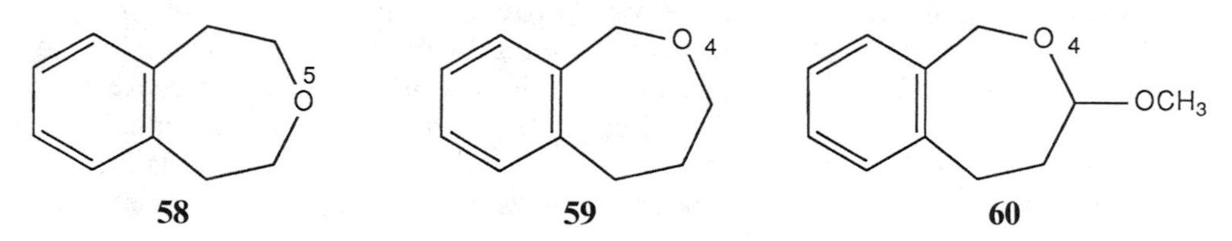

58 **59** **60**

In the benzo derivative of 4-oxacycloheptene (1,2,3,4-tetrahydro-2-benzoxepin), **59**, the presence of a methoxy group adjacent to the ring oxygen causes both the chair and twist–boat to be observable in **60**, whereas only the chair occurs in **59** itself. The ring inversion barrier ($\Delta G^{\ddagger}$) in **59** is 9.4 kcal/mol. The complex ^{1}H NMR spectrum of **60** at -120°C has been analyzed with the help of two-dimensional COSY experiments and corresponds to three distinct forms: chair (equatorial methoxy), chair (axial methoxy), and twist–boat (methoxy antiparallel to one electron pair of the ring oxygen) in the ratios of 12:3:5 and 2:1:1, with dimethyl ether and CHF_2Cl, respectively, as solvents. Only the latter two forms are stabilized by an anomeric effect, and, effectively, the chairs are destabilized either by a sterically hindered axial methoxy group or by an unfavorable geometry at the anomeric site, thus reducing the chair/twist–boat energy difference sufficiently for both types of conformations to be observed. The barrier ($\Delta G^{\ddagger} = 8.8$ kcal/mol) for the change of the twist–boat to the axial chair is significantly higher than that ($\Delta G^{\ddagger} = 11.1$ kcal/mol) for the twist–boat changing to the equatorial chair. The molecule, **60**, is particularly interesting because it shows that in medium rings, unlike in six-membered rings, a single substituent (CH_3O) at an anomeric site is sufficient to cause a qualitative difference in the ring conformation.

2.6.2.2. Rings with Two Oxygens. Monounsaturated heterocycles containing two ring oxygens, and their benzo analogues, exist in chairs or twist–boats, or as mixtures of these conformations. Proton and ^{13}C NMR, vibrational spectroscopy (IR and Raman), X-ray diffraction, and dipole moment and Kerr constant measurements have proved informative, especially when these techniques are used in combination. Compounds in this series with the two oxygens in a 1,3 relationship are easily made, since they are acetals or ketals. The only other compounds that have been studied are cyclic ethers of catechol. The oxygen atoms in all these molecules are symmetrically disposed with respect to the double bond or the benzene ring.

Friebolin et al.[169] were the first to discuss in detail the conformational isomerism in seven-membered rings with one strong torsional constraint. Although they were interested in the conformation of cycloheptene, the complexity of the 60-MHz ^{1}H NMR spectrum of this compound led them to study analogous heterocyclic molecules containing either two oxygens or two sulfurs in a 1,3 relationship. The sulfur-containing heterocycles exist either as chairs or as mixtures of chairs and twist–boats (called just "twist forms" by the above workers). Fusion to a benzene ring increases the conformational barriers, which are of the order of 10 kcal/mol, by 2.4–4 kcal/mol. The oxygen heterocycles have given less definite results because clear dynamic NMR effects have only been observed for **61**, where the methylene protons show an *AB* pattern at low temperatures ($\Delta G^{\ddagger} = 9.7$ kcal/mol at -76°C). The methyl signal of **61** broadens but does not split down to -127°C. Because of the large steric hindrance of the axial methyl group in the chair, a twist–boat was assigned to this molecule, the relatively high barrier arising because pseudorotation is hindered by the axial methyl group in the boat form.

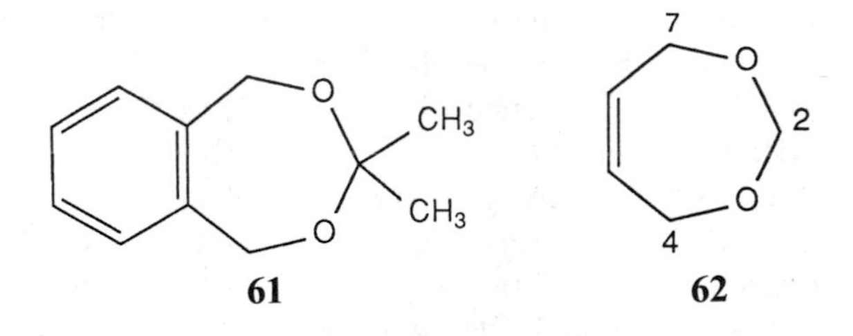

61 **62**

Friebolin et al.[169] were unable to detect any dynamic ^{1}H NMR effect in 1,3-dioxacyclohept-5-ene (**62**) at 60 MHz down to -140°C. Twist–boats have been assigned to the *cis*- and *trans*-4,7-dimethyl derivatives of **62** on the basis only of room temperature ^{13}C NMR,[170] but this method, which has also been used to study 2-substituted derivatives of **62** by other workers,[171] is insensitive to the presence of minor conformations. In the *r*-2-*tert*-butyl-*cis*-4,*cis*-7-dimethyl derivative of **62**, the twist–boat has a strong repulsion between the *tert*-butyl and one methyl group, and this compound undoubtedly exists in the chair.[170] A twist–boat was originally assigned to the diastereomeric 2-*tert*-butyl-*trans*-4-*cis*-7-dimethyl derivative, but a mixture of chair and twist–boat conformations appears more likely (see below for the corresponding benzo derivatives).[172] Investigations of **62** and related molecules by Arbuzov, Klimovitskii, and coworkers, using a variety of methods, have proved that two forms are present.[173]

Dipole moment measurements on **62** show a moment (1.39*D*) intermediate between moments expected of the chair (2.6*D*) and the twist–boat (0.14*D*) and indicating 28% of the chair.[173] The presence of two conformations is also supported by the disappearance of some bands from the IR spectrum of liquid **62** upon crystallization. Furthermore, the temperature and solvent-polarity dependence of the relative intensities of the IR bands in solution indicates that the less polar form—that is, the twist–boat—is the more stable conformation ($\Delta H°$ favors the twist–boat by 1.4 kcal/mol in CCl_4). These workers suggested that the anomeric effect stabilizes the twist–boat, and when this effect is absent, as in cycloheptene, only the chair is observed.

Molecular mechanics, dipole moment, and Kerr constant measurements on **62** have been reported by Pitea et al.,[174] who apparently were unaware of Arbuzov's work. They find that the twist–boat is 3.2 kcal/mole more stable than the chair form, but the force field used tends to give too low an energy for twist–boat geometries in other molecules, and solvation of the polar chair is expected to reduce the above calculated energy difference. Their experimental dipole moment is 1.29*D*, compared with calculated values for the twist–boat and chair of 0.14 and 1.83*D*, respectively, and this leads to equal populations of these two forms, which is only in fair agreement with the conclusions of Arbuzov. Analysis of the Kerr constant data gives a 10:1 ratio favoring the twist–boat. A careful investigation of the high-field ^{1}H and ^{13}C NMR spectra of **62** at low temperatures would probably be fruitful, and could give a good value for the twist–boat/chair equilibrium constant.

Many derivatives of **62** have been studied by Arbuzov, Klimovitskii, and co-workers.[173] For the 2-methyl derivative of **62**, which crystallizes as a twist–boat, and in CS_2 solution $\Delta H°$ favors this conformation over the chair by 0.72 kcal/mol, as shown by the temperature dependence of the IR bands. With the much larger *tert*-butyl group at the same position, it is the more polar (chair) form that crystallizes, and here the chair form is favored by a $\Delta H°$ of 0.3 ± 0.3 kcal/mol. With two substituents at the 2 position in **62**, only the twist–boat conformation is detected, whereas with a single substituent at the 2 position, the twist–boat has an unfavorable gauche interaction across an oxygen atom and is therefore destabilized relative to the chair, where the substituent can be equatorial. When two substituents are present, however, one of them has to occupy the very crowded axial site in the chair, and thus the twist–boat becomes even more favorable than it is in the parent molecule. The series with the olefinic hydrogens replaced by chlorines gives rise to similar effects, and in one case (**63**), the twist–boat geometry was proved by an X-ray diffraction study.[175] Although the ring puckering can be discussed in terms of the

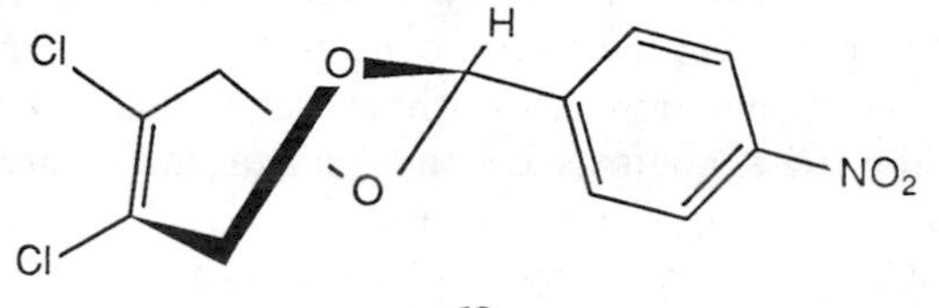

63

Cremer–Pople parameters[31] (or modifications of these parameters[34]), it is easier to describe the conformation as having carbons 2, 4, 5, 6, and 7 coplanar within 0.03 Å, and with the ring oxygens lying above (0.96 Å) and below (0.77 Å) this plane. Thus the twist–boat is slightly pseudorotated from the ideal C_2 symmetry, which requires equality of the oxygen out-of-plane displacements. The chlorine atoms are also slightly twisted from the plane of the double bond. The phenyl ring takes up an arrangement that minimizes nonbonded repulsions with the CH groups in the seven-membered ring.

The barrier to pseudorotation in the twist–boat form of **62** has been estimated to be less than 5 kcal/mol,[176] but the presence of gem dimethyl groups at the 2 position increases the barrier to 6.8 kcal/mol, and also forces the molecule to be entirely in the twist–boat conformation. Although the pseudorotation barrier is not measurable for the 2-methyl derivative, the 2-methoxy derivative has a conformational barrier of 5.2 kcal/mol. Both compounds are twist–boats, as shown by the numbers of different ^1H and ^{13}C chemical shifts, with no other conformation visible at low temperatures, but the chair conformation, at least for the methyl derivative, must be somewhat populated at room temperature to accommodate the IR results of Arbuzov.[173]

The cyclopropane (**64**) and epoxide (**65**) analogues of **62** have been studied by several groups. No dynamic NMR effect occurs for **65** down to $-160°C$ but **64** gives rise at $-90°C$ to two separate species that interconvert with a $\Delta G^{\ddagger}$ of 10.7 kcal/mol (the equilibrium constant was not given, but it appears to be close to 1).[177] The NMR of the exo and endo diastereomers of the 2-isopropyl derivatives of **64** and **65** have also been analyzed, and from the proton vicinal coupling constants it has been deduced that both epoxides have twist–boat conformations whereas both cyclopropanes have chair conformations. The parent molecules, **64** and **65**, have been suggested as having the same conformations as the related isopropyl derivatives, that is, chair and twist–boat, respectively.

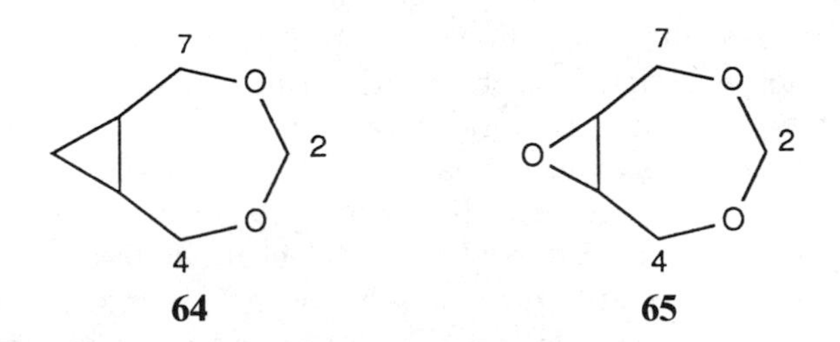

Other workers have proposed a chair conformation for the epoxide and its 2 derivatives, except that the twist–boat becomes favored when there are two substituents at the 2 positions.[171] An X-ray structure of **65** reveals a chair conformation with the epoxide in the endo position (**66**),[178] in agreement with the gauche effect previously mentioned. On the other hand, molecular mechanics calculations give the twist–boats as 2.7 and 1.6 kcal/mol more stable than the chairs for **64** and **65**, respectively.[179] Most recently, the dipole moment and Kerr constant of **65** have been interpreted in terms of an equal mixture of the twist–boat and chair; the same kind of data show that **64** is mostly in a chair

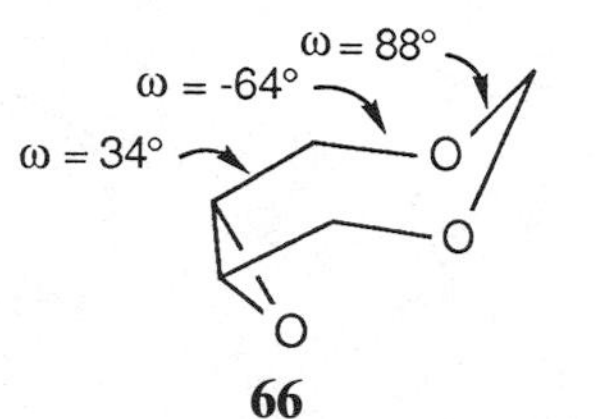

66

form.[174] As with **62**, measurements of high-field ^{1}H and ^{13}C NMR spectra of **65** at low temperatures would probably be useful, as also would a comparison of the IR spectra in the liquid and solid phases.

The 8,8-dichloro-2-phenyl derivative of **64** in the crystalline state has the cyclopropane ring fused exo to a seven-membered ring chair and *trans* to the equatorial phenyl group (misleadingly called the syn isomer in the original paper).[180] A comparison of the NMR spectra of this compound and of the 8,8-dichloro and 8,8-dibromo derivatives of **64** show that they all have similar conformations.[181] In contrast, 8-chloro and 8-bromo derivatives of **64** with the halogen on the cyclopropane ring *trans* to the seven-membered ring have different vicinal coupling constants which are consistent with an endo cyclopropane ring.[181] The introduction of two substituents at the 2 position leads, as expected from previous examples, to twist–boats.[171,181]

The fusion of four- and five-membered rings to the dioxacycloheptane system gives rise to chairs, or if a substituent at the 2 position would be forced to be axial in the chair, to twist–boats, as shown by NMR chemical shifts and coupling constants.[171,182]

Benzo-fused derivatives of **62** with various substituents at the 2 position (**67**) have been investigated by ^{1}H and ^{13}C dynamic NMR,[183] and shown to possess chair and/or twist–boat conformations, depending on the nature of the substituent(s), as shown in Table 2-3, which also gives a list of observed conformational barriers. The twist–boat is favored over the chair in the parent compound (**68**) by 0.4 kcal/mol, according to molecular mechanics calculations,[184] in good agreement with the experimental data. Derivatives of **68** have also been investigated by IR and dipole moment measurements, and the more polar chair form of **67** has been shown to have about 0.3 kcal/mol lower enthalpy than the less

TABLE 2-3. Conformational Equilibria and Barriers[a] of **67** and **68**

R^1	R^2	Chair, %	Twist–Boat, %	$\Delta G^{\ddagger}$, kcal/mol
H	H	79	21	8.0
CH_3	H	100	0	
CH_3	CH_3	0	100	10.0
$(CH_2)_4$		0	100	9.0
$(CH_2)_5$		0	100	9.9
OCH_3	H	0	100	6.7

[a] Data from Ref. 183, for CHF_2Cl solutions at −100° to −150°C.

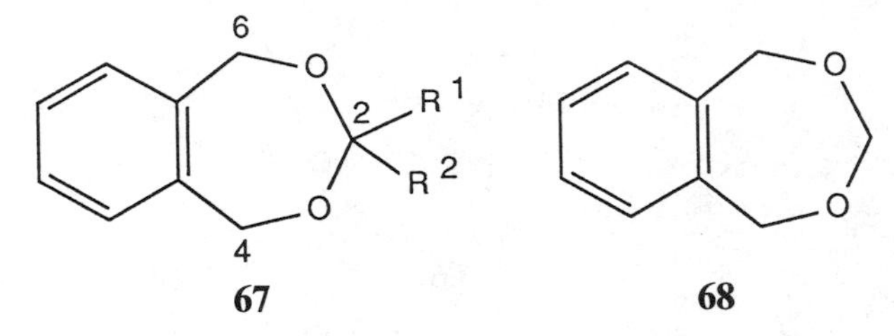

67 **68**

polar twist–boat in CCl_4, in good agreement with the above ^{13}C NMR data ($\Delta G° = 0.38$ kcal/mol at $-130°C$ in $CHFCl_2$).[185] In the case of 2-*p*-substituted phenyl derivatives, the chairs and twist–boats exist in ratios of 2:3 at room temperature.[186] For the 2-phenyl derivative at $-115°C$, the ratios are 56:44 and 77:23 in dimethyl ether and CHF_2Cl, respectively.[187] The fusion of a benzene ring to **62** can be seen to increase the population of the chair form, the energy effect being about 1 kcal/mol. The 2-*tert*-butyl derivative of **68** exists exclusively in the chair form,[185,187] whereas the 2-methoxy derivative is exclusively in the twist–boat because of an anomeric effect involving the methoxy group, and these differences are similar to those in the analogous molecules without fused benzene rings. The *cis*- and *trans*-2-*tert*-butyl-4-phenyl derivatives of **68** exist in chair forms, whith the *trans* isomer having an axial phenyl group and being less stable than the *cis* isomer.[188]

Arbuzov, Klimovitskii, and co-workers have found that fusion to a benzene ring shifts the chair/twist–boat equilibria considerably toward the more polar chair in diastereomeric 4,7-dimethyl- and 2-*tert*-butyl-4,7-dimethyl-1,3-dioxacyclohept-5-enes previously studied by Gianni et al. and discussed above.[172] The crystal structure of *r*-2,*cis*-4,*cis*-7-trimethyl-1,3-dioxa-5,6,-benzocycloheptene (**69**) shows a chair with all the methyl groups equatorial; the geometric features of this molecule have been compared to those of the twist–boat in **63** and of related six- and eight-membered chair conformations.[189]

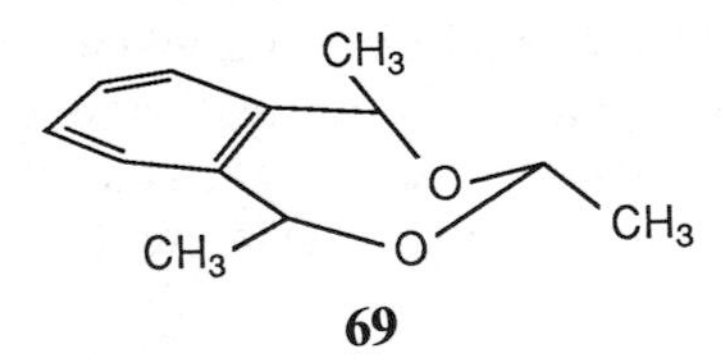

69

Ultraviolet spectroscopy on dihydro-1,5-benzodioxepin **70** and various derivatives, which are cyclic ethers of catechol, lead to rather tentative conclusions indicating a predominance of a chair for the parent molecule, with increasing populations of twist–boats upon substitution at the 3 position.[146] X-ray structures of related compounds (e.g., **71**) that have two substituents at the 3 position show that these molecules exist as twist–boats with approximate C_2 symmetry,[190,191] but minor amounts of chair forms probably occur in solution, as discussed below. St. Jacques and co-workers have reinvestigated the conformations of nine-membered cyclic catechol ethers by dynamic NMR methods in two different kinds of solvent, namely, CHF_2Cl, which is weakly hydrogen bonding and has a moderate dielectric constant, and CH_3OCH_3, which is

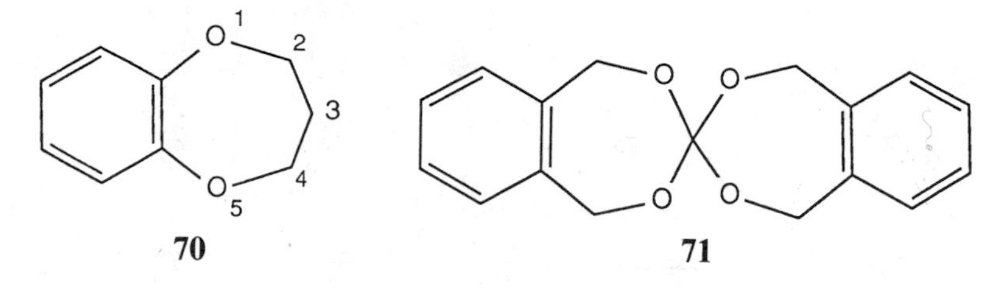

nonpolar and has a low dielectric constant.[192] Chairs and twist–boats in the ratio of about 4:1 are found in the parent molecule at −147°C, and presumably this ratio decreases at higher temperatures. The free-energy barrier for the chair to twist-boat conversion is 6.5 kcal/mol. In this chair conformation of **70**, an axial 3-methyl group is much less hindered than in isomeric molecules discussed previously because that group has gauche interactions with oxygen rather than with methylene moieties (as in the axial conformer of 5-methyl-1,3-dioxane), and thus the 3,3-dimethyl derivative of **70** also exists as a mixture of chair and twist–boat conformations, but in the ratio of about 3:7. Free-energy barriers in this molecule for the chair to twist–boat conversion (7.5 kcal/mol), as well as for the rather hindered twist–boat pseudorotation (6.8 kcal/mol), have been obtained. In both **70** and its 3,3-dimethyl derivative, the twist–boat is slightly favored by a low solvent polarity.

The 3-methyl derivative of **70** exists in the twist–boat, *ax*-chair, and *eq*-chair, whereas the 3-methoxy derivative is in the twist–boat and *ax*-chair, with the latter unobservably populated in dimethyl ether.[193] Particular attention has been paid to the conformational equilibria in 3-halogenated derivatives of **70** as a function of the solvent polarity and the electronegativity of the halogen.[194] Some of the complex spectra obtained at low temperatures have been analyzed with the help of two-dimensional ^{1}H COSY experiments and up to three forms have thus been detected. The conformational equilibria have been interpreted in terms of dipolar interactions and gauche effects, which are attractive when the X substituent in the O—C—C—X moiety is F or OCH_3, but repulsive when X is Cl, Br, or I. The twist–boat is always significantly populated, and the chair with an equatorial substituent, which is the dominant conformation only for X = I, is absent for X = F and X = OCH_3. This ring system is particularly valuable as a testing ground for substituent effects on conformational equilibria,[194] but molecular mechanics calculations, which should prove valuable, have not as yet been applied to this system.

2.6.3. Rings with Two or More Strong Torsional Constraints

Singly and doubly bridged biphenyls containing one or two seven-membered oxygen heterocycles are chiral and racemize more or less slowly at room temperature.[195] The racemization activation energy in **72** is 22 kcal/mol and the half-life at room temperature is about 10 minutes. In the case of **73**, the racemization barrier (E_{act} = 11.2 kcal/mol) is too small to allow resolution at room temperature, but the barrier can easily be measured by dynamic NMR.[196] A similar

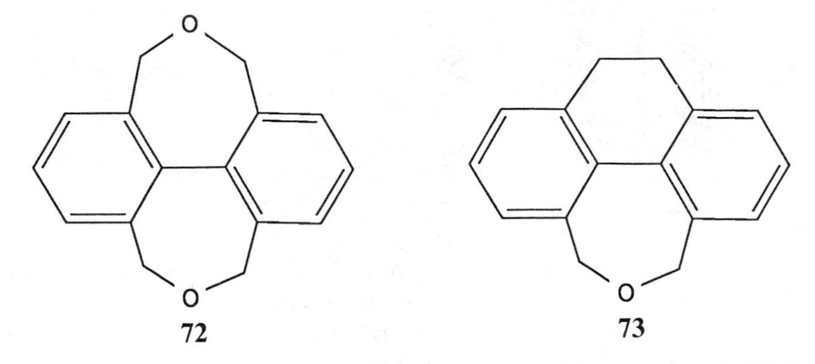

72 73

barrier (E_{act} = 11.2 kcal/mol) is found in the monobridged molecule, **74**, but the tetramethyl derivative, **75**, has a somewhat larger barrier to ring inversion (E_{act} = 13.3 kcal/mol, log A = 12.1).[197]

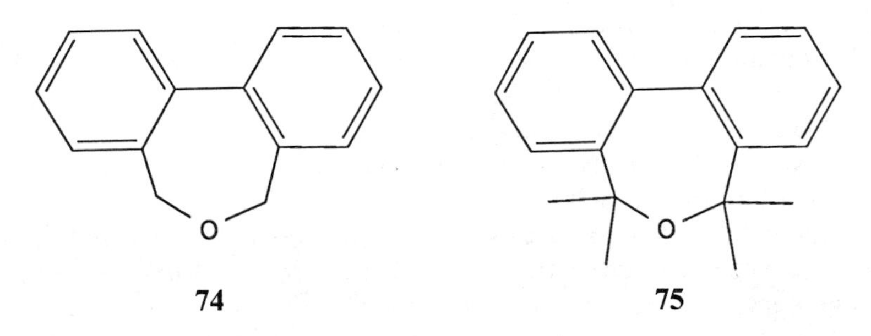

74 75

1,3-Dioxa-4,6-cycloheptadiene has been investigated by molecular mechanics (MMP2) and *ab initio* methods. The energy of the conformation with C_s symmetry (i.e., a flattened boat, **76**) is calculated to be lower than that with C_2 symmetry by 10.4 and 5.7 kcal/mol, respectively.[198] The planar form is calculated by the above methods to lie 12.3 and 7.2 kcal/mol above the boat, respectively. Thus the ring inversion barrier in this molecule should be high enough for measurement by dynamic NMR, but attempts to do so have not been reported. The ultraviolet photoelectron spectrum shows a strong interaction of the π electrons with the oxygen lone pairs, and this interaction is sensitive to the geometry of the molecule.

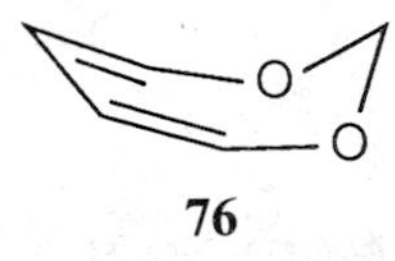

76

Oxepins, which are the most unsaturated rings in this section, are fluxional molecules that can exist in equilibrium with the corresponding benzene oxides. In the parent system, the bicyclic form is favored by a ΔH° of 1.7 kcal/mol.[199] Although this equilibrium has been studied by *ab initio* calculations, the energy surface for ring inversion in the boat-shaped oxepin has not been explored.[14] Substitution at the 2 and 7 positions shifts the equilibrium toward the oxepin form, and such compounds have been used for X-ray diffraction and dynamic

NMR studies. In **77** the free-energy barrier for ring inversion (6.5 ± 0.2 kcal/mol at −138°C) can be obtained because the methylene protons are diastereotopic (*AB* pattern with the CH_3 decoupled) when the inversion is slow on the NMR time scale.[200] The analogous ethyl ester gives complex spectra at low temperatures as a result of restricted carbonyl-ring rotation ($\Delta G^{\ddagger}$ = 7.5 kcal/mol at −113°C) as well as slow ring inversion. The simplest oxepin whose structure is known from X-ray diffraction data is the 2-*tert*-butoxycarbonyl derivative (**78**) which has fold angles of 56.6° (θ_1) and 26° (θ_2).

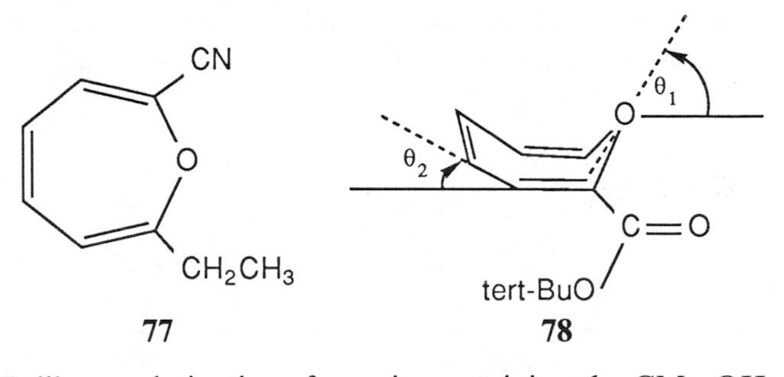

A 2,3:6,7 dibenzo derivative of oxepin containing the CMe_2OH group as a probe (**79**) has been studied by dynamic NMR. The methyl groups are diastereotopic at low temperatures, and thus the free-energy barrier for ring inversion (10.3 kcal/mol at −69°C) can be obtained.[77]

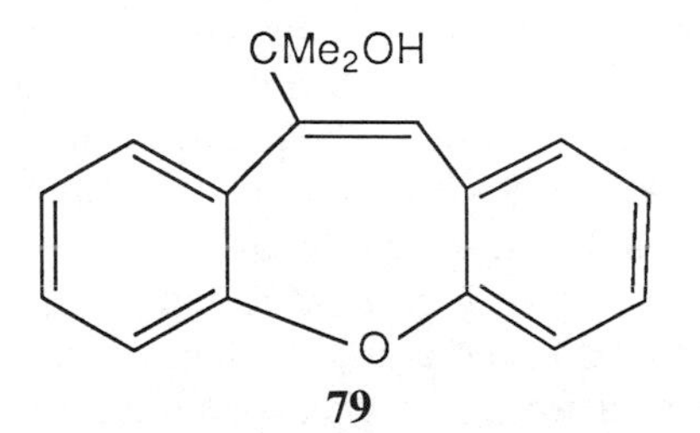

2.7. Eight-membered Oxygen Heterocyclic Rings

2.7.1. Rings with No Strong Torsional Constraint

The conformational properties of oxocane (oxacyclooctane) are similar to those of cyclooctane discussed in section 2.4.2. According to NMR data, the predominant form is a boat–chair with a free-energy barrier to ring inversion of 7.4 kcal/mol,[201] but the position of the oxygen atom is not well defined, and it is possible that several boat–chairs are populated. A crown-family conformation is observed as a minor component at −100°C (ΔG° = 1 kcal/mol), and this type of conformation is slightly more favored than in the case of cyclooctane or azacane.[54] Molecular mechanics calculations show that the chiral *BC*-3 boat–chair (**80**) and its mirror image (*BC*-7) are of lowest energy. The achiral *BC*-1 and the crown (**81**) lie 0.4 and 2.2 kcal/mol above these forms, respectively.[202,203] The boat–boats are the least favorable conformations of oxocane.

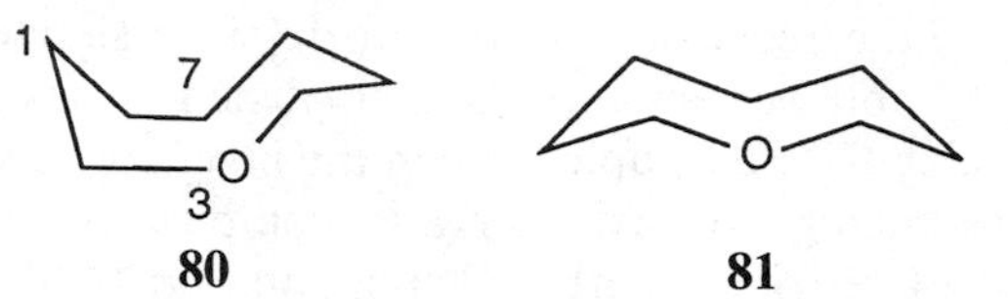

80 **81**

Substituted oxocanes occur as marine natural products and the X-ray structure of one of these compounds, laurencienyne, **82**, shows a *BC*-3 type of conformation (actually, the oxygen is at the enantiomeric 7 position in this chiral molecule) and one of the chlorine substituents is in the relatively unhindered 5 axial position; all other substituents are in equatorial or pseudoaxial positions.[204,205]

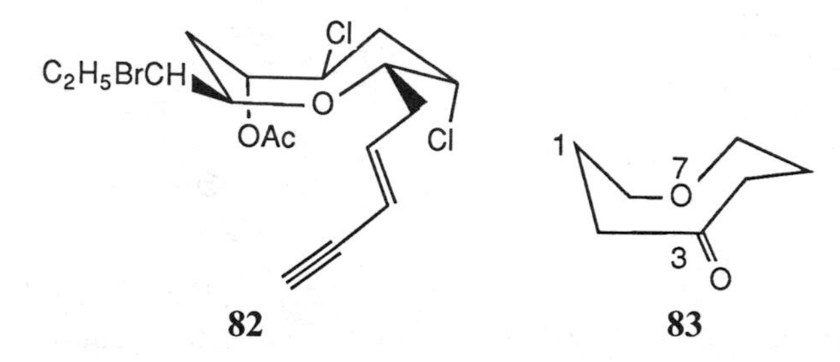

82 **83**

5-Oxocanone has the chiral *BC*-3,7 boat–chair (**83**),[206] as expected from its relationship to cyclooctanone and oxocane. This conformation also allows an attractive transannular oxygen–carbonyl interaction to take place. The presence of relatively high barriers for pseudorotation ($\Delta G^{\ddagger} = 7.9$ kcal/mol) and for ring inversion ($\Delta G^{\ddagger} = 9.0$ kcal/mol) is consistent with this interaction, together with the pseudorotation taking place via a twist–boat–chair rather than a boat–boat. Neither the crown family nor the boat–boat is observably populated. 4-Oxocanones can be prepared in unexpectedly high yields by the $TiCl_4$-catalyzed intramolecular directed aldol condensations of silyl enol ethers, but this is probably due to a template effect and not to a particularly favorable conformation for the eight-membered rings[207]; transannular interactions are unfavorable in this system because a four-membered ring would be formed.

Much work has been done on the conformations of 1,3-dioxocanes, 1,3,5-trioxocane, and 1,3,5,7-tetroxocanes, but little seems to be known about simple 1,2-, 1,4-, and 1,5-dioxocanes. 1,3-Dioxocane and derivatives with *gem*-dimethyl groups at the 2 and/or 6 positions exist exclusively in *BC*-1,3 boat–chairs (**84**) that minimize nonbonded and dipole–dipole repulsions and that are favorable for the anomeric effect, as shown by dynamic NMR and dipole moment data.[201,202,208] The lowest-energy process in 1,3-dioxocane is the pseudorotation of **85** to its enantiomer, **86**, via the plane symmetrical form, **87**. The pseudorotation is greatly hindered by the presence of *gem*-dimethyl groups at the 2 position in **84**, because **87** with such substituents becomes very crowded. The lowest-energy process in the 2,2-dimethyl derivative is the conversion of **88a** to **88b** via the *BB*-1,3 boat–boat intermediate of C_2 symmetry. The methyl groups *A* and *B* in **88** thus exchange sites and this change is actually similar to the high-energy conformational process in 1,3-dioxocane itself. This crossover of mechanisms is shown diagrammatically in Fig. 2-1. Molecular mechanics cal-

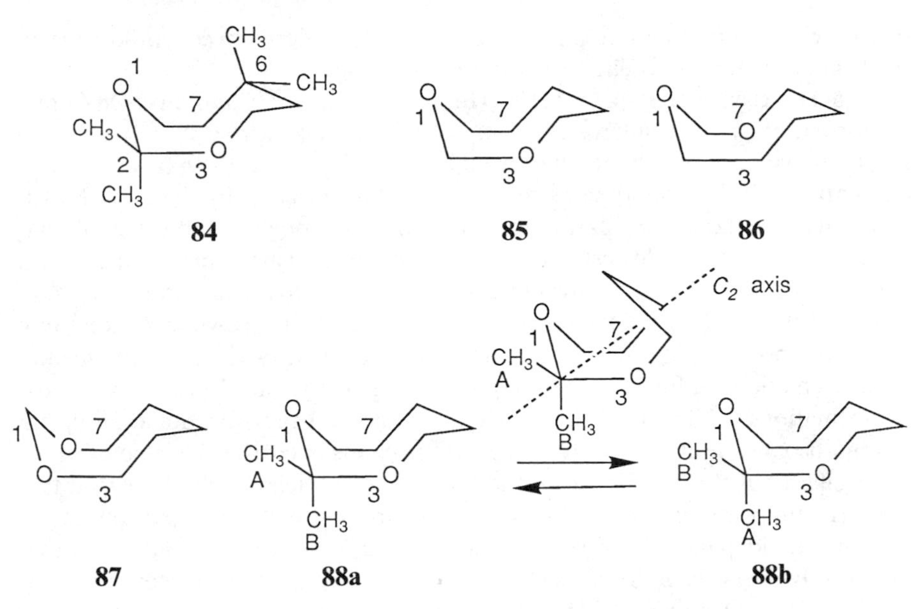

culations show that the *BC*-1,3 is of lowest energy and that the *BB*-1,3 form is the next most favorable ($\Delta E = 1.2$ kcal/mol), where the latter is stabilized by the oxygens being in positions to relieve 1,5 transannular nonbonded interactions. Even the presence of the somewhat sterically crowded axial methyl group in the 6,6-dimethyl derivatives is unable to make the boat–boat

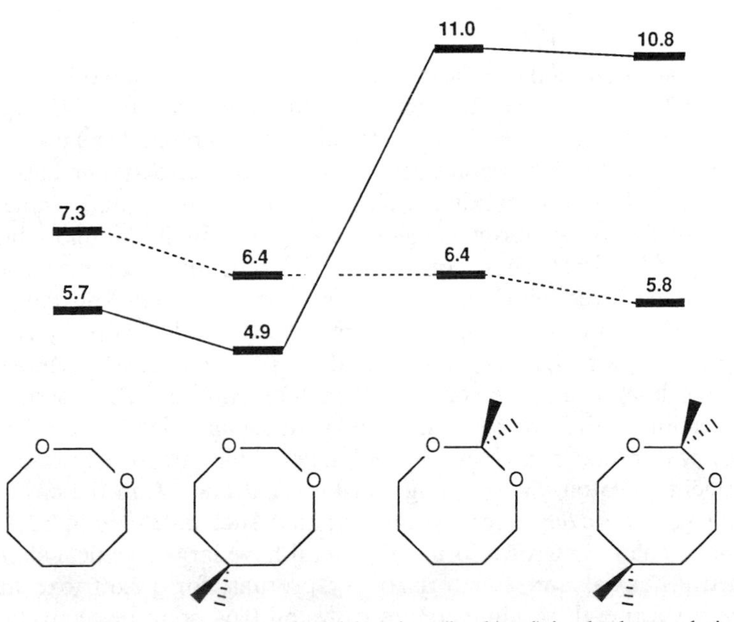

Figure 2-1. Experimental conformational barriers (kcal/mol) in the boat–chair conformations of 1,3-dioxocane and its symmetrical *gem*-dimethyl derivatives. The twist–boat and boat–boat processes introduce C_s and C_2 time-averaged symmetries and the corresponding transition states are labeled by full and dotted lines, respectively.

observable.[202] Originally it was thought that these derivatives would exist in boat–boats in order to relieve this steric crowding.[208]

1,3,6-Trioxocane exists as a 1:1 mixture of a boat–chair and a crown-family conformation with an interconversion barrier of 8.6 kcal/mol.[201,208] The ring inversion barrier in the boat–chair itself is 6.7 kcal/mol but the barrier to pseudorotation in this form appears to be too low to measure by dynamic NMR. Molecular mechanics calculations on this molecule done by Burkert show that the *BC*-1,3,6 (**89**) is of lowest energy and that the next best form is the crown (**90**) ($\Delta E = 1.0$ kcal/mol).[203] Strauss and co-workers have done similar calculations but with a different force field and report that the crown is 4.2 kcal/mol above the boat–chair, which is not in good agreement with the experimental data, even allowing for likely entropy solvation energy differences between the conformations.[209] These workers have also analyzed the vibrational spectrum of 1,3,6-trioxocane and they have assigned the bands in terms of a single *BC*-1,3,6 boat–chair, but this conflicts with the NMR data, which clearly show that two conformations are present.[45] The heats of formation of 1,3,6-trioxocane in the gas and liquid phases have been measured and, together with data on other cyclic polyethers, have been used to determine apparent strain contributions of O—CH_2—O and O—CH_2—CH_2—O units in conformationally similar molecules.[129]

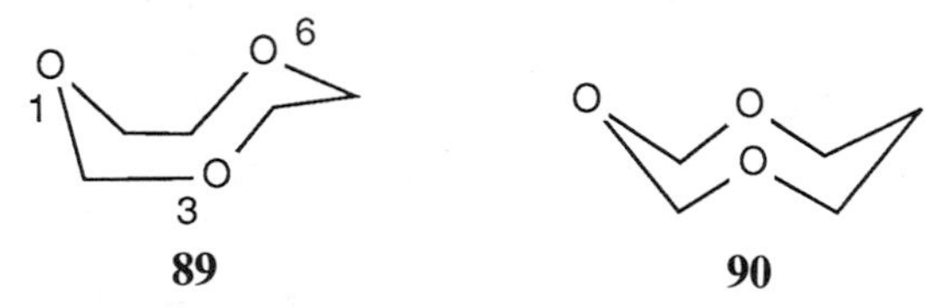

The conformations of the cyclic tetramer of formaldehyde and other aldehydes have been extensively investigated by X-ray diffraction, dynamic NMR, IR and Raman spectroscopy, dipole moment measurements, and molecular mechanics. In the crystalline state, the parent compound, 1,3,5,7-tetroxocane, has a crown conformation (**91**) with approximate C_{4v} (exact C_{2v}) point group symmetry,[210] but there are clearly two forms in solution, as shown by NMR and vibrational spectroscopy.[201,208,211] These forms have different dipole moments and entropies so that the equilibrium constant is strongly solvent and temperature dependent. The crown has a high dipole moment but a low entropy compared with the boat–chair (**92**) ($\Delta S^\circ = 6$ eu) and it is favored at low temperatures, whereas the boat–chair is favored at high temperatures. The barrier to the interconversion of the two conformations is unusually high (12 kcal/mol), but processes within the boat–chair have barriers that are too low for NMR measurements. Proton-^{13}C coupling constants (^1J and ^3J) in the crown show interesting geometric dependences; unfortunately such data cannot be obtained for the boat–chair.[212] Crown forms in general have large chemical shift differences between axial and equatorial CH_2 protons; for 1,3,6-trioxocane and 1,3,5,7-tetroxocane, these numbers are 0.88 and 0.64 ppm, respectively.[201,208] In contrast, this shift difference is only 0.07 ppm in the boat–chair of 1,3,6-trioxocane.

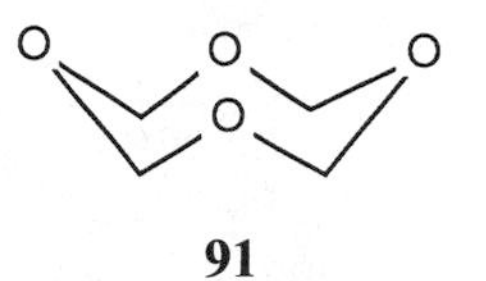

91

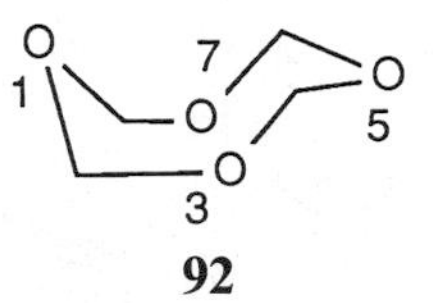

92

Molecular mechanics calculations by Burkert on 1,3,5,7-tetroxocane show that the crown is 2.6 kcal/mol more stable than the BC-1,3,5,7 form (**92**) in the gas phase,[203] and the difference should be even larger in solution as the crown is more highly solvated than the boat–chair. These molecular mechanics calculations do not reproduce the experimental data at all well, but it should be remembered that the strain energy contributions in the two conformations are quite different and thus 1,3,5,7-tetroxocane is a stringent test of a force field. A "valence force field" for the crown has been developed to reproduce IR and Raman vibrational data in the liquid and crystalline phases of this compound and then used to show that the other conformation is the *BC*-1,3,5,7 boat–chair.[211] The temperature dependence of the liquid-phase IR spectra gives an enthalpy difference of 1.0–1.4 kcal/mol, with the crown form the more stable. Lattice frequencies in crystalline 1,3,5,7-tetroxocane have been interpreted in terms of the above intramolecular valence force field, together with additional intermolecular nonbonded (Lennard-Jones) parameters.[213]

Metaldehyde (all-*cis*-2,4,6,8-tetramethyl-1,3,5,7-tetroxocane) exists in the crown form in the crystalline phase with all the methyl groups equatorial, as shown by a very early X-ray study.[214] The analogous metachloral[215] and 2,6-bis-(trichloromethyl)-1,3,5,7-tetroxocane[212] have been shown to exist in crown forms in solution by NMR and dipole moment measurements. It is interesting that crown conformations are also found in several nitrogen- and sulfur-containing heterocycles related to 1,3,5,7-tetroxocane, although 1,3,5,7-tetrathiocane is entirely in the boat–chair.[54]

2.7.2. Rings with One Strong Torsional Constraint

X-ray structures have been obtained on a number of marine natural products containing a single ring oxygen atom and where the torsional constraint is a *cis* double bond, as in laurenyne[216] (**93**) and laurencin[217] (**82**), or a *cis* epoxide group, as in poiteol[218] (**94**), 3-*Z*-epoxyvenustin[219] (**95**), and epoxyrhodophytin[220] (**96**), or a 1,3 bridge, as in chlorofucin[218] (**97**). These molecules have distorted boat–chair conformations as in *cis*-cyclooctene (section 2.4.2), except

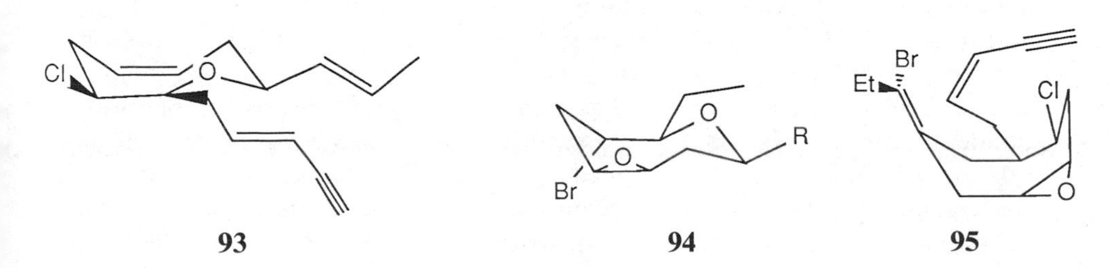

93 **94** **95**

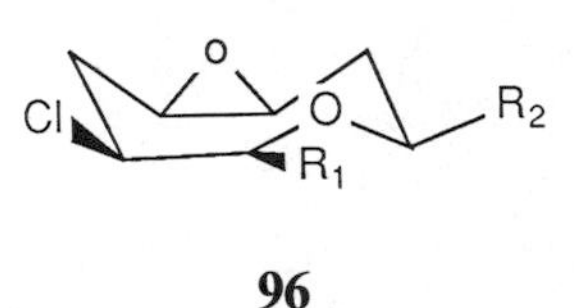

96

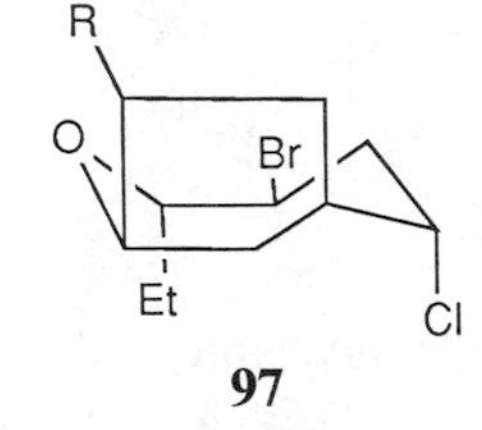

97

for 3-*Z*-epoxyvenustin, which can be considered to be midway between a boat–chair and a boat–boat. These distortions are needed to accommodate an eclipsed ring bond. The solution conformations of these compounds have not been established and need not necessarily correspond to the crystal structures. Extremely complicated fused cyclic polyethers are found in dinoflagellate toxins, such as the brevetoxins.[221–226] These molecules contain oxacyclooctane or oxacyclooctene rings fused to two other heterocyclic rings, which are sometimes also eight membered, but details of their conformations will not be discussed here, except to mention that the lone unsaturated eight-membered ring in brevetoxin B,[221] as well as in a molecule prepared as a tricyclic model for this toxin,[227] have *cis*-cyclooctene-like conformations.

2.7.3. Rings with More Than One Strong Torsional Constraint

Little is known concerning the conformations of eight-membered oxygen heterocycles with two separated torsional constraints—that is, heterocyclic analogues of the *cis-cis*-1,3-, 1,4-, and 1,5-cyclooctadienes and their fused benzo derivatives—although related nitrogen and sulfur compounds have been extensively investigated.[54] An unusual eight-membered unsaturated peroxylactone, isolated from fresh brown seaweed, has two torsional constraints (apart from the peroxide bond) and exists in the crystalline state as a twist–boat conformation related to that of *cis-cis*-1,5-cyclooctadiene.[228]

Peri-fused naphthalene compounds containing an eight-membered carbocyclic or heterocyclic ring have two special torsional constraints (section 2.4.2) and have interesting conformational properties. The cyclic ether, **98**, has been assigned the boat conformation in analogy with its carbocyclic analogue (**99**), and the ring inversion barriers in the two molecules are similar and high enough (13.7 and 14.0 kcal/mol) that broad NMR spectra are obtained at room temperature.[229] (A more detailed study of **98** has subsequently been carried out and a $\Delta G^{\ddagger}$ of 14.5 kcal/mol obtained[230]; related compounds have also been studied.[231–233]) The chair form, which is rigid, is disfavored by eclipsing strains, while the boat, which is flexible, has staggered bonds but considerable nonbonded strain. A twist–boat of C_2 symmetry has some eclipsing as well as nonbonded strains and is an intermediate in the ring inversion of the chair. In the cyclic acetal, **100**, studied by Arbuzov, Klimovitskii, and co-workers, the balance between the boat and chair is closer than in **98** or **99** because the boat

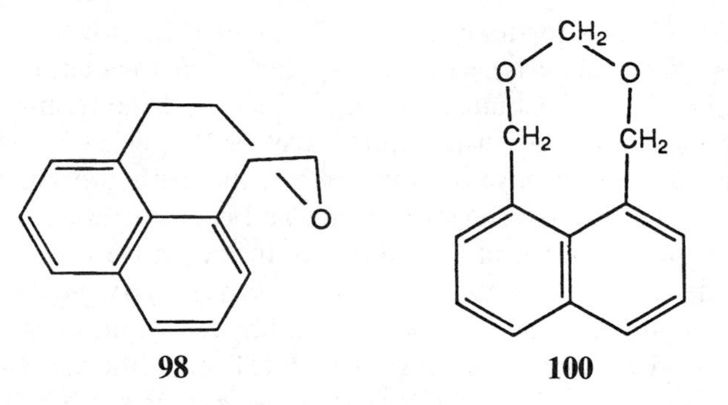

is more puckered and nonbonded strains are larger when two oxygen atoms are present in the ring.[234–236] The *tert*-butyl derivative, **101** in the crystalline state has the chair conformation with the substituent in the equatorial position,[234] but in solution there is a small fraction of the boat conformation (7% at 0°C) in addition to the chair.[236] The twist–boat form is unfavorable in this molecule because of a strong gauche interaction of the *tert*-butyl with a ring methylene group. The parent compound, **100**, which also exists as a mixture of a boat (22% at 0°C) and a chair, has a ring inversion barrier for the chair of 10.8 kcal/mol.[236] The dipole moment of **100** is 2.4*D*, compared with calculated values of 2.6, 1.8, and 0.3*D* for the chair, boat, and twist–boat, respectively.[236] The presence of an additional minor conformation in solution over that present in the crystalline state is shown by IR. The minor form is most populated in nonpolar solvents, and must have a low dipole moment, in agreement with the data given above. Although the major form of **100** is certainly the chair, there is some uncertainty as to whether the minor form is actually the boat rather than the twist–boat. Molecular mechanics calculations indicate that the boat is 1.0 kcal/mol higher in energy than the chair, with the twist–boat considerably less stable than the boat.

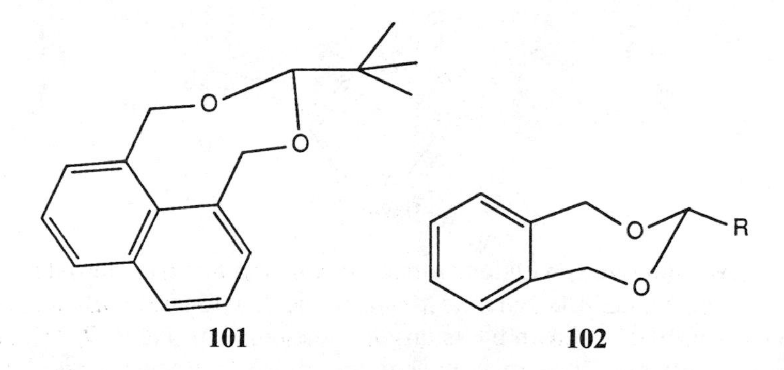

The ring puckering in the chair form of **100**, as measured by the dihedral angle (60.4°) between the O—C—O plane and the plane passing through these two oxygens and their adjacent benzylic carbons, is similar to that in the seven-membered cyclic acetal, **102**, but greater than in 1,3-dioxane, where the dihedral angle is about 70°.[234] Accompanying this difference in puckering is a difference in the C—O—C′—C″ torsional angles (179.1 and 163.5° for the six- and eight-membered rings, respectively), where C″ is a carbon of an equatorial substituent on the acetal carbon.[235] Comparisons of ^{13}C shifts of the terminal carbon (C) in the substituted with the unsubstituted acetals (γ_{anti} effects) as a function of ring size and substituent have been made. For the methyl group, γ_{anti} is -0.1, -1.58, and -0.98 ppm for the six-, seven-, and eight-membered rings, respectively, and these values reflect the difference in the above torsional angles.[235]

A trimer derived from biacetyl has been shown by two-dimensional ^{13}C-^{13}C COSY (60-hour data acquisition!),[237] and by X-ray diffraction to have two mutually bridged 1,3-dioxocane rings (**103**).[238] The addition of methyl lithium gives a plane symmetrical derivative that shows a dynamic NMR effect in the vicinity of room temperature, thus revealing the presence of a fairly high conformational barrier (about 15 kcal/mol). There are two different conformers (**104** and **105**), which differ in the equatorial and axial placements of the OH and CH_3 substituents on the eight-membered ring.

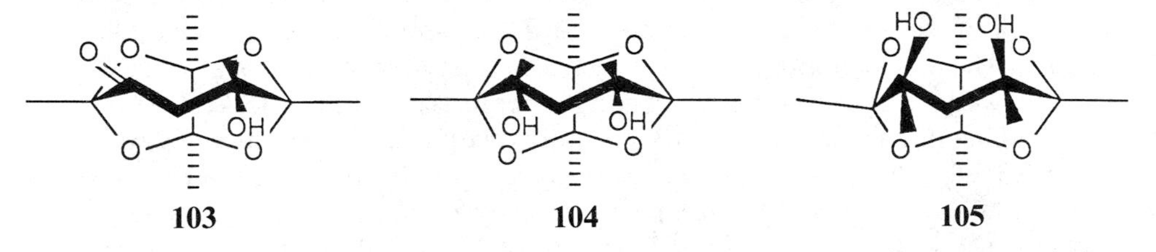

The hydrooxocine, **106**, which has three torsional constraints, has a conformation somewhat similar to 1,3,6-cyclooctatriene,[84] according to NMR data.[239] The 1,4-dioxocine ring also has three torsional constraints and is not aromatic, despite the potential presence of a 10π-electron system. A derivative of 1,4-dioxocine containing a pyrone substituent is nonplanar and has a "twist–boat–chair" conformation,[240] but a chlorocarbonyl derivative of 1,4-dioxocine is planar.[240]

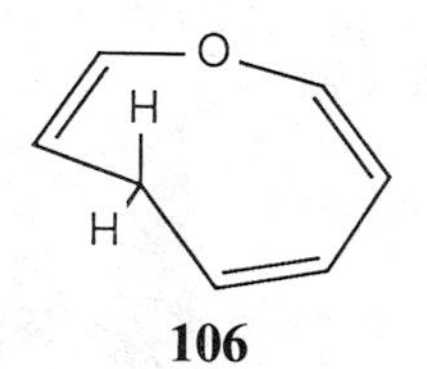

106

Only a few compounds with four torsional constraints have had their conformations examined, and these are mostly aromatic dilactones. Disalicylide exists in a boat form, **107**, as shown by its large dipole moment (6.26*D*)[241]; the chair, which is centrosymmetrical, must have a zero dipole moment and is therefore excluded. The related di-*o*-thymotide is also a boat and has a barrier to ring inversion of 17.7 kcal/mol.[242] In one instance, chair (**108**) and boat (**109**) forms occur and have such a high interconversion barrier that atropisomers can be separated at room temperature, although isomerization does take place at higher temperatures.[243]

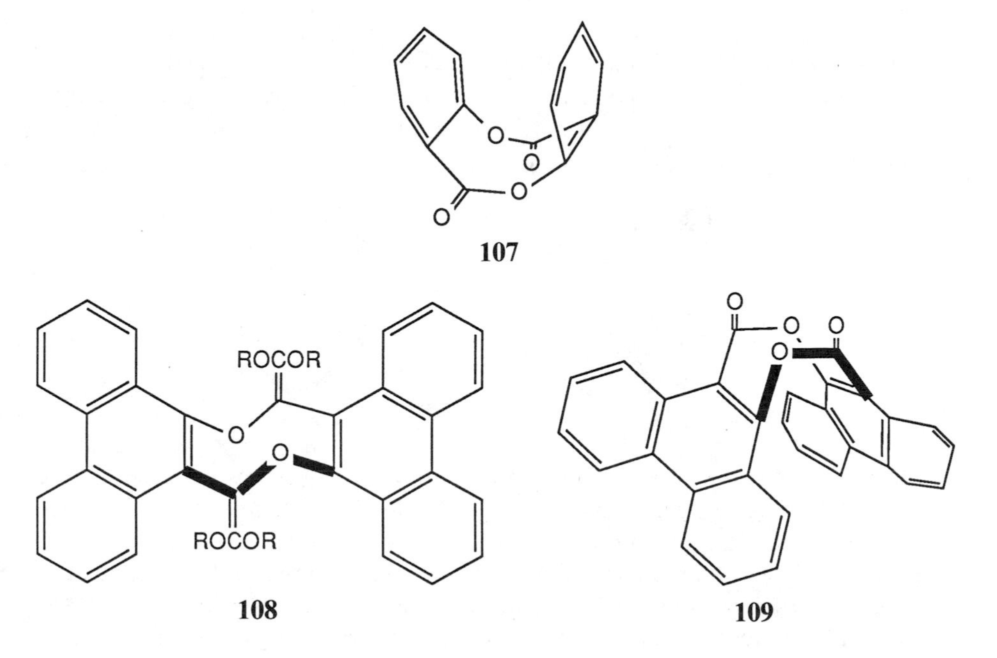

107

108 109

2.8. Nine-membered Oxygen Heterocyclic Rings

The conformations of only a few nine-membered oxygen heterocycles are known, and most of these molecules have several rings or several ring oxygen atoms. 1,4,7-Trioxacyclononane (**110**), which has symmetrically disposed oxygens and is actually a very small crown ether, has the same IR in the crystal as in solution and therefore appears to be conformationally homogeneous.[244] Dynamic ^{13}C NMR shows the presence of two processes ($\Delta G^{\ddagger} = 7.3$ and 7.9 kcal/mol): the spectrum changes from one line at room temperature to two broad lines at about -100°C, and then to five lines in the ratio of 1:1:1:2:1 at -130°C. These data are consistent with a twist–boat–chair (or [234] conformation in Dale's terminology), with the oxygens as shown in **111**, but this form has a high calculated dipole moment in contrast to that observed (1.55 *D*).

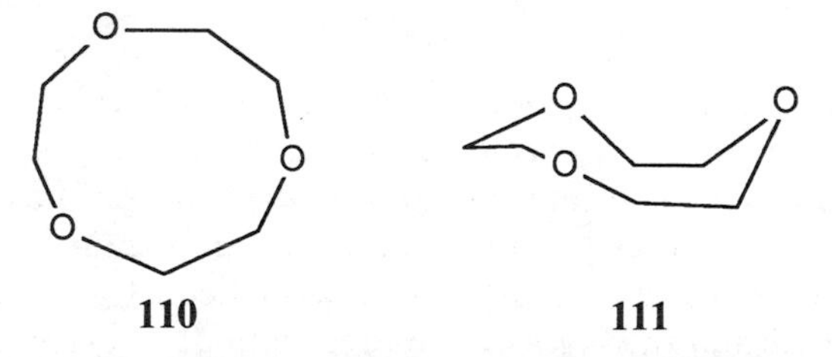

110 111

Molecular mechanics calculations show that the [234] form lies 4.8 kcal/mol above the symmetrical [22221] form (**112**), which is the lowest-energy conformation.[245] The latter form cannot be the sole species because it can only give three ^{13}C chemical shifts. The calculations also reveal that a twist–chair–chair or [9] (also termed [441]) form (**113**), which has the proper features to explain the NMR data, lies only 0,8 kcal/mol above the [22221] form. Further experimental data are needed to determine the conformational features of this interesting molecule.

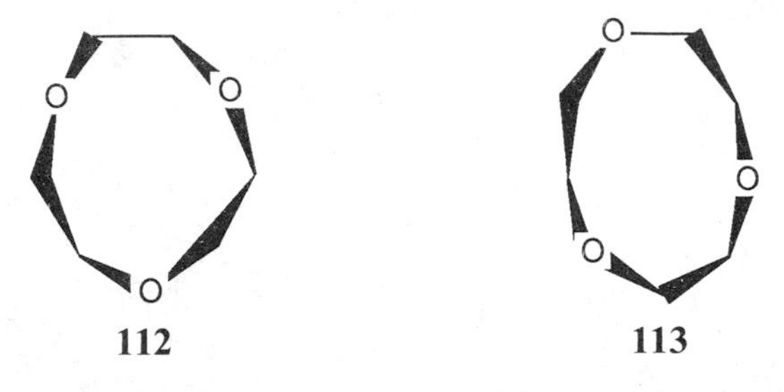

Ketones give rise under appropriate conditions to (explosive) cyclic nine-membered peroxides. The conformation (**114**), [333], of "trimeric acetone peroxide" in the crystal has approximate D_3 symmetry, and the crystals are built up from both enantiomeric molecules.[246] The all-*cis* isomer of the trimeric peroxide derived from chloroacetone shows a sharp *AB* quartet in its ^{1}H NMR spectrum, even at 155°C, and this means that the free-energy barrier for enantiomerization is higher than 24 kcal/mol and that this compound, and presumably also acetone trimeric peroxide, is potentially resolvable at room temperature.[135] However, no report of the resolution of such molecules, which are interesting because they belong to the relatively rare D_3 point group and lack stereogenic carbons, has yet been published.

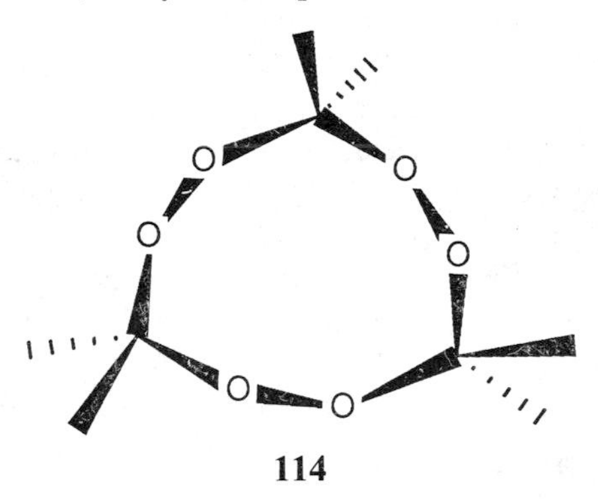

114

The conformations of *cis*-cyclononene and a symmetrical *cis*-oxacyclononene (**115**) have been investigated by molecular mechanics calculations.[247,248] The best conformations of these rings are almost the same as those in two marine natural products, isolaurallene (**116**) and the diastereomeric neolaurallene (**117**). The latter molecules have an additional torsional constraint caused by the fusion of a tetrahydrofuran ring as compared with **115**. Eleven energy minima for **115** have been identified, and their complicated interconversions have been studied by molecular mechanics, but unfortunately there are no experimental data for this molecule.

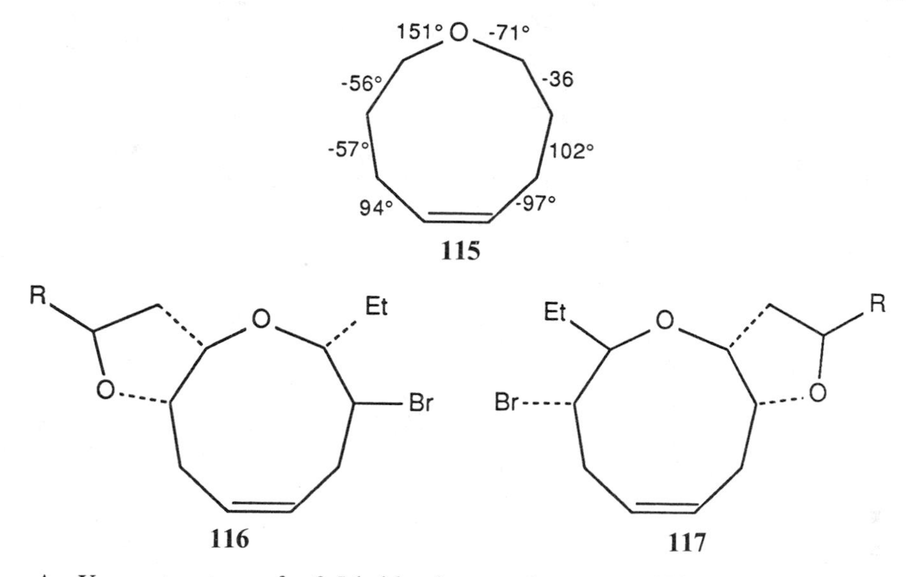

An X-ray structure of a 3,5 bridged oxacyclononane, **118**, shows a distorted boat–chair (or twist–boat–chair), represented by **119**, for the nine-membered ring.[249]

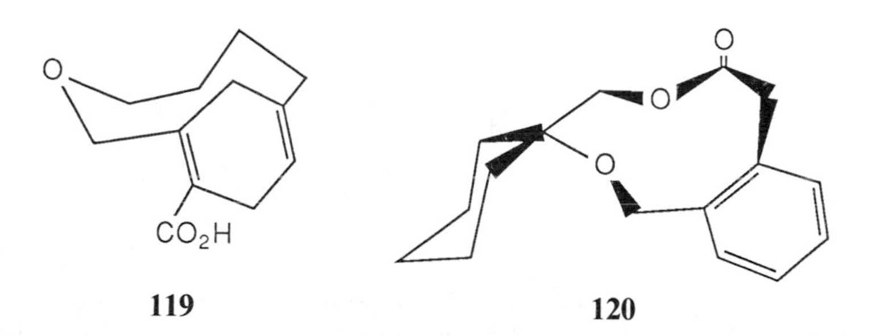

A benzo-fused nine-membered ring containing an ether and a lactone moiety has the structure shown in **120**.[250] The lactone moiety, unlike that in a smaller ring, has the syn arrangement as found in esters (corresponding to a *trans* double bond in the analogous carbocycle). The O—C—Ar—C—C(carbonyl) part of the ring has torsional angles corresponding to an approximate (local) C_s symmetry, while in the other part of the molecule, the torsional angles in the C—O—C(spiro)—C—O—C(carbonyl) chain correspond to an approximate (local) C_2 symmetry. This kind of situation is fairly common in unsymmetrical medium-ring conformations, such as the conformation of *cis*-cyclooctene.

A biphenyl-fused and two related binaphthyl-fused unsaturated 1,5-dioxacyclononanes, (e.g., **121**) have been studied by X-ray diffraction and found to have similar conformations.[251–253] The torsional angles in the nine-membered ring are given in Table 2-4 and show that this ring is quite unsymmetrical. There is an approximate (local) C_s symmetry of the biphenyl or binaphthyl and the attached O—C bonds in the nine-membered ring.

TABLE 2-4. Torsional Angles in the Nine-membered Rings of **121** and Its 1,1-Binaphthyl Analogue[251–253]

Ring Atoms	Torsional Angles (degrees)	
	121	Binaphthyl Analogue
1–2–3–4	50	45
2–3–4–5	−80	−82
3–4–5–6	114	115
4–5–6–7	−120	−117
5–6–7–8	7	5
6–7–8–9	58	64
7–8–9–1	8	4
8–9–1–2	−134	−130
9–1–2–3	64	68

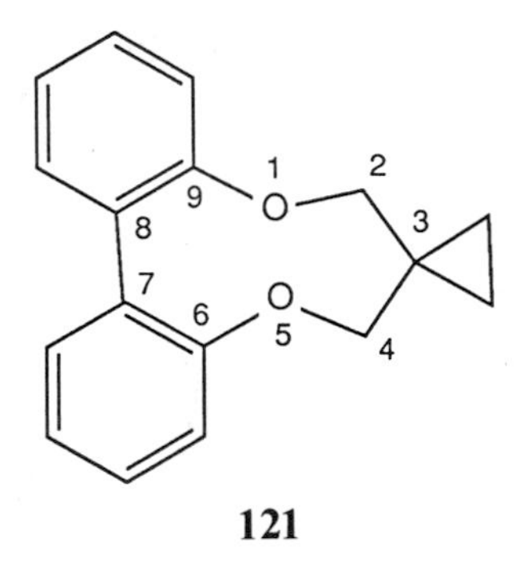

121

The barrier to ring inversion in 1,4,7-trioxonin (**122**),[254] which presumably exists as a crown and is a heterocyclic analogue of 1,4,7-cyclononatriene, cannot be obtained by dynamic NMR, but this would be possible in derivatives of this interesting molecule. The tribenzo derivative of a 3-oxa-1,4,7-cyclononatriene (**123**) has been assigned a crown form on the basis of a large (about 1-ppm) chemical shift difference between the CH_2 protons at −90°C.[255] The low barrier to ring inversion in **123** (estimated to be about 10 kcal/mol; no actual coalescence temperature is given) compared with its carbocyclic analogues, some of which can be resolved at room temperature (about 26 kcal/mol),[95] has been ascribed to the large bond angle (124 ± 5°) expected at the oxygen atom in the diphenyl ether moiety.

122

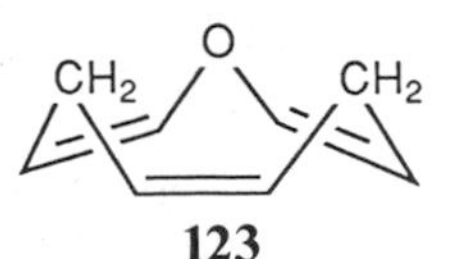

123

2.9. Ten- to 12-Membered Oxygen Heterocyclic Rings

The conformations of only a few oxygen heterocycles that have ring sizes in the range of 10 to 12 members seem to be known. Some of these molecules contain several oxygens and can be considered small crown ethers. The X-ray structure of the 10-membered lactone, 2-oxacyclodecane-1,6-dione, shows a boat–chair–boat (or [2323]) conformation (**124**) with the cyclic ester in a *syn* arrangement and both carbonyl groups relieve transannular repulsions. There is, in fact, a weak transannular attraction between the ring lactone oxygen and the carbon atom of the ketone carbonyl group, as shown by a C . . . O distance of 2.83 Å, compared with the van der Waals distance of 3.1 Å.

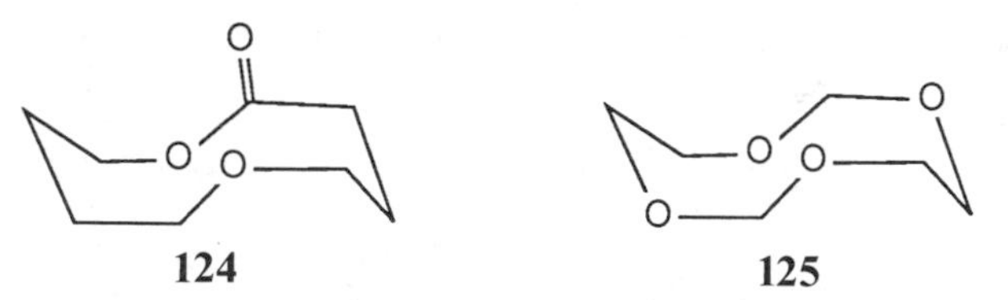

124 **125**

1,3,6,8-Tetraoxacyclodecane in the crystalline phase has a centrosymmetrical diamond–lattice (boat–chair–boat or [2323]) conformation (**125**) similar to the preferred conformation of cyclodecane.[256] The oxygens occupy positions that relieve transannular nonbonded strains and the C—O—CH_2—O—C units have the ideal dimethoxymethane geometry that minimizes electrostatic repulsions and optimizes the anomeric effect (section 2.5). 1,3,5,7,9-Tetraoxacyclodecane, which is a pentamer of formaldehyde, has a slightly lower gaseous enthalpy of formation per CH_2O unit than 1,3,7-tetroxocane or 1,3,5-trioxane.[257]

An X-ray structure of a 3,5 bridged oxacyclodecane homologous to **118** (section 2.8) shows that the 10-membered ring is a distorted boat–chair–chair.[249] A derivative of **125**, constrained by two bridging six-membered rings, gives ^1H–^1H vicinal coupling constants corresponding to the latter rings apparently being in twist–boat forms. The molecule has a C_2-axis, at least on the NMR time scale at room temperature. Molecular mechanics calculations, done without considering electrostatic or anomeric effects, give a conformation that has transannular oxygens very close to one another (2.35 Å). Since the agreement between calculated and observed coupling constants is only fair, the proposed conformation cannot be regarded to be well established and an X-ray structure determination would be desirable.[258] The dilactone, **126**, is constrained in the

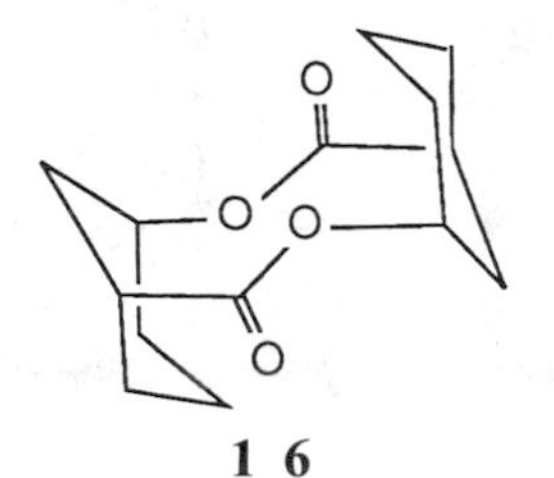

1 6

same way as is the previous compound and it has the [2323] conformation for its 10-membered rings, with a syn ester moiety.[259]

Some fungal metabolites contain 10-membered lactone rings. An X-ray structure has been obtained for one of these compounds in order to establish the relative configurations of its stereocenters, but its conformational features have not been discussed.[260]

The conformations of a few 11-membered-ring oxygen heterocycles have been investigated, but the static conformers suggested for these molecules,[261] which can be regarded as 2:11-bridged oxacycloundecanes (e.g., **127**), are not supported by any real evidence, and alternate forms with less eclipsing strains can be drawn. No X-ray structures are available nor have molecular mechanics calculations been done. However, barriers have been measured by dynamic NMR, and it is clear that these molecules show interesting conformational processes.[261–263]

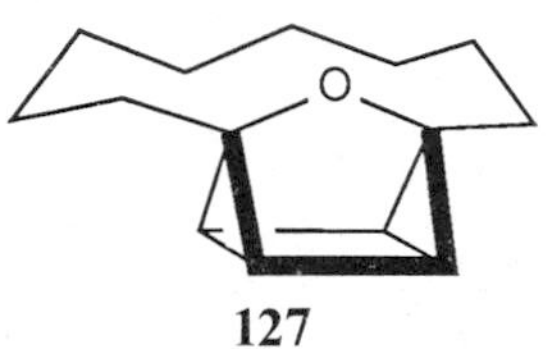

127

The conformation of 12-crown-4 (1,4,7,10-tetraoxacyclododecane), **128**, which is a tetramer of ethylene oxide, and several of its metal complexes have been investigated by X-ray and dynamic NMR methods. The 12-membered ring in the LiCNS complex of **128** has been assigned a [3333] conformation (**129**) with the oxygen atoms on side positions, according to dynamic NMR data.[264] This ring conformation is also found in sodium and calcium salt complexes of **128**, as shown by X-ray structures.[265] The same conformation was originally also assigned to the uncomplexed crown ether, although the dipole moment is not in agreement with this form.[264] In the crystalline state, 12-crown-4 has the [66] geometry (**130**),[266] and more recent work shows that it exists as a mixture of forms in solution. In most solvents, **128** is probably in the [66] form, but the [3333] and [66] forms appear to exist in a ratio of 2:1 in $CHCl_2F$ and $CHClF_2$ at low temperatures.[267] Molecular mechanics calculations give the [3333] form as being the lowest in energy, and the energy of the [66] form is a surprising

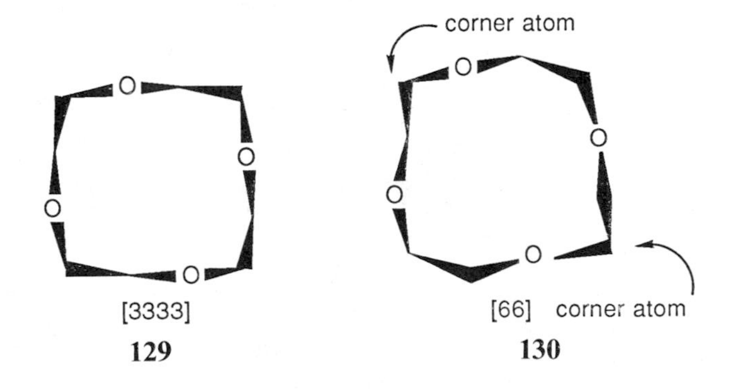

129 **130**

2.9 kcal/mol higher. Also, three other low-energy forms have been found.[245] The possible conformational processes in this molecule have been discussed and the slowness of the exchange between the complexed and uncomplexed crown has been ascribed to the necessity for a major conformational change in this process.[268]

Tri-*o*-thymotide (**131**) undergoes spontaneous resolution and gives rise to clathrates when crystallized from some solvents. In the crystal, **131** has a propeller form (C_3) with all the ester units syn, but in solution it exists as a mixture of the propeller and a "helix form" that lacks symmetry. The exchange between these forms has been studied by dynamic NMR; the racemization barrier is about 20 kcal/mol.[269]

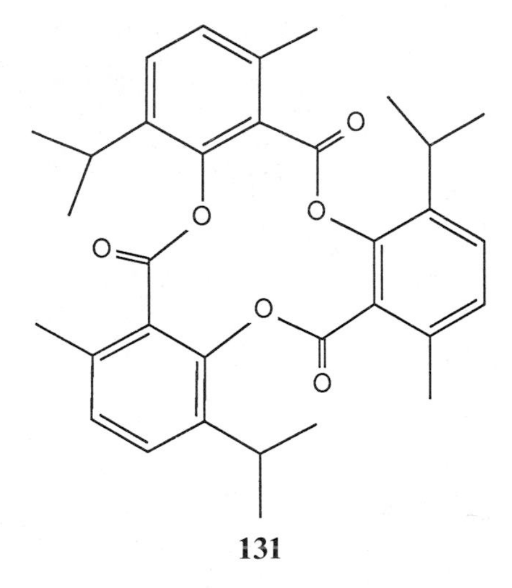

131

2.10. Conclusions

The conformations of medium-ring oxygen heterocycles exhibit a rich variety and are now fairly well understood. The ring sizes nine to 12 are sparsely represented, and further work in this area, particularly dealing with simple cyclic ethers, lactones, cyclic acetals, and ketals, is needed in order to obtain both static and dynamic conformational information. X-ray and dynamic NMR are the most important techniques for such studies, but vibrational spectroscopy and dipole moment measurements can also provide valuable information, especially in not too flexible molecules. Improvements in force fields used in molecular mechanics for oxygen-containing molecules can be anticipated, even though a great deal of work has already been done in this area. The effect of solvent polarity needs to be studied by examining the effect of clusters of solvating molecules on the conformational energies of the ring molecules. It is hoped that the application of the existing conformational knowledge of medium-ring oxygen heterocycles for organic synthetic purposes, as well as in physical organic chemistry research, will be made easier by the present review.

References

1. Prelog, V. *Bull. Soc. Chim. France* **1960**, 1433.
2. Prelog, V. *Pure Appl. Chem.* **1963**, *6*, 545.
3. Sicher, J. *Progr. Stereochem.* **1962**, *3*, 202.
4. Still, W. C.; Galynker, I. *Tetrahedron* **1981**, *23*, 3981.
5. Still, W. C.; MacPherson, L. J.; Harada, T.; Callahan, J. F.; Rheingold, A. L. *Tetrahedron* **1984**, *40*, 2275.
6. Nicolaou, K. C.; McGarry, D. G.; Somers, P. K.; Veale, C. A.; Furst, G. T. *J. Am. Chem. Soc.* **1987**, *109*, 2504.
7. Cope, A. C.; Martin, M. M.; McKervey, M. A. *Q. Rev. Chem. Soc.* **1966**, *20*, 119.
8. Riddell, F. G. "The Conformational Analysis of Heterocyclic Compounds." Academic Press: London, 1980.
9. March, J. "Advanced Organic Chemistry." Wiley-Interscience: New York, 1985.
10. Woolley, R. G. *Isr. J. Chem.* **1980**, *19*, 30.
11. Bader, R. F. W.; Tal, Y.; Anderson, S. G.; Nguyen-Dang, T. T. *Isr. J. Chem.* **1980**, *19*, 8.
12. Boeyens, J. C. A. *Struct. Bonding* **1985**, *63*, 65.
13. Cheng, A. K.; Anet, F. A. L.; Mioduski, J.; Meinwald, J. *J. Am. Chem. Soc.* **1974**, *96*, 2887.
14. Schulman, J. M.; Disch, R. L.; Sabio, M. L. *J. Am. Chem. Soc.* **1984**, *106*, 7696.
15. Mislow, K. *Bull. Soc. Chim. Belg.* **1977**, *86*, 595.
16. Eliel, E. L. *Isr. J. Chem.* **1977**, *15*, 7.
17. Bryant, R. G. *J. Chem. Educ.* **1983**, *60*, 933.
18. Dunitz, J. D.; Waser, J. *J. Am. Chem. Soc.* **1972**, *94*, 5645.
19. Pickett, H. M.; Strauss, H. L. *J. Chem. Phys.* **1970**, *53*, 376.
20. Sachse, H. *Ber.* **1890**, *23*, 1363.
21. Sachse, H. *Z. Physik. Chem. (Leipzig)* **1892**, *10*, 203.
22. Mohr, E. *J. Prakt. Chem. [2]* **1922**, *103*, 316.
23. Hückel, W. *Ann.. Chem.* **1925**, *441*, 1.
24. Eliel, E. L.; Allinger, N. L.; Angyal, S. J.; Morrison, G. A. "Conformational Analysis." Wiley-Interscience: New York, 1965.
25. Hanack, H. "Conformational Theory." Academic Press: New York, 1965.
26. Anet, F. A. L.; Anet, R. In "Dynamic Nuclear Magnetic Resonance Spectroscopy," Jackman, L. M.; Cotton, F. A., Eds. Academic Press: New York, 1975, pp 543–619.
27. Anet, F. A. L.; Kozerski, L. *J. Am. Chem. Soc.* **1973**, *95*, 3407.
28. Kilpatrick, J. E.; Pitzer, K. S.; Spitzer, R. *J. Am. Chem. Soc.* **1947**, *69*, 2483.
29. Altona, C.; Sundaralingam, M. *J. Am. Chem. Soc.* **1972**, *95*, 8205.
30. Dunitz, J. D. *Tetrahedron* **1972**, *28*, 5459.
31. Cremer, D.; Pople, J. A. *J. Am. Chem. Soc.* **1975**, *97*, 1354.
32. Cremer, D. *Isr. J. Chem.* **1979**, *20*, 12.
33. Miller, R. W.; McPhail, A. T. *J. Chem. Res. (M)* **1979**, 3122.
34. Zefirov, N. S.; Palyulin, V. A. *Proc. Acad. Sci. U.S.S.R., Chem. Ser. (Engl. transl.)* **1980**, *252*, 207.
35. Dale, J. *Top. Stereochem.* **1976**, *9*, 199.
36. Rao, S. T.; Westhof, E.; Sundaralingam, M. *Acta Crystallogr. Sect. A* **1981**, *37*, 421.
37. Petit, G. H.; Dillen, J.; Geise, H. J. *Acta Crystallogr. Sect. B* **1983**, *39*, 648.
38. Cremer, D. *Acta Crystallogr. Sect. B* **1984**, *40*, 498.
39. Essén, H.; Cremer, D. *Acta Crystallogr. Sect. B* **1984**, *40*, 418.
40. Pickett, H. M.; Strauss, H. L. *J. Am. Chem. Soc.* **1970**, *92*, 7281.
41. Bocian, D. F.; Pickett, H. M.; Rounds, T. C.; Strauss, H. L. *J. Am. Chem. Soc.* **1974**, *97*, 687.
42. Bocian, D. F.; Strauss, H. L. *J. Am. Chem. Soc.* **1977**, *99*, 2866.
43. Bocian, D. F.; Strauss, H. L. *J. Am. Chem. Soc.* **1977**, *99*, 2876.
44. Rounds, T. C.; Strauss, H. L. *J. Chem. Phys.* **1978**, *69*, 268.
45. Pakes, P. W.; Rounds, T. C.; Strauss, H. L. *J. Phys. Chem.* **1981**, *85*, 2476.
46. Graveron-Demilly, D. *J. Chem. Phys.* **1977**, *66*, 2874.
47. Offenbach, J. L.; Strauss, H. L.; Graveron-Demilly, D. *J. Chem. Phys.* **1978**, *69*, 3441.
48. Loz'ach, N. *J. Chem. Inf. Comput. Sci.* **1985**, *25*, 180.
49. Ōki, M. "Applications of Dynamic NMR Spectroscopy to Organic Chemistry." VCH Publishers: Deerfield Beach, Fla., 1985.
50. Hendrickson, J. B. *J. Am. Chem. Soc.* **1967**, *89*, 7047.
51. Boeyens, J. C. A. *J. Cryst. Mol. Struct.* **1978**, *8*, 317.

52. Stoddart, J. F. "Stereochemistry of Carbohydrates." Wiley-Interscience: New York, 1971.
53. Anet, F. A. L. Unpublished work.
54. Moore, J. A.; Anet, F. A. L. In "Comprehensive Heterocyclic Chemistry," Lwowski, W. Ed. Pergamon Press: Oxford, 1984, Vol. 5, pp 653–707.
55. Engler, E. M.; Andose, J. D.; Schleyer, P. v. R. *J. Am. Chem. Soc.* **1973**, *95*, 8005.
56. Hehre, W. J.; Radom, L.; Schleyer, P. v. R.; Pople, J. A. "*Ab Initio* Molecular Orbital Theory." Wiley-Interscience: New York, 1986.
57. Dodziuk, H. *J. Mol. Struct.* **1979**, *55*, 107.
58. Burkert, U.; Allinger, N. L. "Molecular Mechanics." American Chemical Society: Washington, 1982.
59. Rasmussen, K. "Potential Energy Functions in Conformational Analysis." Springer-Verlag: Berlin, 1985.
60. Ermer, O. *Struct. Bonding* **1976**, *27*, 161.
61. Still, W. C. In "Current Trends in Organic Synthesis," Nozaki, H., Ed. Pergamon: Oxford, 1983, p 233.
62. Saunders, M. *J. Am. Chem. Soc.* **1987**, *109*.
63. Diehl, P.; Jokisaari, J. In "Applications of NMR Spectroscopy to Problems in Stereochemistry and Conformational Analysis," Takeuchi, Y.; Marchand, A. P., Eds. VCH Publishers: Deerfield Beach, Fla., 1986, p 41.
64. Haasnoot, C. A. G.; de Leeuw, F. A. A. M.; de Leeuw, H. P. M.; Altona, C. *Recl. Trav. Chim. Pays-Bas* **1979**, *98*, 576.
65. Kessler, H.; Bermel, W. In "Applications of NMR Spectroscopy to Problems in Stereochemistry and Conformational Analysis," Takeuchi, Y.; Marchand, A. P., Eds. VCH Publishers: Deerfield Beach, Fla., 1986, p 179.
66. Lambert, J. B.; Vagenas, A. R. *Org. Magn. Reson.* **1981**, *17*, 265.
67. Dessinges, A.; Castillon, S.; Olesker, A.; Thang, T. T.; Lukacs, G. *J. Am. Chem. Soc.* **1984**, *106*, 450.
68. Sandström, J. "Dynamic NMR Spectroscopy." Academic Press: London, 1982.
69. Esteban, A. L.; Galiano, C.; Diez, E.; Bermejo, F. J. *J. Chem. Soc. Perkin Trans. 2* **1982**, 657.
70. De Clercq, P. J. *J. Org. Chem.* **1981**, *46*, 667.
71. Kabuß, S.; Schmid, H. G.; Friebolin, H.; Faißt, W. *Org. Magn. Reson.* **1969**, *1*, 451.
72. St. Jacques, M.; Vaziri, C. *Org. Magn. Reson.* **1972**, *4*, 77.
73. Grunwald, E.; Price, E. *J. Am. Chem. Soc.* **1965**, *87*, 3139.
74. v. Bredow, K.; Jaeschke, A.; Schmid, H. G.; Friebolin, H.; Kabuß, S. *Org. Magn. Reson.* **1970**, *2*, 543.
75. Saebø, S.; Boggs, J. E. *J. Mol. Struct.* **1982**, *87*, 365.
76. Favini, G.; Maggi, A.; Todeschini, R. *J. Mol. Struct.* **1983**, *105*, 17.
77. Nógrádi, M.; Ollis, W. D.; Sutherland, I. O. *J. Chem. Soc., Chem. Commun.* **1970**, 158.
78. Anet, F. A. L. *Top. Curr. Chem.* **1974**, *45*, 169.
79. Anet, F. A. L. In "Conformational Analysis, Scope and Present Limitations," Chiurdoglu, G., Ed. Academic Press: New York, 1971, pp 15–29.
80. Servis, K. L.; Noe, E. A. *J. Am. Chem. Soc.* **1973**, *95*, 171.
81. Ermer, O. *J. Am. Chem. Soc.* **1979**, *98*, 3964.
82. Allinger, N. L.; Sprague, J. T. *Tetrahedron* **1975**, *31*, 21.
83. Anet, F. A. L.; Yavari, I. *J. Am. Chem. Soc.* **1977**, *99*, 6986.
84. Anet, F. A. L.; Yavari, I. *Tetrahedron Lett.* **1975**, 4221.
85. Fletschinger, M.; Zipperer, B.; Fritz, H.; Prinzbach, H. *Tetrahedron Lett.* **1987**, *28*, 2517.
86. Anet, F. A. L.; Krane, J. *Isr. J. Chem.* **1980**, *20*, 72.
87. Groth, P. *Acta Chem. Scand. Ser. A* **1976**, *30*, 667.
88. Borgen, G.; Dale, J. *Acta Chem. Scand. Ser. B* **1976**, *30*, 733.
89. Cope, A. C.; Banholzer, K.; Keller, H.; Pawson, B. A.; Whang, J. J.; Winkler, H. J. S. *J. Am. Chem. Soc.* **1965**, *87*, 3644.
90. Anet, F. A. L.; Yavari, I. *J. Am. Chem. Soc.* **1977**, *99*, 7640.
91. Anet, F. A. L.; Yavari, I. *J. Am. Chem. Soc.* **1977**, *99*, 6496.
92. Luche, J. L.; Damiano, J. C.; Crabbé, P.; Cohen-Addad, C.; Lajzerowicz, J. *Tetrahedron* **1977**, *33*, 961.
93. Anet, F. A. L.; Ghiaci, M. *J. Am. Chem. Soc.* **1979**, *103*, 2528.
94. Cookson, R. C.; Halton, B.; Stevens, I. D. R. *J. Chem. Soc. (B)* **1968**, 767.
95. Collet, A.; Gabard, J. *J. Org. Chem.* **1980**, *45*, 5400.
96. Anet, F. A. L.; Cheng, A. K.; Wagner, J. J. *J. Am. Chem. Soc.* **1972**, *94*, 9250.
97. Hilderbrandt, R. L.; Wieser, J. D.; Montgomery, L. K. *J. Am. Chem. Soc.* **1973**, *95*, 8598.
98. Anet, F. A. L.; Cheng, A. K.; Krane, J. *J. Am. Chem. Soc.* **1973**, *95*, 7877.

99. Alvik, T.; Borgen, G.; Dale, J. *Acta Chem. Scand.* **1972**, *26*, 1805.
100. Anet, F. A. L.; St. Jacques, M.; Henrichs, P. M.; Cheng, A. K.; Krane, J.; Wong, L. *Tetrahedron* **1974**, *30*, 1629.
101. White, D. N. J.; Bovill, M. J. *Tetrahedron Lett.* **1975**, 2239.
102. Vicković, I.; Watson, W. H.; Silva, M.; Garvez, M. E.; Hoeneisen, M. *Acta Crystallogr. Sect. C* **1985**, *41*, 436.
103. Fischer, N. H.; Malcolm, A. J.; Olivier, E. J.; Fronczek, F. R.; Delord, T. J.; Watkins, S. F. *J. Chem. Soc., Chem. Commun.* **1982**, 1243.
104. Anet, F. A. L.; Rawdah, T. N. *J. Am. Chem. Soc.* **1978**, *100*, 7810.
105. Groth, P. *Acta Chem. Scand. Ser. A* **1974**, *28*, 294.
106. Dunitz, J. D.; Shearer, H. M. M. *Helv. Chim. Acta* **1960**, *43*, 18.
107. Anet, F. A. L.; Rawdah, T. N. *J. Am. Chem. Soc.* **1978**, *100*, 7166.
108. Groth, P. *Acta Chem. Scand. Ser. A* **1979**, *33*, 203.
109. Wolf, R. E., Jr.; Hartman, J. R.; Storey, J. M. E.; Foxman, B. M.; Cooper, S. R. *J. Am. Chem. Soc.* **1987**, *109*, 4328.
110. Zefirov, N. S. *Tetrahedron* **1977**, *33*, 3193.
111. Lide, D. R., Jr. *J. Chem. Phys.* **1960**, *33*, 1514.
112. Blukis, V.; Kasal, P. H.; Myers, R. J. *J. Chem. Phys.* **1963**, *38*, 2753.
113. Hoyland, J. R. *J. Chem. Phys.* **1968**, *49*, 1908.
114. Allinger, N. L.; Chang, S. H.-M.; Glaser, D. H.; Hönig, H. *Isr. J. Chem.* **1979**, *20*, 51.
115. Eisenstein, O.; Anh, N. T.; Jean, Y.; Cantacuzène, J.; Salem, L. *Tetrahedron* **1974**, *30*, 1717.
116. Allan, A.; McKean, D. C.; Perchard, J.-P.; Josien, M.-L. *Spectrochim. Acta, A* **1971**, *27*, 1409.
117. Edward, J. T., *Chem. Ind. (London)* **1955**, 1102.
118. Kirby, A. J. "The Anomeric Effect and Related Stereoelectronic Effects at Oxygen." Springer-Verlag: Berlin, 1983.
119. Jørgensen, F. S.; Nørskov-Lauritsen, L. *Tetrahedron Lett.* **1982**, *23*, 5221.
120. Anderson, J. E.; Heki, K.; Hirota, M.; Jørgensen, F. S. *J. Chem. Soc., Chem. Commun.* **1987**, 554.
121. Anet, F. A. L.; Yavari, I. *J. Am. Chem. Soc.* **1977**, *99*, 6752.
122. Briggs, A. J.; Glenn, R.; Jones, P. G.; Kirby, A. J.; Ramaswamy, P. *J. Am. Chem. Soc.* **1984**, *106*, 6200.
123. Allen, F. A.; Kirby, A. H. *J. Am. Chem. Soc.* **1984**, *106*, 6197.
124. Jones, P. G.; Kirby, A. J. *J. Am. Chem. Soc.* **1984**, *106*, 6207.
125. Nørskov-Lauritsen, L.; Allinger, N. L. *J. Comput. Chem.* **1984**, *5*, 326.
126. Aped, P.; Apeloig, Y.; Ellencweig, A.; Fuchs, B.; Goldberg, I.; Karni, M.; Tartakovsky, E. *J. Am. Chem. Soc.* **1987**, *109*, 1486.
127. Cossé-Barbi, A.; Dubois, J.-E. *J. Am. Chem. Soc.* **1987**, *109*, 1503.
128. Gittins, V. M.; Wyn-Jones, E.; White, R. F. M. In "Internal Rotation in Molecules," Orville-Thomas, W. J. Ed. Wiley: London, 1974, pp 425–480.
129. Byström, K.; Månsson, M. *J. Chem. Soc. Perkin Trans. 2* **1982**, 565.
130. Wolfe, S. *Acc. Chem. Res.* **1972**, *26*, 1707.
131. Eliel, E. L.; Juaristi, E. In "Anomeric Effect Origins and Consequences," Szarek, W. A.; Horton, D., Eds. American Chemical Society: Washington, 1979, Vol. 87, p 95.
132. Lowe, J. P. *Progr. Phys. Org. Chem.* **1968**, *6*, 1.
133. Cremer, D. *J. Chem. Phys.* **1978**, *69*, 4440.
134. Salomon, M. F.; Salomon, R. G. *J. Am. Chem. Soc.* **1977**, *99*, 3500.
135. Anet, F. A. L.; Yavari, I. *Tetrahedron Lett.* **1976**, 3787.
136. Rushkin, S.; Bauer, S. H. *J. Phys. Chem.* **1980**, *84*, 3061.
137. Blom, C. E.; Günthard, Hs. H. *Chem. Phys. Lett.* **1981**, *84*, 267.
138. Wiberg, K. B.; Laidig, K. E. *J. Am. Chem. Soc.* **1987**, *109*, 5935.
139. Grindley, T. B. *Tetrahedron Lett.* **1982**, *23*, 1757.
140. Philip, T.; Cook, R. L.; Malloy, T. B., Jr.; Allinger, N. L.; Chang, S.; Yuh, Y., *J. Am. Chem. Soc.* **1981**, *103*, 2151.
141. Thomas, S. A. *J. Cryst. Spectr. Res.* **1985**, *15*, 115.
142. Clark, A. H. In "Internal Rotation in Molecules," Orville-Thomas, W. J., Ed. Wiley: London, 1974, pp 325–384.
143. Garcia, G.; Grunwald, E. *J. Am. Chem. Soc.* **1980**, *102*, 6407.
144. Allen, G.; Fewster, S. In "Internal Rotation in Molecules," Orville-Thomas, W. J., Ed. Wiley: London, 1974, pp 255–424.
145. Anderson, G. M., III; Kollman, P. A.; Domelsmith, L. N.; Houk, K. N. *J. Am. Chem. Soc.* **1979**, *101*, 2344.

146. Archer, A. W.; Claret, P. A.; Hayman, D. F. *J. Chem. Soc. (B)* **1971**, 1231.
147. Rounds, T. C.; Strauss, H. L. *Vibr. Spectr. Struct.* **1978**, *7*, 237.
148. Maple, S. R.; Allerhand, A. *J. Am. Chem. Soc.* **1987**, *109*, 3168.
149. James, V. J.; Stevens, J. D. *Carbohydr. Res.* **1980**, *82*, 167.
150. Beale, J. P.; Stephenson, N. C.; Stevens, J. D. *J. Chem. Soc., Chem. Commun.* **1971**, 484.
151. McConnell, J. F.; Stevens, J. D. *Cryst. Struct. Commun.* **1973**, *2*, 619.
152. Craig, D. C.; Stevens, J. D. *Cryst. Struct. Commun.* **1979**, *8*, 225.
153. Choong, W.; McConnell, J. F.; Stephenson, N. C.; Stevens, J. D. *Aust. J. Chem.* **1980**, *33*, 979.
154. McConnell, J. F.; Stevens, J. D. *J. Chem. Soc. Perkin Trans. 2* **1974**, 345.
155. Jackobs, J.; Reno, M. A.; Sundaralingam, M. *Carbohydr. Res.* **1973**, *28*, 75.
156. Grainger, C. T.; Rukvichai, S.; Stevens, J. D. *Cryst. Struct. Commun.* **1982**, *11*, 1939.
157. James, V. J.; Stevens, J. D. *Cryst. Struct. Commun.* **1982**, *11*, 79.
158. Yavari, I. *Org. Magn. Res.* **1980**, *14*, 511.
159. Kamalov, G. L.; Luk'yanenko, N. G.; Samitov, Y. Y.; Bogatskii, A. V. *J. Org. Chem. U.S.S.R. (Engl. transl.)* **1977**, *13*, 1005.
160. Sauriol-Lord, F.; Grindley, R. B. *J. Am. Chem. Soc.* **1981**, *103*, 936.
161. Burkert, U. *J. Comput. Chem.* **1980**, *1*, 192.
162. Cameron, T. S.; Cordes, R. E.; Grindley, T. B. *Acta Crystallogr. Sect. B* **1977**, *33*, 3718.
163. Noe, E. A.; Roberts, J. D. *J. Am. Chem. Soc.* **1971**, *93*, 7261.
164. Majchrzak, M. W.; Kotelko, A.; Guryn, R.; Lambert, J.; Wharray, S. M. *Tetrahedron* **1981**, *37*, 1075.
165. Allinger, N. L. *Pure Appl. Chem.* **1982**, *54*, 2515.
166. Gingerich, S. B.; Campbell, W. H.; Bricca, C. E.; Jennings, P. W. *J. Org. Chem.* **1981**, *46*, 2589.
167. Canuel, L.; St. Jacques, M. *Can. J. Chem.* **1974**, *52*, 3581.
168. Anet, F. A. L; Kopelevich, M. *J. Am. Chem. Soc.* **1986**, *108*, 2109.
169. Friebolin, H.; Mecke, R.; Kabuß S.; Lüttringhous, A. *Tetrahedron Lett.* **1964**, 1929.
170. Gianni, M. H.; Adams, M.; Kuivila, H. G.; Wursthorn, K. *J. Org. Chem.* **1975**, *40*, 450.
171. Soulier, J.; Farines, M.; Laguerre, A.; Bonafos-Bastouill, A. *Bull. Soc. Chim. France* **1976**, 307.
172. Arbuzov, B. A.; Klimovitskii, E. N.; Sergeeva, G. N. *J. Gen. Chem. U.S.S.R. (Engl. transl.)* **1983**, *52*, 2139.
173. Arbuzov, B. A.; Klimovitskii, E. N.; Remizov, A. B; Timirbaev, M. B. *Bull. Acad. Sci. U.S.S.R., Chem. Ser. (Engl. transl.)* **1981**, 794.
174. Pitea, G.; Moro, G.; Favini, G. *J. Chem. Soc. Perkin Trans 2* **1987**, 313.
175. Petrov, V. N.; Litvinov, I. A.; Struchkov, Y. T.; Klimovitskii, E. N.; Timirbaev, M. B.; Arbuzov, B. A. *J. Gen. Chem. U.S.S.R. (Engl transl.)* **1983**, 112.
176. St. Amour, R.; St. Jacques, M. *Can. J. Chem.* **1981**, *59*, 2281.
177. Gianni, M. H.; Cody, R.; Asthana, M. R.; Wursthorn, K.; Patanode, P.; Kuivila, H. G. *J. Org. Chem.* **1977**, *42*, 365.
178. Pilati, T.; Simonetta, M. *Cryst. Struct. Commun.* **1981**, *10*, 265.
179. Favini, G.; Pitea, D.; Sottocornola, M.; Todeschini, R. *J. Mol. Struct.* **1982**, *87*, 53.
180. Clark, G. R.; Palenik, G. J. *J. Chem. Soc. Perkin Trans. 2* **1973**, *194*, 197.
181. Taylor, K. G.; Chaney, J.; Deck, J. C. *J. Am. Chem. Soc.* **1976**, *98*, 4163.
182. Gianni, M. H.; Prezzavento, B.; Shea, K. *J. Org. Chem.* **1985**, *50*, 1554.
183. Blanchette, A.; Sauriol-Lord, F.; St. Jacques, M. *J. Am. Chem. Soc.* **1978**, *100*, 4055.
184. Plyamovatyi, A. K.; Klimovitskii, E. N. *Proc. Acad. Sci. U.S.S.R., Phys. Chem. Sect. (Engl. transl.)* **1982**, *268*, 107.
185. Arbuzov, B. A.; Klimovitskii, E. N.; Remizov, A. B.; Sergeeva, G. N. *Bull. Acad. Sci. U.S.S.R., Chem. Ser. (Engl. transl.)* **1980**, 211.
186. Arbuzov, B. A.; Klimovitskii, E. N.; Remizov, A. B.; Sergeeva, G. N. *Bull. Acad. Sci. U.S.S.R., Chem. Ser. (Engl. transl.)* **1979**, 1873.
187. St. Amour, R.; St. Jacques, M. *Can. J. Chem.* **1983**, *61*, 109.
188. Arbuzov, B. A.; Klimovitskii, E. N.; Sergeeva, G. N. *J. Gen. Chem. U.S.S.R. (Engl. transl.)* **1983**, *52*, 2144.
189. Litvinov, I. A.; Struchkov, Y. T.; Klimovitskii, E. N.; Timirbaev, M. B.; Arbuzov, B. A. *J. Gen. Chem. (Engl. transl.)* **1986**, *56*, 135.
190. Stadnicka, K.; Lebioda, Ł. *Acta Crystallogr. Sect. B* **1979**, *35*, 1517.
191. Stadnicka, K.; Lebioda, Ł.; Grochowski, J. *Acta Crystallogr. Sect. B* **1979**, *35*, 2763.
192. Ménard, D.; St. Jacques, M. *Can. J. Chem.* **1981**, *59*, 1160.
193. Ménard, D.; St. Jacques, M. *J. Am. Chem. Soc.* **1984**, *106*, 2055.
194. Dionne, P.: St. Jacques, M. *J. Am. Chem. Soc.* **1987**, *109*, 2616.
195. Mislow, K.; Glass, M., A., W. *J. Am. Chem. Soc.* **1961**, *83*, 2781.

196. Ōki, M,; Iwamura, H.; Hayakawa, N. *Bull. Chem. Soc. Jpn.* **1963**, *36*, 1542.
197. Sutherland, I. O.; Ramsey, M. V. J. *Tetrahedron* **1965**, *21*, 3401.
198. Jørgensen, F. S.; Gajhede, M.; Frei, B. *Helv. Chim. Acta* **1985**, *68*, 2148.
199. Vogel, E.; Günther, H. *Angew. Chem. Int. Ed. Engl.* **1967**, *6*, 385.
200. Jennings, W. B.; Rutherford, M.; Agarwal, S. K.; Boyd, D. R.; Malone, J. F.; Kennedy, D. A. *J. Chem. Soc., Chem. Commun.* **1986**, 970.
201. Anet, F. A. L.; Degen, P. J. *J. Am. Chem. Soc.* **1972**, *95*, 1390.
202. Anet, F. A. L.; Degen, P. J.; Krane, J. *J. Am. Chem. Soc.* **1976**, *98*, 2059.
203. Burkert, U. *Z. Naturforsch., Teil B* **1980**, *35*, 1479.
204. Caccamese, S.; Azzolina, R.; Duesler, E. N.; Paul, I. C.; Rinehart, K. L., Jr. *Tetrahedron Lett.* **1980**, *21*, 2299.
205. Duesler, E. N.; Rinehart, K. L., Jr.; Paul, I. C.; Caccamese, S.; Azzolina, R. *Cryst. Struct. Commun.* **1980**, *9*, 777.
206. Anet, F. A. L.; Degen, P. J. *Tetrahedron Lett.* **1972**, 3613.
207. Cockerill, G. S.; Kocienski, P. *J. Chem. Soc., Chem. Commun.* **1983**, 705.
208. Dale, J.; Ekeland, T.; Krane, J. *J. Am. Chem. Soc.* **1972**, *95*, 1389.
209. Pakes, P. W.; Rounds, T. C.; Strauss, H. L. *J. Phys. Chem.* **1981**, *85*, 2469.
210. Chatani, Y.; Yamauchi, T.; Miyake, Y. *Bull. Chem. Soc. Jpn.* **1974**, *47*, 583.
211. Kobayashi, M.; Kawabata, S. *Spectrochim. Acta A* **1977**, *33*, 549.
212. Ladd, J. A. *J. Mol. Struct.* **1977**, *36*, 329.
213. Kobayashi, M. *J. Chem. Phys.* **1977**, *66*, 32.
214. Pauling, L.; Carpenter, D. C. *J. Am. Chem. Soc.* **1936**, *58*, 1274.
215. Barón, M.; de Mandirola, O. B.; Westerkamp, J. F. *Can. J. Chem.* **1963**, *41*, 1893.
216. Falshaw, C. P.; King, T. J.; Imre, S.; Islimeyi, S.; Thomson, R. H. *Tetrahedron Lett.* **1980**, *21*, 4951.
217. Forbes, C. A.; Cheung, K. K.; Ferguson, G.; Robertson, J. M. *J. Chem. Soc. (B)* **1969**, 559.
218. Howard, B. M.; Schulte, G. R.; Fenical, W.; Solheim, B.; Clardy, J. *Tetrahedron* **1980**, *36*, 1747.
219. Suzuki, M.; Kurosawa, E.; Furusaki, A.; Matsumoto, T. *Chem. Lett.* **1983**, 779.
220. Howard, B. M.; Fenical, W.; Hirotsu, K.; Solheim, B.; Clardy, J. *Tetrahedron* **1980**, *36*, 171.
221. Lin, Y.-Y.; Risk, M.; Ray, S. M.; Van Engen, D.; Clardy, J.; Golik, J.; James, J. C.; Nakanishi, K. *J. Am. Chem. Soc.* **1981**, *103*, 6773.
222. Chou, H.-N.; Shimizu, Y.; Van Duyne, G.; Clardy, J. *Tetrahedron Lett.* **1985**, *26*, 2865.
223. Shimizu, Y.; Chou, H.-N.; Bando, H.; Van Duyne, G.; Clardy, J. C. *J. Am. Chem. Soc.* **1986**, *108*, 514.
224. Shimizu, Y.; Bando, H.; Chou, H.-N.; Van Duyne, G.; Clardy, J. C. *J. Chem. Soc., Chem. Commun.* **1986**, 1656.
225. Lee, M. S.; Repeta, D. J.; Nakanishi, K.; Zagorski, M. G. *J. Am. Chem. Soc.* **1986**, *108*, 7855.
226. Pawlak, J.; Tempesta, M. S.; Golik, J.; Zagorski, M. G.; Lee, M. S.; Nakanishi, K.; Iwashita, T.; Gross, M. L.; Tomer, K. B. *J. Am. Chem. Soc.* **1987**, *109*, 1144.
227. Nicolaou, K. C.; Duggan, M. E.; Hwang, C.-K. *J. Am. Chem. Soc.* **1986**, *108*, 2468.
228. Gonzalez, A. G.; Martin, J. D.; Perez, C.; Rovirosa, J.; Tagle, B.; Clardy, J. *Chem. Lett.* **1984**, 1649.
229. Nelsen, S. F.; Gillespie, J. P. *J. Am. Chem. Soc.* **1973**, *95*, 2940.
230. Kamada, T.; Yamamoto, O. *Bull. Chem. Soc. Jpn.* **1979**, *52*, 1159.
231. Kamada, T.; Yamamoto, O. *Chem. Lett.* **1976**, 843.
232. Kamada, T.; Yamamoto, O. *Tetrahedron Lett.* **1977**, 1341.
233. Kamada, T.; Yamamoto, O. *Bull. Chem. Soc. Jpn.* **1980**, *53*, 994.
234. Litvinov, I. A.; Klimovitskii, E. N.; Yufit, D. S.; Sergeeva, G. N.; Struchkov, Y. T.; Arbuzov, B. A. *Proc. Acad. Sci. U.S.S.R., Chem. Ser. (Engl. transl.)* **1981**, 115.
235. Arbuzov, B. A.; Klimovitskii, E. N.; Sergeeva, G. N.; Remizov, A. B.; Chernov, P. P. *J. Gen. Chem. U.S.S.R. (Engl. transl.)* **1983**, 2497.
236. Arbuzov, B. A.; Klimovitskii, E. N.; Klochkov, V. V.; Aganov, A. V. *Izv. Akad. Nauk S.S.S.R. Ser. Khim.* **1985**, 1685.
237. Hudec, J.; Turner, D. L. *J. Chem. Soc. Perkin Trans. 2* **1982**, 951.
238. Poje, M.; Perina, I.; Vickovic, I.; Bruvo, M. *Tetrahedron* **1985**, *41*, 1985.
239. Zipperer, B.; Fletschinger, M.; Hunkler, D.; Prinzbach, H. *Tetrahedron Lett.* **1987**, *28*, 2513.
240. Altenbach, H.-J.; Lex, J.; Linkenheil, D.; Voss, B.; Vogel, E. *Angew. Chem. Int. Ed. Engl.* **1984**, *23*, 966.
241. Crossley, R.; Downing, A. P.; Nógrádi, M.; Braga de Oliveira, A.; Ollis, W. D.; Sutherland, I. O. *J. Chem. Soc. Perkin Trans. 1* **1973**, 205.

242. Ollis, W. D.; Stoddart, J. F. *J. Chem. Soc., Chem. Commun.* **1973**, 571.
243. Saalfrank, R. W. *Angew. Chem. Int. Ed. Engl.* **1977**, *16*, 185.
244. Borgen, G.; Dale, J.; Anet, F. A. L.; Krane, J. *J. Chem. Soc., Chem. Commun.* **1974**, 243.
245. Bovill, M. J.; Chadwick, D. J.; Sutherland, I. O.; Watkin, D. *J. Chem. Soc. Perkin Trams. 2* **1980**, 1529.
246. Groth, P. *Acta Chem. Scand.* **1969**, *23*, 1311.
247. Furusaki, A.; Katsuragi, S.-I.; Suehiro, K.; Matsumoto, T. *Bull. Chem. Soc. Jpn.* **1985**, *58*, 803.
248. Suzuki, M.; Kurosawa, E.; Furusaki, A.; Katsuragi, S.-I.; Matsumoto, T. *Chem. Lett.* **1984**, 1033.
249. Shea, K.; Burke, L. D.; Doedens, R. J. *Tetrahedron* **1986**, *42*, 1844.
250. Kerr, B.; McCullough, K. J. *J. Chem. Soc., Chem. Commun.* **1985**, 590.
251. Karle, I. L.; Grochowski, J. *Acta Crystallogr. Sect. A* **1978**, *34*, S146.
252. Stadnicka, K. *Acta Crystallogr. Sect. B* **1979**, *35*, 2757.
253. Stadnicka, K.; Lebioda, Ł. *Acta Crystallogr. Sect. B* **1979**, *35*, 2760.
254. Vogel, E.; Altenbach, H.-J.; Sommerfeld, C. D. *Angew. Chem. Int. Ed. Engl.* **1972**, *11*, 939.
255. Sato, T.; Uno, K. *J. Chem. Soc. Perkin Trans. 1* **1973**, 895.
256. Bassi, I. W.; Scordamaglia, R.; Fiore, L. *J. Chem. Soc. Perkin Trans. 2* **1975**, 1129.
257. Månsson, M. *Acta Chem. Scand. Ser. B* **1974**, *28*, 895.
258. Gagnaire, D.; Tran, Y.; Vignon, M. *J. Chem. Soc., Chem. Commun.* **1976**, 6.
259. Tanaka, I.; Tajima, I.; Hayakawa, I.; Okada, M.; Bitoh, M.; Ashida, T.; Sumimoto, H. *J. Am. Chem. Soc.* **1980**, *102*, 7873.
260. Ackland, M. J.; Hanson, J. R.; Hitchcock, P. B.; Mabelis, R. P.; Ratcliffe, A. H. *J. Chem. Soc. Perkin Trans. 1* **1984**, 2755.
261. Hogeveen, H.; Nusse, B. J. *Tetrahedron Lett.* **1976**, 699.
262. Helder, R.; Wynberg, H. *Tetrahedron Lett.* **1973**, 4321.
263. Nozaki, H.; Koyama, T.; Mori, T. *Tetrahedron* **1969**, *25*, 5357.
264. Anet, F. A. L.; Krane, J.; Dale, J.; Daasvatn, K.; Kristiansen, P. O. *Acta Chem. Scand.* **1973**, *27*, 3395.
265. Van Remoortere, F. P.; Boer, F. P. *Inorg. Chem.* **1974**, *13*, 2071.
266. Groth, P. *Acta Chem. Scand. Ser. A* **1978**, *32*, 279.
267. Borgen, G.; Dale, J.; Daasvatn, K.; Krane, J. *Acta Chem. Scand. Ser. B* **1980**, *34*, 249.
268. Krane, J.; Amble, E.; Dale, J.; Daasvatn, K. *Acta Chem. Scand. Ser. B* **1980**, *34*, 255.
269. Downing, A. P.; Ollis, W. D.; Sutherland, I. O. *J. Chem. Soc. (B)* **1970**, 24.

Nitrogen Heterocycles

Roger W. Alder and Jonathan M. White
University of Bristol,
Bristol, England

3.1. Introduction

This chapter discusses current knowledge of the structure and conformational properties of seven- to 12-membered rings containing one or more nitrogen atoms. The subject has been reviewed several times. Riddell[1] gives a brief introduction, but the most extensive previous review was that by Armarego.[2] There are also useful discussions of seven-, eight, and nine-membered rings in "*Comprehensive Heterocyclic Chemistry*"; the section on the conformations of eight-membered rings is particularly good.[3] Other useful reviews are those by Anet[4] and by Dale.[5] The reader is also referred to the *Specialist Periodical Reviews*, published by the Royal Society of Chemistry, which cover this area.[6] In this review of nitrogen heterocycles, we include compounds that also have ring oxygen atoms, but heterocycles containing sulfur and other elements are excluded. Monocyclic systems are described first, in order of increasing ring size. Conformations in these compounds are often strongly influenced by the presence of rigid elements that impose torsional constraints, for example, double bonds and benzo-ring fusions (we include amide bonds in this category, although rotation about these bonds may be possible). For each ring size, compounds with no strong torsional constraints are discussed first, followed by compounds with one, two, or three constraints, and so on.

It is assumed that the reader is familiar with the trivial (and often rather silly) names given to the common conformations found for seven- to 12-membered rings; these were discussed earlier in this book. We shall refer to chair, twist–chair, boat, and twist–boat conformations for seven-membered rings; to BB (boat–boat), BC, CC, and crown for eight-membered rings; to the D_3 or [333] conformation of nine-membered rings; to the BCB or [3232] conformation of 10-membered rings; and to the [3333] of 12-membered rings. There is an almost exponential decay in knowledge as we progress from seven- to 12-membered rings. In particular, information on saturated nine-membered and all types of 11-membered rings is almost nonexistent.

Following this survey of ring conformations, the influence of ring size on nitrogen geometry and properties is briefly discussed; what emerges is that, in general, the effects are remarkably small. There is, of course, much more evidence for important transannular interactions involving nitrogen in these rings, and this is discussed next.[7–11]

The discussion of bicyclic compounds is confined to structures built entirely from seven- to 12-membered rings. There is, of course, extensive work on bicyclo-[3.3.1]nonanes and similar ring systems, which contain at least one medium-sized ring. However, the structures of these compounds are determined by the common-sized rings they contain, and they are hardly typical medium-ring compounds. Where the results seem particularly important, we discuss these compounds under the smallest appropriate monocyclic medium-ring system. Bicyclic compounds built entirely from medium rings are often much more strained[12] than their monocyclic counterparts, and, in particular, bridgehead nitrogens in these systems are strongly distorted from standard geometry, leading to unusual properties. There are also interesting chemical consequences of interactions between nitrogens at both bridgeheads. The ring conformations in these compounds are still imperfectly understood; it seems obvious that conformational possibilities are limited by the presence of the second bridge, as compared with monocyclic counterparts, but defining these restrictions is no simple matter. We shall discuss an approach based on looking at the conformations of the bridges individually rather than at the conformations of the constituent rings.

Polycyclic compounds are considered only briefly, since very few systems consisting entirely of medium rings are known. There are, of course, many more polycyclic systems containing common-sized rings or macrorings in addition to medium rings, but we discuss these under the smallest appropriate monocycle. Some of these are cyclophanes and these have been extensively reviewed.[13] Other examples are the bridged hetero[10]annulenes prepared by Vogel and his co-workers, while some cryptands contain 12-membered rings.

3.2. Seven-membered Rings

3.2.1. Seven-membered Rings without Torsional Constraint

In most cases, saturated seven-membered rings are very mobile, so that their conformations are not very amenable to study by dynamic nuclear magnetic resonance (NMR) methods. Most of the work described in this section is therefore based on crystallographic investigations.

3.2.1.1. Compounds with One Nitrogen Atom. Two independent crystallographic studies on salts of azacycloheptane have been reported.[14,15] The structures of the *p*-chlorobenzoate (**1a**) and the dimorphs of the *p*-bromobenzoate (**1b**) were determined. The structures of the cation are virtually identical in the three cases, and assume twist–chair conformations. The hydrochloride salt (**1c**)[15] similarly adopts a twist–chair conformation. Twist–chair conformations are also observed for the azepine (**2**),[16] spirans (**3**[17] and **4**[18]), and the tricycles (**5**) and (**6**).[19] The structure of the indole (**7**)[20] was described as a twist–chair conformation with some distortion toward a chair form: it was suggested that this distortion serves to alleviate nonbonded transannular interactions between hydrogen atoms bonded to C(2) and C(5a).

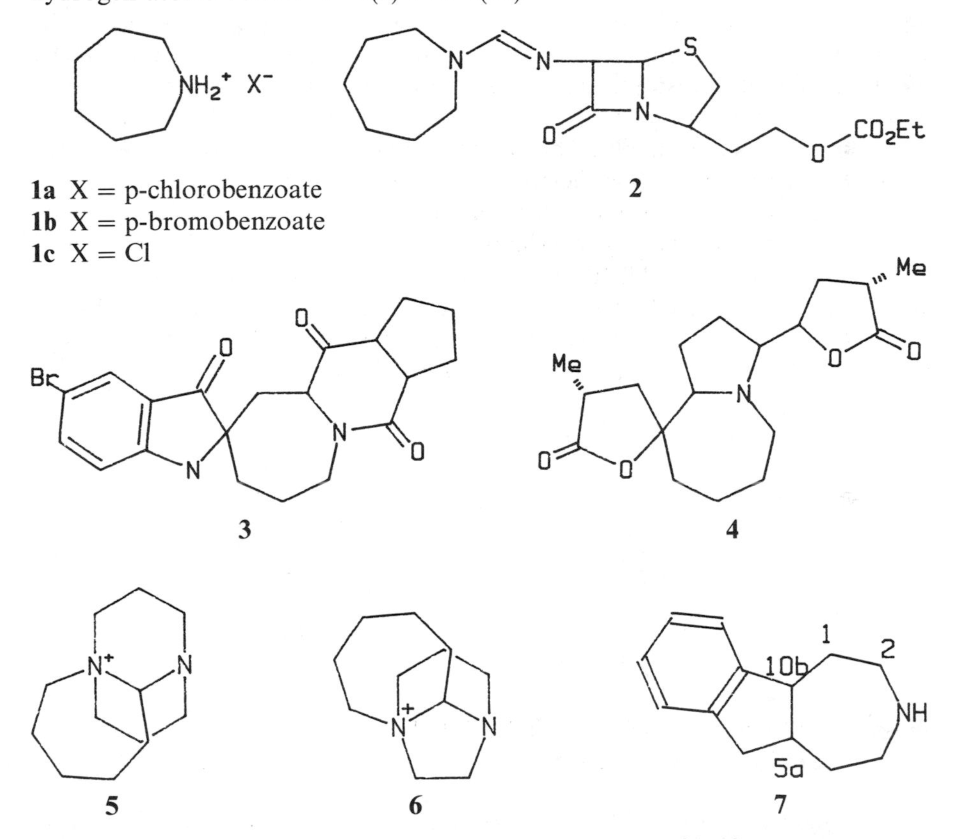

1a X = p-chlorobenzoate
1b X = p-bromobenzoate
1c X = Cl

2

3 **4**

5 **6** **7**

Many structures have been reported in the literature,[21–23] with the seven-membered ring showing various degrees of distortion between a "pure" twist–chair form and a "pure" chair form. Considering the low barrier to pseudorotation of twist forms (via chairs) to twist forms, these observed conformations may well be influenced by crystal packing forces. The structure of the N-triazole derivative (**8**)[24] demonstrates the "softness" of cycloheptane conformational processes. From an analysis of the torsion angles, the ring is best described as a slightly distorted chair. However, the direction and magnitude of the thermal elipsoids of the seven-membered ring atoms suggest that this conformation is

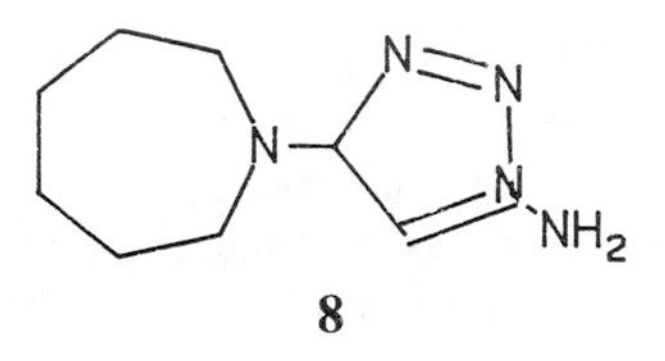

8

an "averaged" one, possibly between two rapidly interconverting twist–chair forms.

3.2.1.2. Compounds with More Than One Heteroatom. The conformational properties of rings having more than one heteroatom present are likely to be dependent upon the relative positions of the two heteroatoms. For example, conformations of 1,4-dihetero compounds may be influenced by the gauche effect and 1,3-diheterosubstituted rings by the anomeric effect.[25] The greatest effect on conformational properties is seen, however, with 1,2-hetero substitution. This is a result of the high barrier to rotation of heteroatom–heteroatom bonds compared with carbon–heteroatom or carbon–carbon bonds; barriers to rotation about the heteroatom–heteroatom bond have been calculated for hydrazine (46 kJ/mol) and hydroxylamine (41 kJ/mol).[26] This is clearly demonstrated by the dynamic NMR studies on the hydroxylamine derivative (**9**),[27] which has an observed barrier to ring inversion of 80 kJ/mol; this is about 42 kJ/mol larger than that observed for the ketone (**10**).[28] The inversion process involves rotation about the N—O bond.

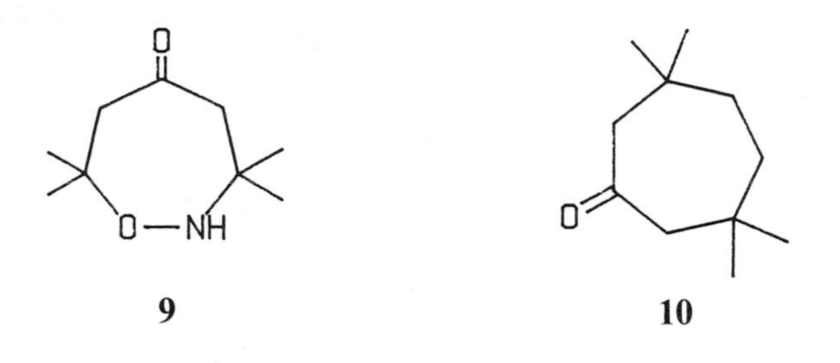

9 **10**

3.2.2. Seven-membered Rings with One Torsional Constraint

Seven-membered rings with one nitrogen atom that is part of a rigid group—an amide group, for example—are likely to have conformations similiar to that of cycloheptene. In a series of structural papers on medium-ring lactams,[29] the structures of ϵ-caprolactam (**11**) and its protonated salt (**12**) were described. The conformations of both structures in the solid state were close to the geometry of chair cycloheptene calculated by force field methods.[30] Similar conformations were observed for tetrazole (**13**),[31] and amide (**14**).[32] ^{19}F dynamic NMR studies on the 4,4-difluoro caprolactam (**15**)[33] indicate a likely chair conformation for the seven-membered ring. The barrier to inversion in this lactam is observed to be 44 kJ/mol. Since the barrier to rotation about the amide C—N bond is likely

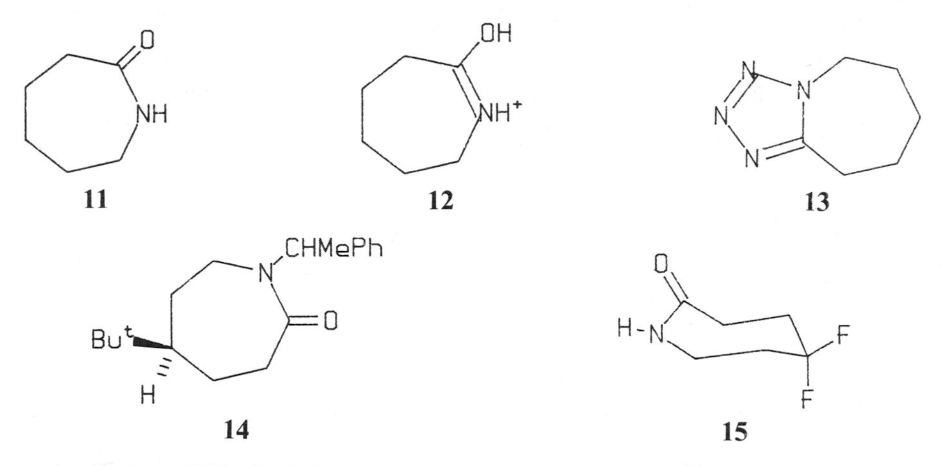

to be close to 76 kJ/mol (observed for dimethylacetamide[34]), it seems unlikely that pseudorotation is responsible for the exchange of axial and equatorial positions of the fluorine atoms.

A series of benzo-fused 1,5-diamines (**16a–e**) were studied[35] by ^{1}H NMR; the 2-methyl (**16b**), the 2,4-dimethyl (**16d**), and the 2,2,4-trimethyl (**16e**) derivatives showed static spectra at room temperature and the conformations were assigned as chairs. The most interesting result was the conformational behavior of the unsubstituted diamine (**16a**) and the 2,2-dimethyl derivative (**16c**). Both of these compounds gave "time-averaged" spectra as low as $-90°C$. Given that cycloheptene freezes out into one conformation at $-80°C$, this suggests that the introduction of the nitrogen atoms into the ring decreases the barrier to inversion of the ring.

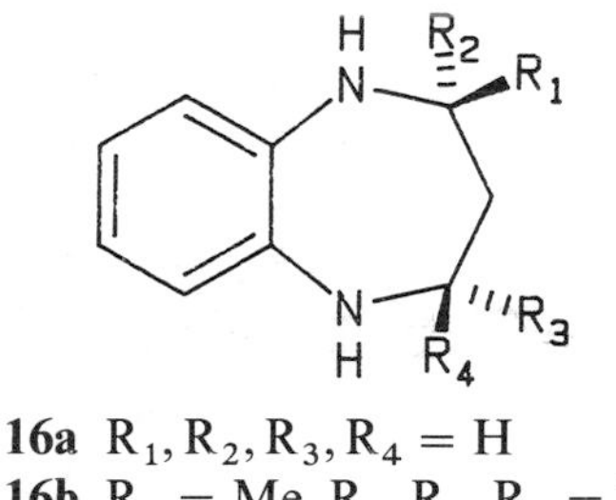

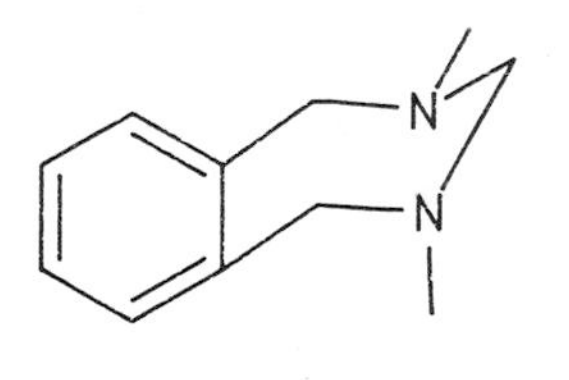

16a $R_1, R_2, R_3, R_4 = H$
16b $R_1 = Me, R_2, R_3, R_4 = H$
16c $R_1, R_2 = Me, R_3, R_4 = H$
16d $R_1, R_4 = Me, R_2, R_3 = H$
16e $R_1, R_3, R_4 = Me, R_2 = H$

17

Studies on the benzoaminal (**17**) by St. Jacques et al.[36] nicely show the influence that the anomeric effect can have on conformation. The seven-membered ring assumes a chair conformation, and barriers to ring inversion (46 kJ/mol) and nitrogen inversion (26 kJ/mol) were determined by dynamic NMR. The interesting aspect of the structure is the axial, equatorial arrangement of the N-methyl groups, which allows the lone pair on the nitrogen with the axial methyl to be antiperiplanar to the other C—N bond. It was pointed out that a twist–boat conformation with an equatorial, axial arrangement of N-methyl groups (**18**) would also allows an antiperiplanar disposition for one of the lone pairs and a C—N bond. This conformation was ruled out, however, as it seemed unlikely that such a conformation would have a barrier to ring inversion as high as that observed.

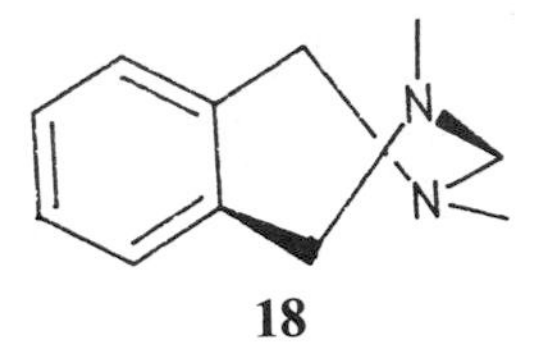

18

3.2.3. Seven-membered Rings with Two Torsional Constraints

This section will be divided into two broad classes—those rings with the rigid elements in 1,3 positions and those with them in 1,4 positions.

3.2.3.1. 1,3-Diconstrained Systems. Studies of a number of derivatives of the cyclic azine (**19**) were carried out using ^{1}H NMR spectroscopy.[37] The azines (**19a**) and (**19b**) were shown to exist in solution as pairs of enantiomeric twist–boat conformations. The signals for the ring methylenes for (**19a**) showed a well-resolved AB splitting pattern at −36°C, which coalesced at 7°C and gave a sharp line at 62°C; this was consistent with an inversion barrier of 59 kJ/mol. The triazepam (**19b**) was studied in the temperature range −59 to +34°C and a barrier to inversion of 51 kJ/mol was calculated; there was no evidence of slowing of nitrogen inversion in this temperature range.

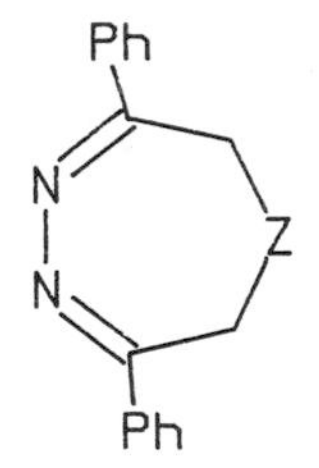

19a Z = CMe_2
19b Z = N-benzyl

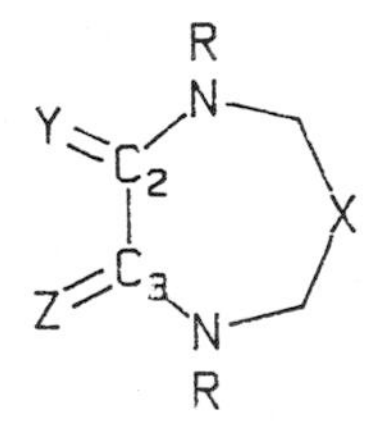

20a Y = Z = O, X = CH_2, R = $CHMe_2$
20b Y = Z = O, X = CH_2, R = benzyl
20c Y = Z = O, Z = CHMe, R = benzyl
20d Y = O, Z = S, X = CH_2, R = $CHMe_2$
20e Y = O, Z = S, X = CH_2, R = benzyl
20f Y = Z = S, X = CH_2, R = $CHMe_2$
20g Y = Z = S, X = CH_2, R = benzyl
20h Y = Z = S, X = $CHMe_2$, R = benzyl

The series of oxamide derivatives (**20a–e**)[38] illustrate the dependence of the inversion barrier on the nature of the elements *Y* and *Z*. The conformations of these compounds were deduced from a combination of ^{1}H NMR spectroscopy

TABLE 3-1. Barriers to Ring Inversion for Cyclic Oxamides (**20a–h**)

Compound	Solvent	T_c (°K)	$\Delta G^{\ddagger}$ (kJ/mol)
20a	$CHCl_2F$	208.8	46
20b	$CDCl_3$	202.9	45
20c	$CDCl_3$	215.1	50
20d	$CDCl_3$	298.6	72
20e	$CDCl_3$	284.4	66
20f	Isoquinoline	447.6	101
20g	*o*-Dichloro-benzene	431.5	97
20h	Isoquinoline	409.4	97

and molecular mechanics to be twisted boats, with the twofold axis through group *X* and bisecting the 2,3-bond. Kinetic studies on these derivatives illustrate the relatively small dependence of the ring inversion barrier on groups *R* and *X*, see Table 3-1 when compared with the large dependence of the barrier on atoms *Y* and *Z*. For $Y = Z = 0$ barriers, it is about 41 kJ/mol; for $Y = 0$, $Z = S$, about 67 kJ/mol; and for $Y = Z = S$, about 100 kJ/mol. This is compatible with a transition state involving eclipsing of the $C(2) = Y$ and $C(3) = Z$ bonds. Support for this is provided by the observed barriers to rotation about the oxamide bond in the acyclic oxamide analogues (**21a**), (**21b**), and (**21c**) of 41, 75, and >100 kJ/mol respectively.[39]

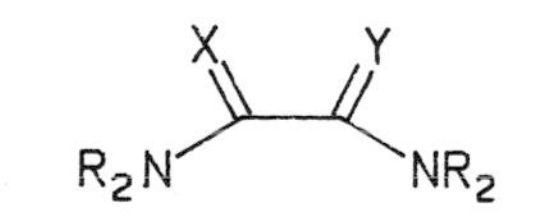

21a X = Y = O, R = benzyl
21b X = O, Y = S, R = benzyl
21c X = Y = S, R = benzyl

Much of the literature on these 1,3 constrained systems concerns the *o,o'*-substituted derivatives of biphenyl.[40] The preferred conformation is one in which the phenyl groups are twisted out of plane with respect to one another. For example, the degree of twist in the azepine (**22**)[41] is calculated to be 45.8°. The barrier to inversion in the simple derivatives is of the order of 41 kJ/mol.

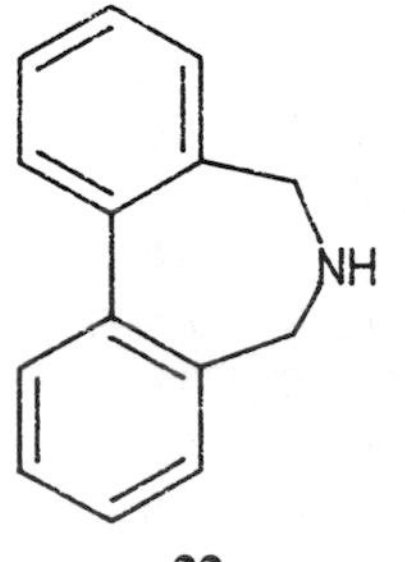

22

This barrier is raised significantly by the addition of groups at the *o,o′* positions of the phenyl rings; for example, the dinitro (**23**) and dimethoxy (**24**) derivatives are rigid, and if prepared optically pure, they remain optically active for several hours at room temperature.

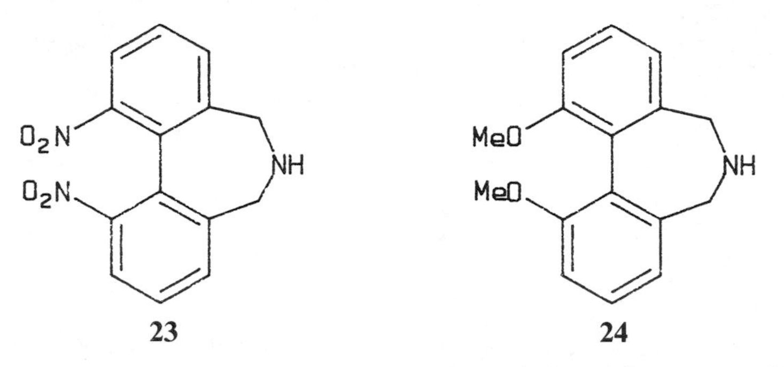

The protonated azepam system (**25**) has been deduced from a combination of ^{1}H NMR, ultraviolet (UV) spectroscopy, and X-ray crystallography to exist in a "half-chair" conformation (**26**).[42,43] The planar arrangement of atoms N(1), N(4) and C(5)–C(7) was deduced from the observed *W* coupling of 1.8 Hz between the N—H and C(6)–H protons. This requires that the N—H bond be coplanar with the C(6)—H bond, an arrangement that cannot be satisfied by the alternative boat geometry (**27**). Further evidence of this conformation was provided by the UV spectrum; this was essentially identical to that of the fused

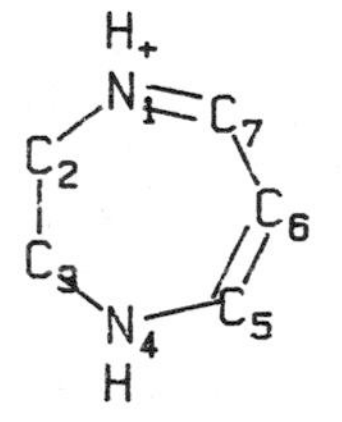

25

26

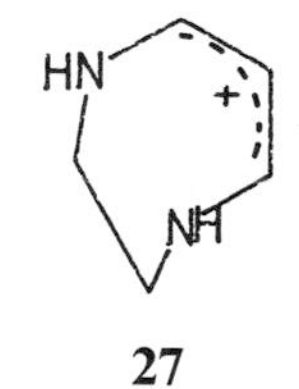

27

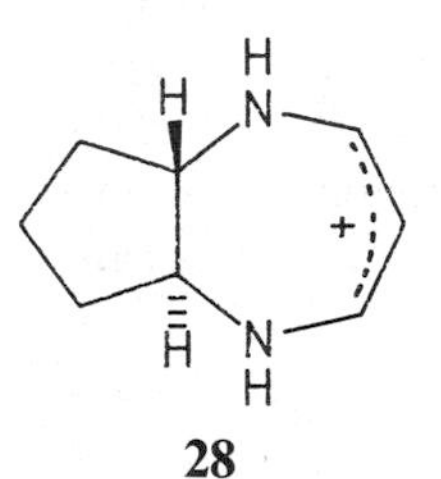

28

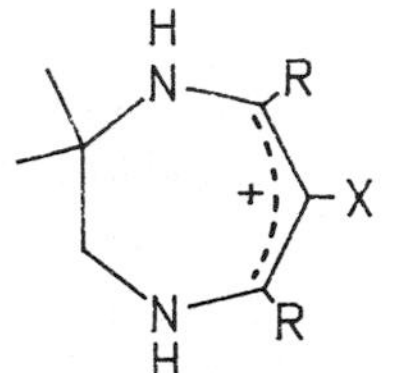

29a R = Me, X = H
29b R = Me, X = NO_2
29c R = Me, X = OMe
29d R = Me, X = Cl
29e R = Me, X = Br
29f R = Me, X = I
29g R = Me, X = Me
29h R = Ph, X = H

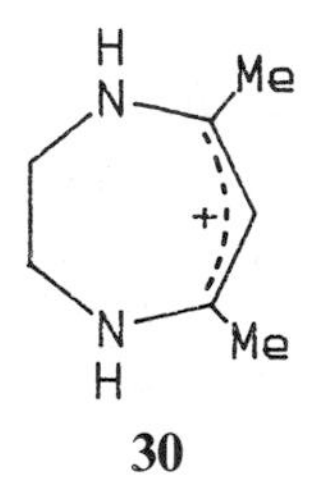

30

TABLE 3-2. BARRIERS TO RING INVERSION IN DIAZEPINIUM SALTS (**29a–h**)

Compound	T_c (°C)	$\Delta G^{\ddagger}$ (kJ/mol)
29a	4	58.7
29b	−27	51.8
29c	−25	52.0
29d	−23	52.6
29e	−25	52.5
29f	−32	50.8
29g	−49	47.1
29h	−2	56.9

cyclopentane derivative (**28**),[44] which must have a "zigzag" structure for N(1)–N(4). Kinetic studies on the 2,2-dimethyl derivatives (**29a–h**) gave barriers to ring inversion ranging from 58 kJ/mol for (**29a**) down to 47 kJ/mol for (**29g**); see Table 3-2. An X-ray study on the dimethylazepine salt (**30**)[45] revealed a conformation essentially the same as that believed to exist in solution, although in the solid state the atoms N(1), N(4), C(5)–C(7) are slightly distorted from coplanarity, to form a partial helix.

3.2.3.2. 1,4-Diconstrained Systems. Although many heterocyclic derivatives having this basic framework have been studied with respect to their synthesis and pharmaceutical properties, few conformational studies have been reported. Compounds based on the fused dibenzyl system, such as imipramine (**31a**), are important antidepressant drugs. A detailed analysis of the dynamic NMR properties of these derivatives was carried out by Abraham et al.[46] It was observed that if *R* is an *n*-alkyl group, or dimethylaminopropyl as in the case of imipramine (**31a**), then the ring shows an averaged ^1H NMR spectrum as low as −100°C. The replacement of *n*-alkyl with an acyl function, however (**31b**), leads to a drastic change in the conformational behavior. The seven-membered ring, which was identified as having the conformation (**32**) with a fold in the molecule through the nitrogen and the C(4) methylene, undergoes three dynamic processes: (1) twisting of the ethano fragment, which results in a coalescence of an ABCD splitting pattern to an AA′BB′ splitting at about 7°C; (2) rotation of the amide

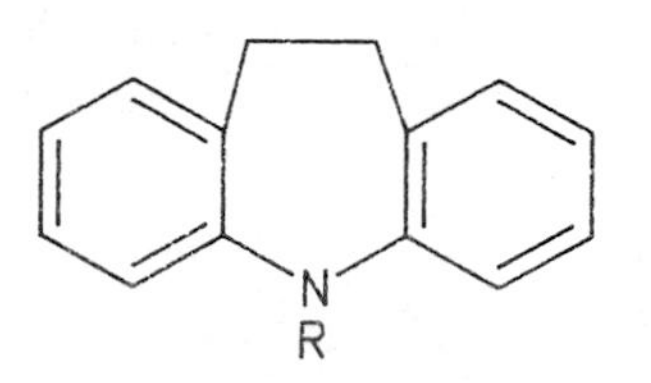

31a R = $CH_2CH_2CH_2NMe_2$
31b R = acyl

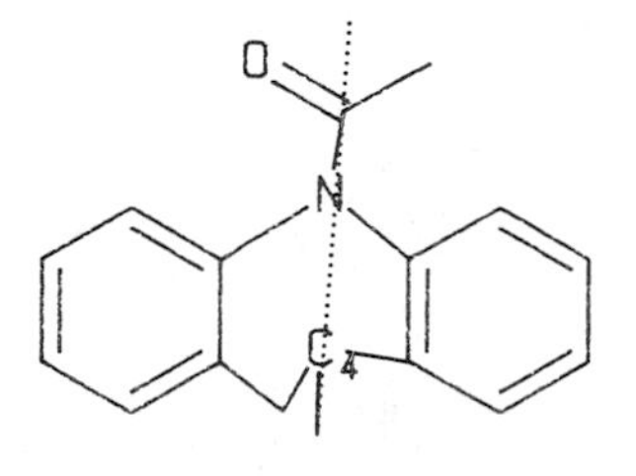

32

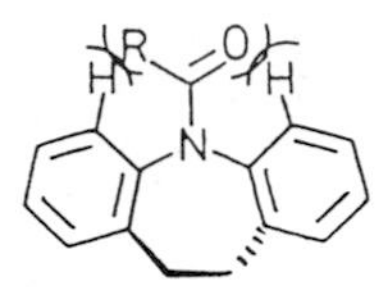

Figure 3-1. Nonbonded interactions in the transition state for ring inversion of **32b**.

bond; (3) ring inversion, which results in coalescence of the ethano protons to a singlet at about 120°C.

The barriers to inversion of the seven-membered ring with various acyl groups correspond well to the barriers to rotation of the corresponding acyclic amides. This was interpreted as follows: The presence of a planar amide function hinders ring inversion (see Fig. 3-1) as this would involve severe interactions between the acyl group and the *o*-protons at the transition state. As the temperature is raised, the amide rotational barrier is overcome, and ring inversion is allowed to occur. Similar behavior was observed for the cyano derivatives (**33**).[47] X-ray structure analyses have been carried out on imipramine (**31a**) and a number of its derivatives.[48–50] These seem to confirm the conformations deduced from solution studies, although there are variations in the degree of twist in the ring.

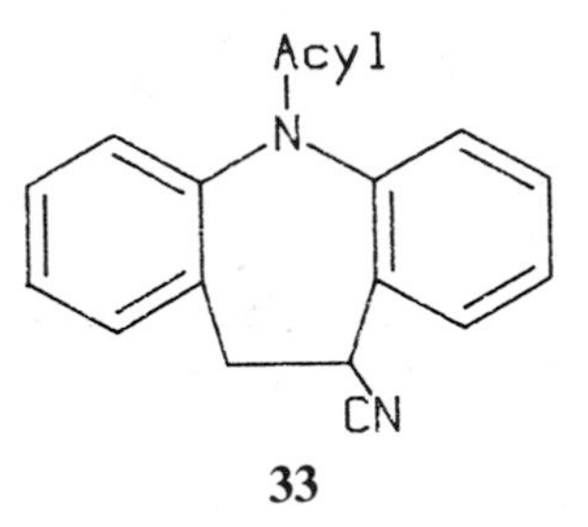

3.2.4. Seven-membered Rings with Three Torsional Constraints

The only reasonable conformation of a seven-membered ring with three torsion angles fixed at zero is the boat conformation. The effect of heteroatom substitution in this ring does not alter this basic conformation to any great extent but does have varying effects on the ring inversion barriers. The simple azepine (**34**) and N-alkyl derivatives are unstable in solution[51]; therefore, conformational studies are difficult. The N-*p*-bromobenzenesulphonyl derivative (**35**), however, is stable, and has been shown by X-ray crystallographic studies to exist in the solid state in a shallow boat conformation.[52] NMR experiments

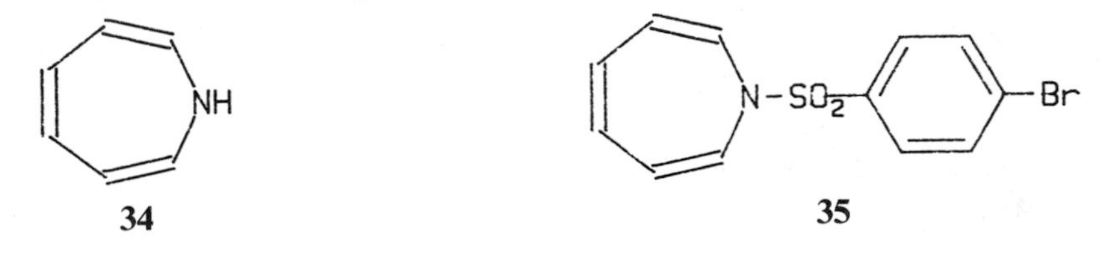

suggest that the barrier to inversion in this azepine is less than 21 kJ/mol. This is at least 8 kJ/mol less than the barrier observed for cycloheptatriene,[53] and is presumably related to the nearly planar nature of the nitrogen atom, a characteristic of sulfonamides, which would suffer less strain at a planar transition state.

The incorporation of amide or amidine functions into the ring has the effect of raising the inversion barrier[54]; for example, the barrier to inversion in the amide (**36**) was estimated to be between 36 and 40 kJ/mol, and in the amidine (**37**), 43 kJ/mol. The most striking effect that heteroatom substitution can have on the inversion barriers is observed for the 1,2-azepines of the structural formula **38**.[55–57] For example, the triphenyl-1,2-azepine (**39**)[57] has a barrier of 72 kJ/mol. This high barrier is a result of the necessary eclipsing of the nitrogen lone pairs at the transition state to inversion, which is disfavored; consistent with this is the greatly reduced barrier in the monoprotonated derivative (**40**) of 43 kJ/mol, for which this repulsion is removed.

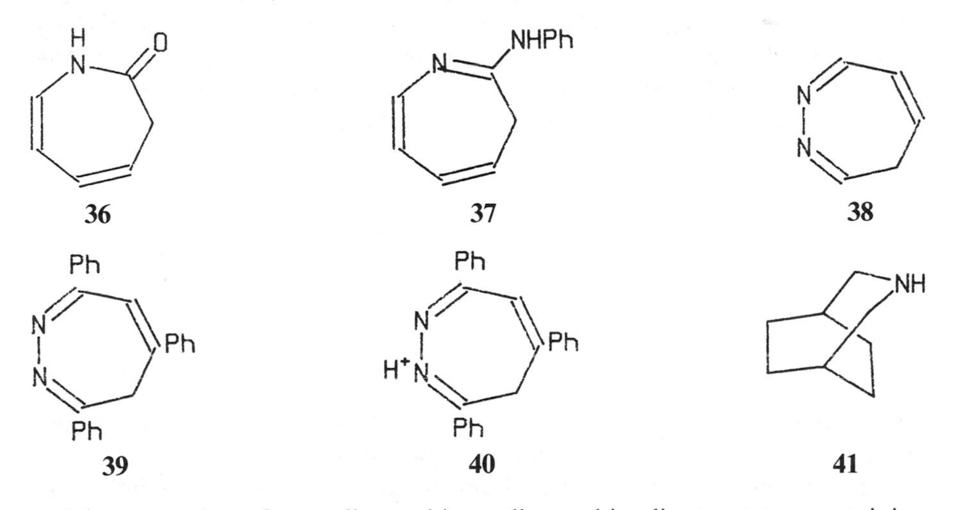

A large number of crystallographic studies on bicyclic structures containing a seven-membered ring as the smallest medium ring have been reported. In many cases, these are large complex molecules and will not be described here in any detail. For a list of references to these structures, see Ref. 58. The 2-azabicyclo[3.2.2]nonane (**41**)[59] shows a dynamic effect in the ^{1}H NMR spectrum, which was believed to be due to inversion of the seven-membered ring; a barrier of 25 kJ/mol for this process was measured.

3.3. Eight-membered Rings

3.3.1. Eight-membered Rings without Torsional Constraint

Solution studies[60] on azocane (**42a**) by dynamic ^{13}C NMR show that this compound exists as a 97:3 mixture of the boat–chair and a crown family conformation. The free energy difference between these two was calculated to be

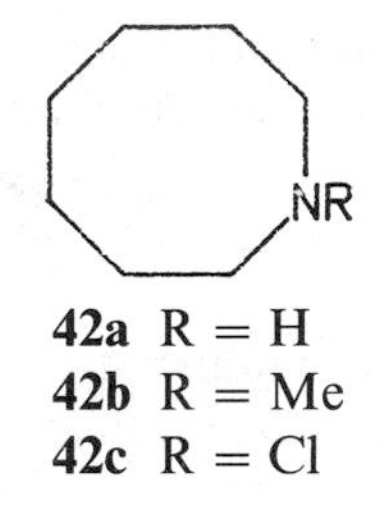

42a R = H
42b R = Me
42c R = Cl

5 kJ/mol, with a barrier to their interconversion of 44 kJ/mol. A dynamic effect was also observed in the ^{1}H NMR spectrum, which was attributed to ring inversion in the boat–chair, with a barrier of 31 kJ/mol. No dynamic effects due to slowing of pseudorotation or nitrogen inversion were observed, as the barriers involved in the processes are too low. The position of the nitrogen atom in the boat–chair or crown was not experimentally determined, however, likely positions are the crown (**43**) and the boat–chairs (**44** and **45**). The N-methyl (**42b**) and N-chloro (**42c**) derivatives of azocane[61] have also been studied by dynamic NMR, and also exist in solution in the boat–chair conformation. The X-ray structure of the tosylate salt of azocane has been determined.[15] It shows that the protonated salt is in a boat–chair conformation very similar to that calculated for cyclooctane.

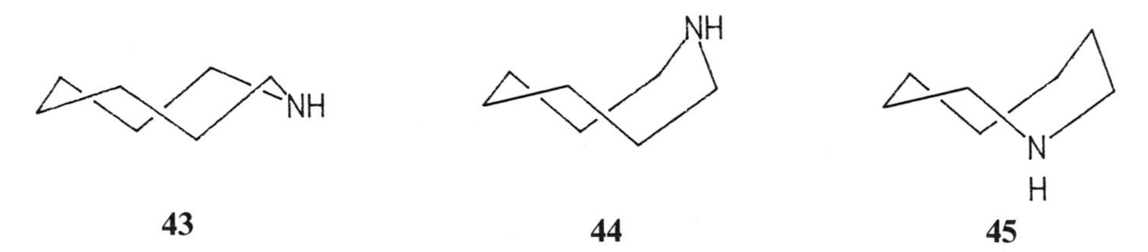

While the boat–chair seems to be the preferred conformation of cyclooctane and azocane, the introduction of further heteroatoms into the ring results in other conformations being preferred. For example, X-ray structural studies carried out on three of the four polymorphs of the explosive HMX (**46**) show that the eight-membered ring exists in a chair–chair (or extended-crown) conformation.[62–64] Crown family conformations were also observed in the X-ray structures of the tetraamide (**47**),[64] azide (**48**),[65] and tetracyclopropyl (**49**).[66] In contrast, the conformation observed in the beta form of HMX[67] was a slightly twisted chair with the molecule possessing a center of inversion.[68] Similar conformations were also observed for the diamide (**50**)[69] and three derivatives of the diaza compound (**51a–c**).[70]

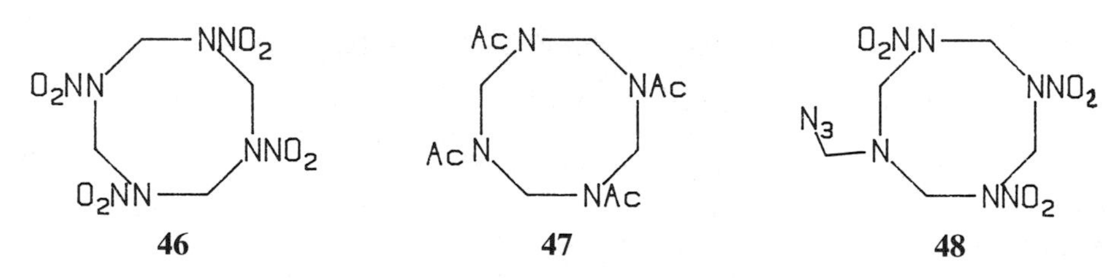

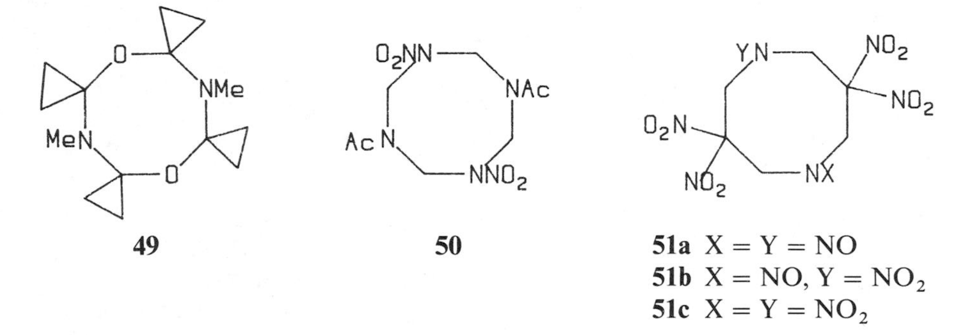

49 **50** **51a** X = Y = NO
51b X = NO, Y = NO_2
51c X = Y = NO_2

The boat conformation of cyclooctane, although having the least torsional strain of all the possible conformations, suffers from severe transannular non-bonded interactions. This conformation is possible for eight-membered rings when favorable transannular interactions are present; compounds in this class are discussed in section 3.8.

3.3.2. Eight-membered Rings with One Torsional Constraint

The structures of enantholactam (**52**)[71] and its protonated salt (**53**)[72] were determined by X-ray crystallography. The torsion angles of the amide and *o*-protonated amide groups are close to zero. The ring conformations are very similar, and are best described as being derived from the boat–chair conformation of cyclooctane, with the amide function at the position of the smallest torsion angle. This is similar to the most stable conformation calculated for *cis*-cyclooctene.[30] This geometry is close to the transition state in the boat–chair to twist–boat–chair pseudorotation in cyclooctane. A similar conformation is also observed for both rings in the natural product homaline (**54**).[73]

The eight-membered ring is the smallest ring system that can accommodate a stable *trans*-double bond. The azine (**55**)[74] has been shown by UV, dipole moment, and europium shift data to contain a *trans*-aza linkage; details of the ring conformation, however, were not discussed. A number of derivatives of this

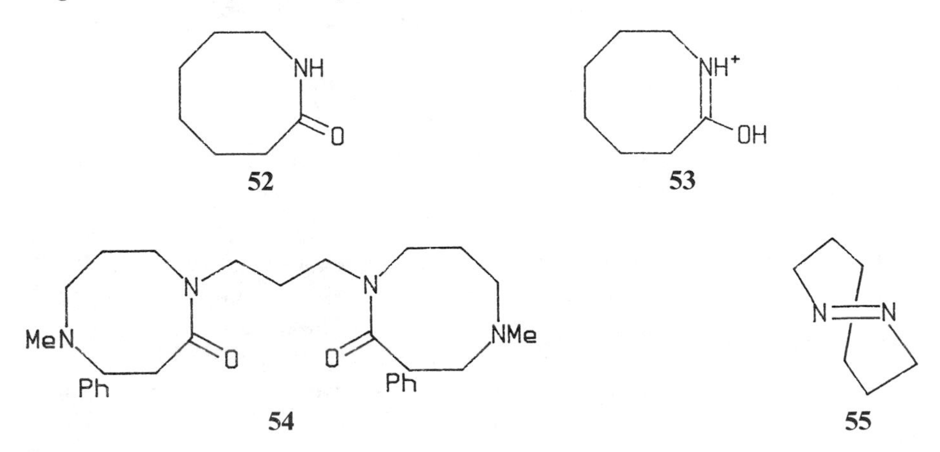

52 **53**

54 **55**

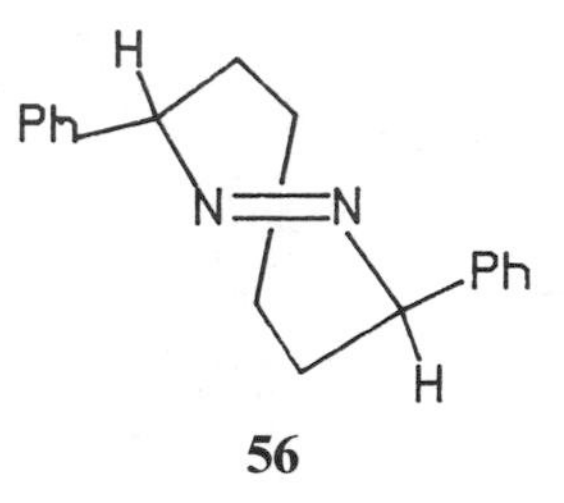

56

azine were later studied, and the X-ray structure of **56**[75] was determined. The eight-membered ring of **56** exists in a twisted-chair conformation, with an approximate C_2 axis of symmetry bisecting the N=N double bond. The torsion angle of the double bond of 155.7° shows a large deviation from planarity.

3.3.3. Eight-membered Rings with Two Torsional Constraints

We shall discuss these under the subheadings of 1,3-, 1,4-, and 1,5-diconstrained systems.

3.3.3.1. 1,3-Diconstrained Systems. There are two possible conformations of an eight-membered ring having two double bonds in a 1,3-relationship—the twist–boat–chair and the twist–boat.[76] In both of these conformations, the diene system is nonplanar, with torsion angles of about 54° in TBC and about 42° in TB. The amide (**57**)[77] exists in solution, as a mixture of these two conformations in the ratio TBC:TB of 10.8:1. These two conformers are stable and separable by chromatography. The barrier to interconversion between the two, measured in both directions, was about 105 kJ/mol. An X-ray structure of the major conformer confirmed the twist–boat–chair geometry. An X-ray crystal structure of the imine (**58**)[78] showed that the eight-membered ring exists in the solid state as a twist–boat–chair.

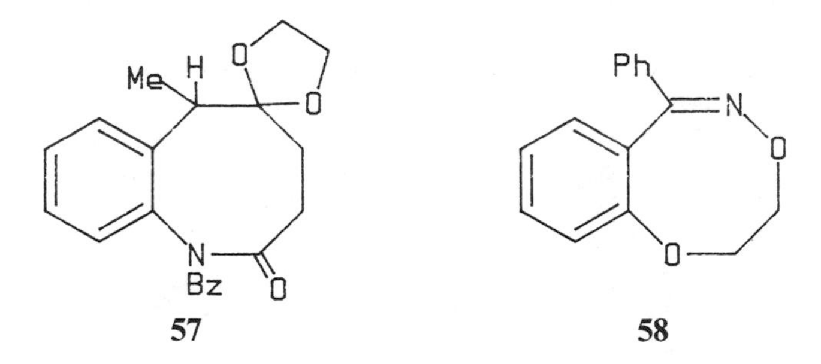

57 **58**

3.3.3.2. 1,4-Diconstrained Systems. Eight-membered rings with two double bonds derived from a 1,4-diene have two conformational minima, the "rigid" boat–chair and the flexible twist–boat. Studies of these systems have been restricted to compounds based on the fused dibenzo skeleton (**59**). X-ray structure analysis of the N-methyl azocine (**59b**)[79a] shows that this compound exists in the solid state as the rigid boat–chair, in contrast to the N-t-butyl azocine

TABLE 3-3. CONFORMATIONAL PROPERTIES OF 1,4-DIBENZO HETEROCYCLES (**59a–f**)

Compound	Process	ΔG°	$\Delta G^\ddagger$ (kJ/mol)
59a	BC = BC	0	67
59b	BC = TB	6.7	63–69
59c	BC = TB	−4.6	
	TB = TB*	0	<46
59d	BC = TB	3.4	>75
59e	BC = TB	1.5	61.5
60	Boat = boat	0	<33

(**59c**),[79b] which exists in the solid state as enantiomeric pairs of twist–boats. This is in accord with solution studies using NMR methods,[80] which show the N-methyl azocine (**59b**) to exist in solution as a mixture of predominantly boat–chair (96.5%) with a small amount of twist–boat (3.5%) and the t-butyl azocine (**59c**) as predominantly twist–boat (95%), with the rest boat–chair.[81] Studies on a number of derivatives (**59a–e**)[80–82] by dynamic NMR and force-field calculations are summarized in Table 3-3. In all these derivatives, apart from the N-t-butyl (**59b**), the boat–chair is the preferred conformation. The minor conformation is the twist–boat, which, in some cases, can only be detected by NMR line-shape methods. For the heterocycles (**59a–d**), the twist–boat was observed to have higher entropy than the boat–chair, due to the greater flexibility of the twist–boat conformation. The ketoamine (**60**)[82] is postulated to exist as the boat–boat conformation rather than the twist–boat, due to a transannular interaction between the nitrogen lone pair and the carbonyl group.

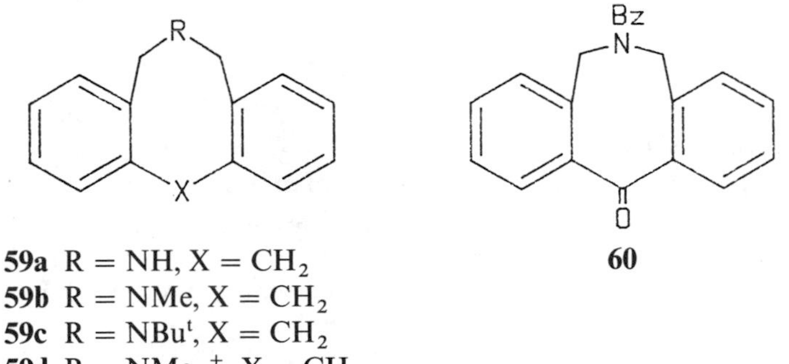

59a R = NH, X = CH_2
59b R = NMe, X = CH_2
59c R = NBu^t, X = CH_2
59d R = NMe_2^+, X = CH_2
59e R = NBenzyl, X = O

60

3.3.3.3. 1,5-Diconstrained Systems. There are three conformational minima for eight-membered rings derived from a 1,5-diene—the rigid chair, the flexible boat, and a skew form, which is sometimes referred to as a twist–boat. With the exception of the diamide (**61**), conformational studies are mostly restricted to compounds with the dibenzo skeleton (**62**). The diamide (**61**) was shown by X-ray crystallography[83] to exist in the solid state as a twisted boat with C_2 symmetry. The results of force-field calculations on this diamide,[84] and the fact

TABLE 3-4. CONFORMATIONAL PROPERTIES OF 1,5-DIBENZO HETEROCYCLES **(64c–e, 65b,c, 66b,c)**

Compound	Process	ΔG°	$\Delta G^\ddagger$
64c	Chair = boat	2.0	46
	Boat = boat	0	84
64d	Chair = boat*	−0.5	47
	Boat = boat*	0	74
64e	Chair = boat	2.8	44
	Boat = boat*	0	60
65b	Chair = boat	1	94
	'Inversion'	0	92
65c	Chair = boat	0	87
	'Inversion'	0	88
66b	Chair = boat	−0.5	54
	Boat = boat*	0	71
70c	Chair = boat	0.3	58
	Boat = boat*	0	60

that the vibrational tensors in the crystal structure are inconsistent with significant torsional oscillations about the C3—C4 and C7—C8 bonds, show that this structure is not as flexible as molecular models would suggest.

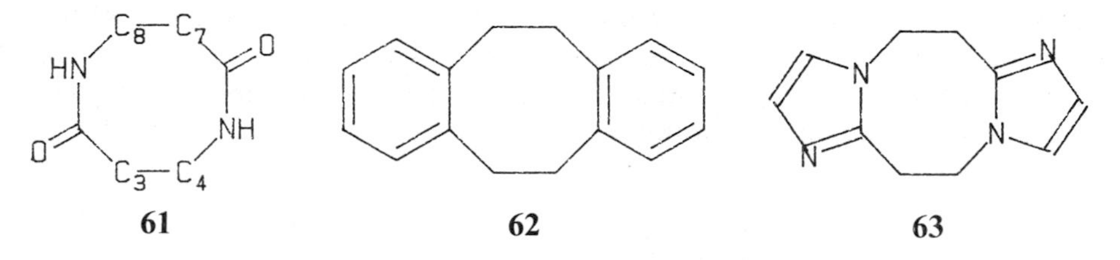

61 **62** **63**

The crystal structure of the bis-imidazole **(63)**[85] shows that the eight-membered ring exists in a chair conformation, with the molecule lying over a crystallographic center of inversion. A series of monoaza and diaza compounds derived from the dibenzo skeleton have been studied[86,87]; the conformational properties of some are summarized in Table 3-4. While the parent hydrocarbon **(62)** shows observable dynamic properties in solution, the introduction of one

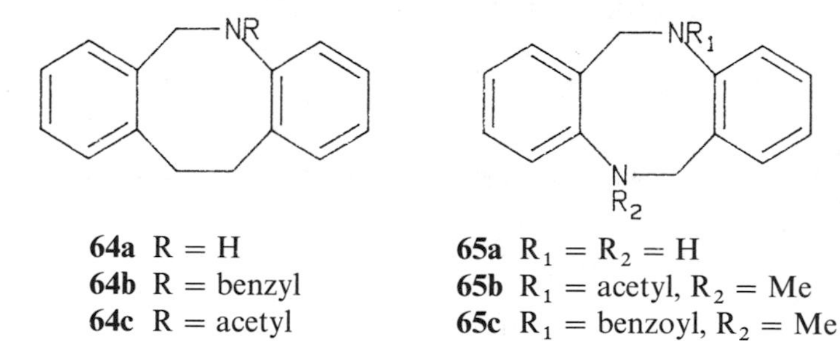

64a R = H
64b R = benzyl
64c R = acetyl
64d R = benzoyl
64e R = tosyl

65a $R_1 = R_2 = H$
65b R_1 = acetyl, R_2 = Me
65c R_1 = benzoyl, R_2 = Me

TABLE 3-5. Barriers to Ring Inversion for Azocinones (**70a–h**)

Compound	Proton	T_c (°C)	$\Delta G^{\ddagger}$
70a	$3CH_2$	−52	43.09
70b	$3CH_2$	−49	43.93
	NCH_2	−49	44.34
70c	$3CH_2$	−46	44.35
	NCH_2	−46	44.35
	N—∗—$(CH_3)_2$	−60	45.59

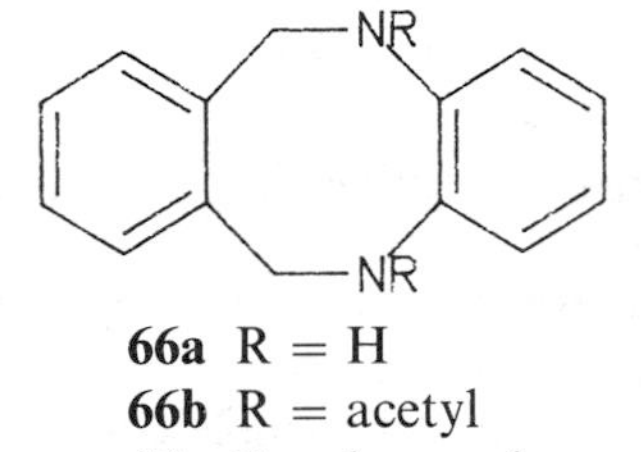

66a R = H
66b R = acetyl
66c R = benzoyl

or more nitrogen atoms into the ring to give **64a,b**, **65a**, and **66a** results in compounds that have no observable dynamic NMR effects. If the nitrogen substituent is an acyl or sulphonyl group as in **64c–e**, **65b,c**, and **66b,c**, then dynamic effects are observed (see Table 3-5). The interconversion pathway for the observed processes is outlined in Fig. 3-2. The boat to chair and boat* to chair* involve the transition state (**67a**) and are referred to as local inversions. The

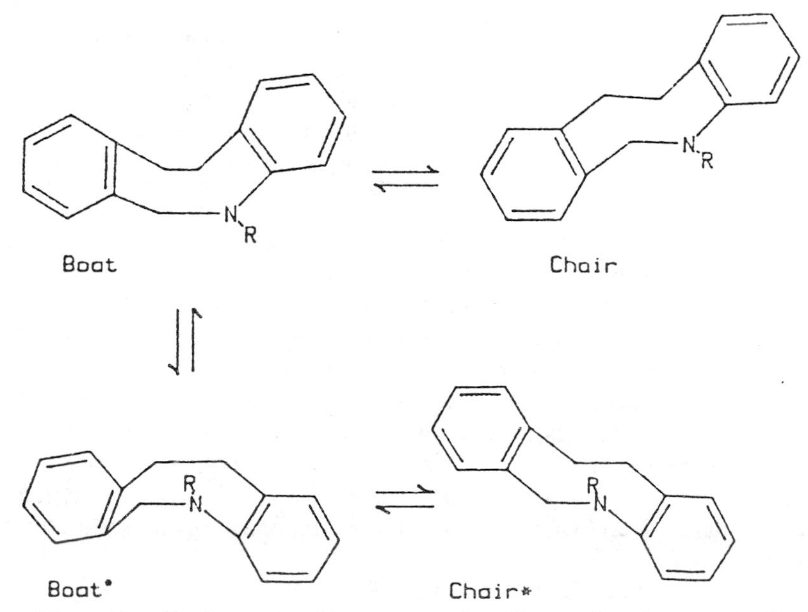

Figure 3-2. Conformational interconversion diagram for compounds **64**.

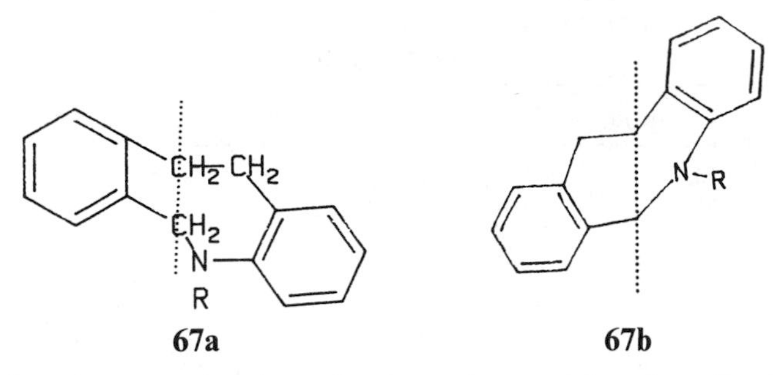

67a **67b**

boat-to-boat* conversion is a ring inversion and involves the transition state (**67b**).

3.3.4. Eight-membered Rings with Three Constraints

There are two possible constitutional isomers of an eight-membered ring with three torsional constraints—those related to the 1,3,5-triene (**68**) and those related to the 1,3,6-triene (**69**). The most stable conformation for **68** is the twist–boat[88] while the most stable conformation for **69** is the twist–boat–chair.

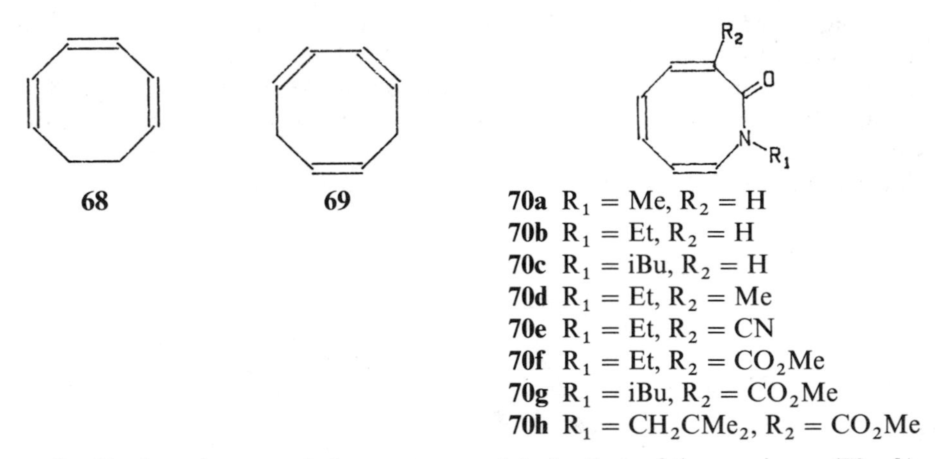

68 **69**

70a R_1 = Me, R_2 = H
70b R_1 = Et, R_2 = H
70c R_1 = iBu, R_2 = H
70d R_1 = Et, R_2 = Me
70e R_1 = Et, R_2 = CN
70f R_1 = Et, R_2 = CO_2Me
70g R_1 = iBu, R_2 = CO_2Me
70h R_1 = CH_2CMe_2, R_2 = CO_2Me

Studies have been carried out on several derivatives of the azocinone (**70a–h**). It has been found that if a bulky substituent is present at the 3-position, as in **70e–h**, then the geminal groups on the exocyclic carbon next to the nitrogen are inequivalent and their chemical shifts are temperature dependent.[89] This was initially thought to be a dynamic process involving rotation of the exocyclic N—C bond, but was later explained as being a result of population changes between different rotamers about this bond at different temperatures.[90] Dynamic effects were later measured for compounds **70a–c**[91] and the observed barriers to ring inversion are summarized in Table 3-5. An X-ray structure of the cyano derivative (**70e**)[92] shows that the ring exists in the solid state in a twist–boat conformation.

The dibenzo amide (**71**) studied by Ollis et al.[93] has a barrier for the ring inversion of 90 kJ/mol.

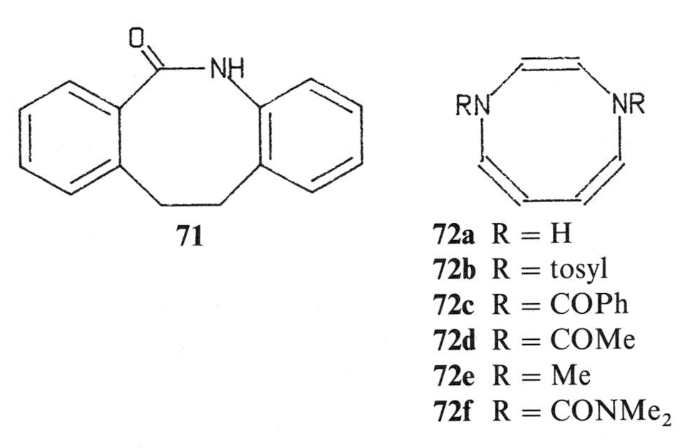

71

72a R = H
72b R = tosyl
72c R = COPh
72d R = COMe
72e R = Me
72f R = $CONMe_2$

Studies on a number of N,N′-disubstituted derivatives of the diamine (**72a–f**)[94] illustrate an interesting dependence of the conformation on the electronic nature of the nitrogen substituents. If the substituents are electron donating, as in the case of **72a,e,f** and the dianion (**73**), then the eight-membered ring is a planar structure with strong 10-electron delocalization whereas those with electron-acceptor substituents (**72b–d**) prefer the twist–boat–chair conformation. Later studies on the cyano derivatives (**74a,b**)[95] show that the anions formed when they are treated with a suitable base are planar and aromatic even when acceptor substituents are present on the nitrogen.

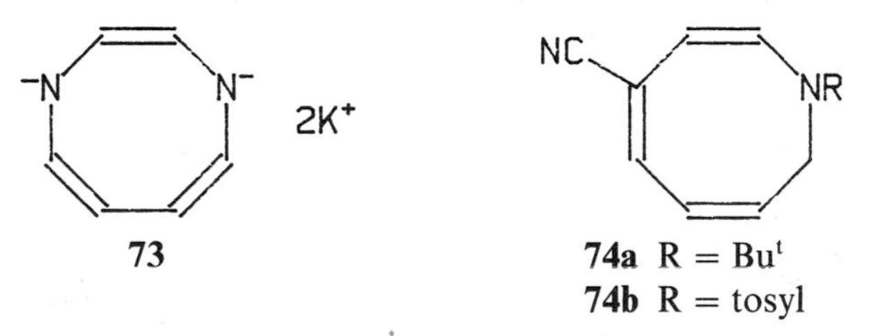

73

74a R = Bu^t
74b R = tosyl

3.3.5. Eight-membered Rings with Four Constraints

Eight-membered rings having four torsional constraints exist as nonplanar tub-shaped conformations that are capable of possessing chirality with appropriate substitution. For example, a number of derivatives of the dibenzo-diimide (**75**) have been prepared[96,97] and the optical isomers of the aldehyde (**76**) separated by conversion to diastereomeric oxazolidinones using 1-ephedrine and

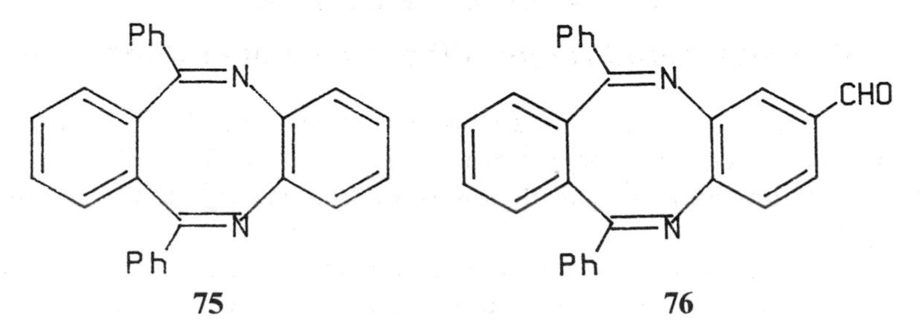

75 **76**

separating by crystallization, followed by regeneration of the aldehyde, which is optically pure. Measurements on the rates of racemization provided a barrier to inversion of 155 kJ/mol.

Dynamic NMR studies on the diamides (**77** and **78**)[93] gave lower limits for the ring inversion in both of 113 kJ/mol.

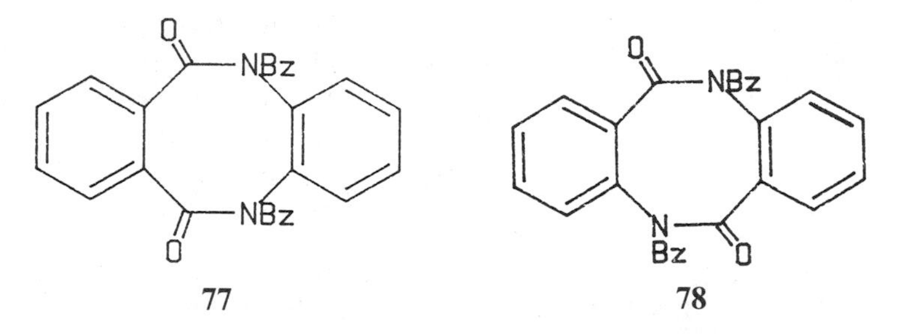

77 **78**

A number of bicyclic structures with eight-membered rings as the smallest medium ring have been studied, for example, thebaine derivative (**79**) containing a [5.3.1] bicyclic structure.[98] Other [5.3.1] structures are given in Ref. 99. The structure of bicyclomycin (**80**)[100] has been determined; the 4-carbon bridge of this [4.2.2] bicycle has a g+g−g+ sequence of torsion angles, and the [4.2.2] skeleton has an approximate twofold axis bisecting the central bond of the 4-carbon bridge and the centroid of the six-membered ring. Other structures of this type are grouped in Ref. 101.

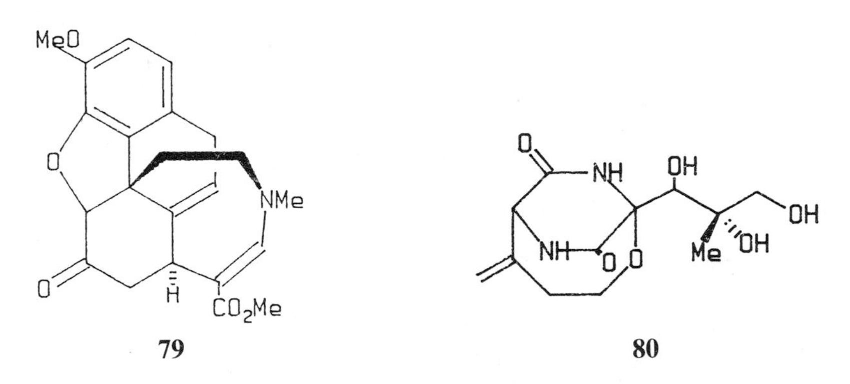

79 **80**

3.4. Nine-membered Rings

3.4.1. Nine-membered Rings without Torsional Constraint

The structure of octamethyleneammonium tosylate has been determined.[15] There is no symmetry, but the conformation is close to a favored nonsymmetrical conformation found by calculation for cyclononane.[102] The mean internal bond angle is 117.6°.

One of the best known unconstrained nine-membered ring nitrogen compound is 1,4,7-triazacyclononane, readily prepared by the methods of Richman and Atkins.[103] However, very little is known about the conformation of these compounds; ring and nitrogen inversion barriers are apparently very low in simple derivatives, but MM2 calculations give the [333] conformation with "side" rather than "corner" nitrogens as the global minimum for the N,N′,N″-trimethyl derivative.[104] On the other hand, there are many X-ray structures of metal complexes.[105] The cyclic tripeptides discussed later in this section are, of course, derivatives of these compounds.

3.4.2. Nine-membered Rings with One Torsional Constraint

The nine-membered ring is close to the point where *cis*- and *trans*-lactams are of equal stability and so this has been carefully studied. In solution, caprylolactam (azacyclononan-2-one) (**81**) is a 4:1 mixture of *cis* and *trans* forms, with the *cis* predominating.[106,107] The structures of caprylolactam,[108,109] its hydrochloride,[29] and the 4,4,7,7-tetramethyl derivative[110] have been reported. The two neutral lactams are *trans* in the solid state, but are quite nonplanar (torsion angles average 147°) as demanded by the ring system, whereas the *o*-protonated lactam is *cis*. The preference for the *trans*-lactam in the solid state is due to a more favorable intermolecular hydrogen-bonding pattern. The greater demand for planarity about the C—N bond in a protonated lactam is probably decisive in making the salt become *cis*.

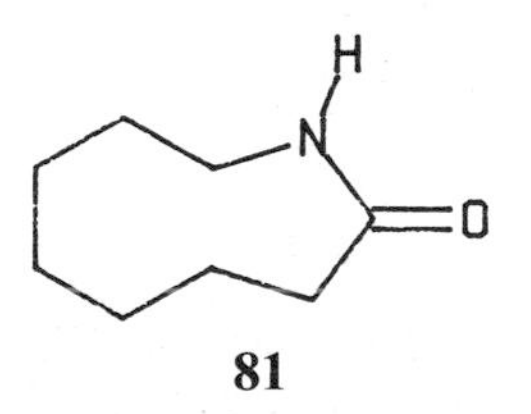

81

1,3-Diazacyclonona-1,2-diene, an interesting cyclic carbodiimide, has been studied by low-temperature ^{1}H NMR and showed[111] a conformational process with a barrier of 28 kJ/mol. This is remarkably close to that observed for diisopropylcarbodiimide,[112] but it is not clear as to what type of process is occurring in the cyclic case, especially in view of the complex behavior of the corresponding allene.[113]

3.4.3. Nine-membered Rings with Two and Three Torsional Constraints

A number of nine-membered ring systems with two fused benzo or a benzo and a naphtho fusion have been studied in considerable detail by Ollis and co-workers. Preliminary communications[114,115] were followed by full papers,[116,117] and the work has been reviewed.[118] Compounds with two benzo fusions but no double bond, for example, **82**, adopt a flexible conformation that was described as a C_2 chair, whereas compounds with an extra double bond exist as a conformational mixture of rigid-chair (**83a**) and flexible-boat (**83b**)

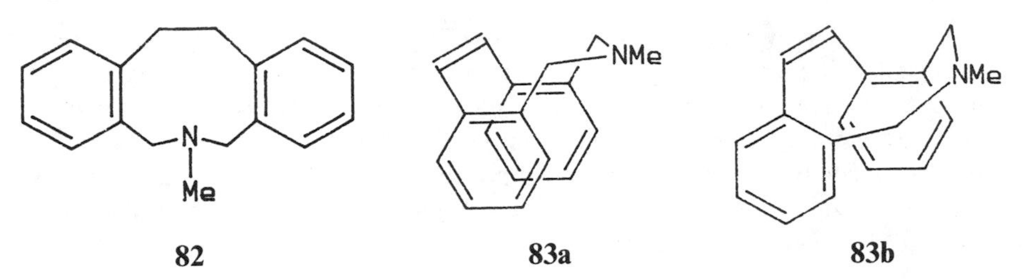

82 **83a** **83b**

conformations; the chair is favored by 4.7 kJ/mol at −22°, and while the boat-to-boat conversion was too fast to follow, the chair-to-boat processes could be followed by dynamic NMR (barrier 68 kJ/mol).[114,116] In another series, compounds with two nitrogens and a naphtho and a benzo fusion exist in a TB conformation (**84**) and a TB = TB* conformational process can be followed by NMR (barrier 59 kJ/mol for the N-Me compound).[115,117] In related work, compound **83** was shown to undergo protonation with transannular reaction, but at a slower rate than an 11-membered ring compound in which the benzo fusions are replaced by 1,8-naphtho fusions.[119]

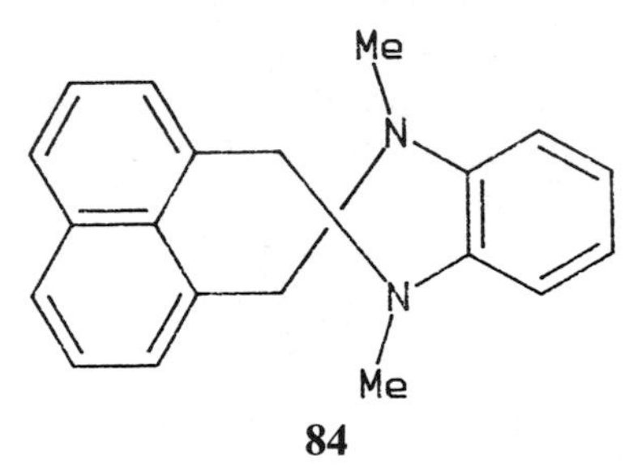

84

In a classic piece of work,[120] Berson showed that thebaine and phenylmagnesium bromide gave a mixture of 4,5-(3-hydroxy-4-methoxybenzo)-6,7-(4-methoxybenzo)-1-methyl-2-phenyl-2,3:8,9-tetrahydroazonines, which were diastereomeric at C-2. This provided the first unambiguous absolute configuration for a biphenyl derivative. The structure of the biphenyl derivative (**85**) shows a strongly twisted structure with an angle of 86.6° between the benzene rings.[121]

The structure of the 3-aza[5](1,7)naphthalenophane (**86**) shows an extremely distorted naphthalene ring; a related 10-membered ring compound was also studied.[122]

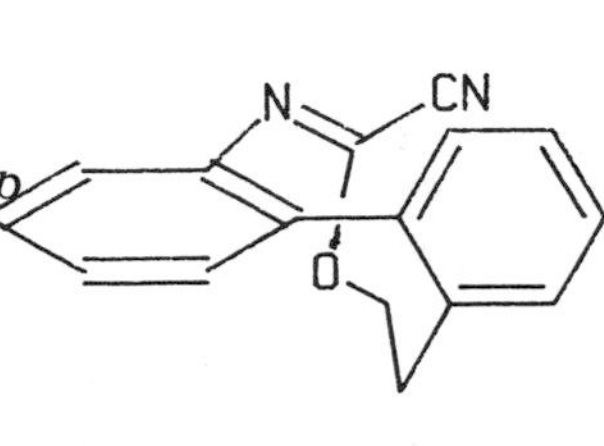

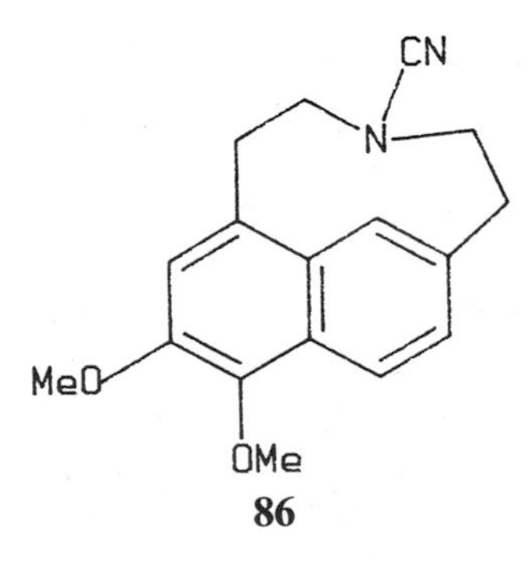

85 **86**

Cyclic tripeptides have been extensively studied. Early work indicated that a C_3 ("crown") conformation was preferred, for example, for cyclo(Pro_3)[123] and cyclo(Sar_3),[124] but subsequent NMR studies by Kessler and his group have shown that a boat conformation can be significantly populated.[125] Thus, cyclo(Pro_2-benzylglycine) exists as a mixture of 90% boat and 10% crown in $CDCl_3$, but the proportion of crown increases with increasing solvent polarity. The equilibrium can be elegantly studied by two-dimensional exchange spectroscopy.[126] Cyclo(di-L-Pro-D-proline) is in a boat conformation in the crystal.[127]

3.4.4. Nine-membered Rings with Four Torsional Constraints

The fully unsaturated azonines have been extensively studied by Anastassiou and others; progress has been reviewed several times.[128–130] The major interest is in the possible aromaticity of these systems, and the conclusion seems to be that only systems in which the nitrogen is very electron rich are aromatic. Thus, compounds with electron-withdrawing substituents have been described as having heavily buckled, heavily localized, molecular frames; they are frequently thermally labile, undergoing electrocyclic transformations to bicyclic isomers. N-(N,N-dimethylcarbamoyl)azonine has a tub-shaped structure with localized double bonds.[131] The ^1H NMR spectrum of N-ethoxycarbonylazonine is temperature dependent, but this is due to restricted rotation in the ethoxycarbonyl group and not to nitrogen inversion.[132]

Alkali metal derivatives of azonines are aromatic, air sensitive, delocalized, and diatropic. N-alkyl derivatives have been described as intermediate, but detailed conformational information is lacking. Several compounds with fused benzo and related heterocyclic rings have been studied, and these undergo a dramatic *cis-trans* isomerization. Thus, the nonaromatic all-*cis* potassium derivative (**87**) isomerizes to the aromatic mono-*trans* isomer (**88**) with an activation energy of 83.7 kJ/mol,[133] but when this is converted into the N-ethoxycarbonyl derivative, it reverts[134] to the all-*cis* isomer when heated (activation energy 136 kJ/mol at 140°).

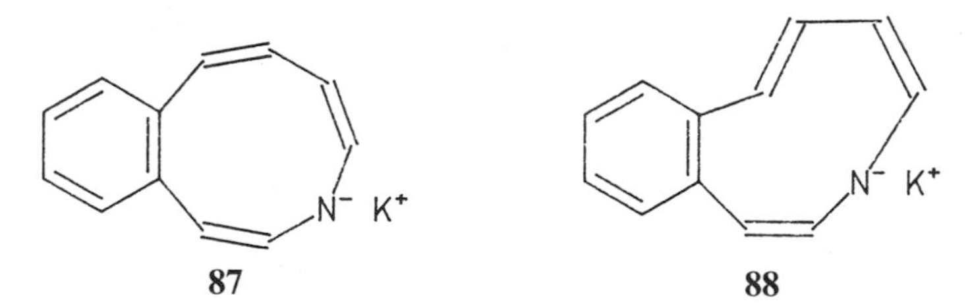

87 **88**

The conformational properties of some [5.2.2] systems, containing a nine-membered ring as the smallest medium ring, have been examined. Substituted 4-azabicyclo[5.2.2]undeca-8,10-dien-3-ones have been studied by NMR and a barrier of 80 kJ/mol is supposedly due to ring inversion.[135] The structure has been reported of the N-phenyltriazolinedione adduct of *cis,trans*-cyclononadiene, an *in,out* [5.2.2] system.[136] The structures of a number of complex alkaloids containing nine-membered rings have been reported (a list is given in Ref. 137).

3.5. Ten-membered Rings

The well-known BCB conformation for 10-membered rings was largely established experimentally from X-ray studies of salts of mono- and diamino derivatives of cyclodecane.[138] However, there are very little data on saturated unconstrained mono- and diaza derivatives of this ring system. We have recently solved the structure of 1,6-ditosyl-1,6-diazacyclodecane.[139] The conformation is BCB, with the completely flat nitrogen atoms occupying sites on the symmetry plane of the BCB structure (**89**). The parent 1,6-diazacyclodecane and its mono- and dialkyl derivatives have first pK_a values over 12 but unusually low second pK_a values (0.4 for the dimethyl compound).[140] The monoprotonated ions are clearly stabilized by a strong transannular hydrogen bond, which is broken by a second protonation. The crystal structure[139] of the monohydroiodide of 1,6-dimethyl-1,6-diazacyclodecane shows a *cis*-decalinlike geometry (C_2 symmetry) (**90**), and the ^{13}C NMR spectrum shows the expected three lines at high and five lines at very low temperature. The apparent ring inversion barrier is <40 kJ/mol, much less than in *cis*-decalin itself, but this could still represent the same type of conformational process with the hydrogen bond remaining intact throughout. It is likely that the diamine itself, as well as its diprotonated ion, adopts BCB conformations, and that part of the pK_a enhancement results from strain relief.

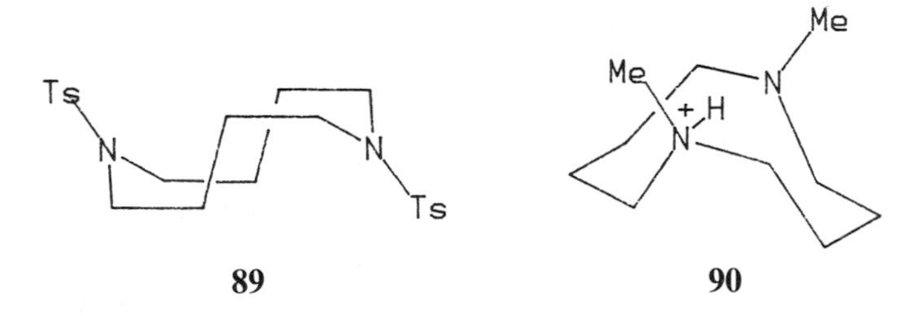

89 **90**

Since the BCB conformation is derived from the diamond lattice, it can be fused to chair cyclohexane rings with little distortion. This is illustrated by the structure (**91**) of 1,4,8,11-tetraazatricyclo[9.3.1.14,8]hexadecane[141]; a hexamethyl derivative adopts the same conformation.[142]

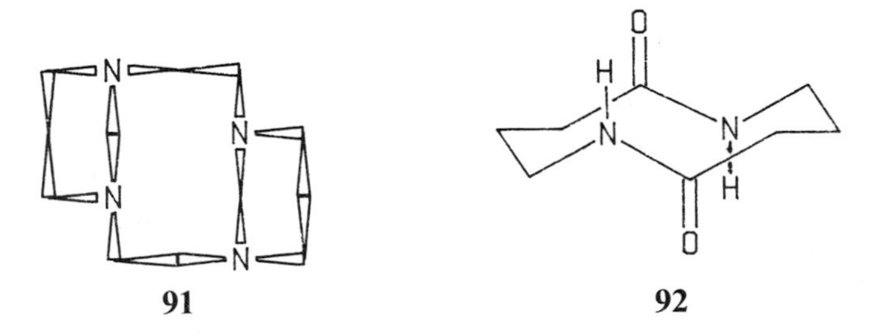

91 **92**

1,6-Diaza-2,7-cyclodecadione has *trans* amides but is CCC, not BCB (**92**).[143] This was deduced from a detailed analysis of the ^1H NMR spectrum, specifically the large (11-Hz) coupling between the NH protons and an adjacent axial hydrogen. In a BCB conformation, there would be a serious interaction between the carbonyl oxygens and the methylene groups lying on the symmetry plane. For dilactam (**93**), both symmetrical *trans,trans*- and *cis,trans*-lactam structures are present in solution, with the latter predominating.[144]

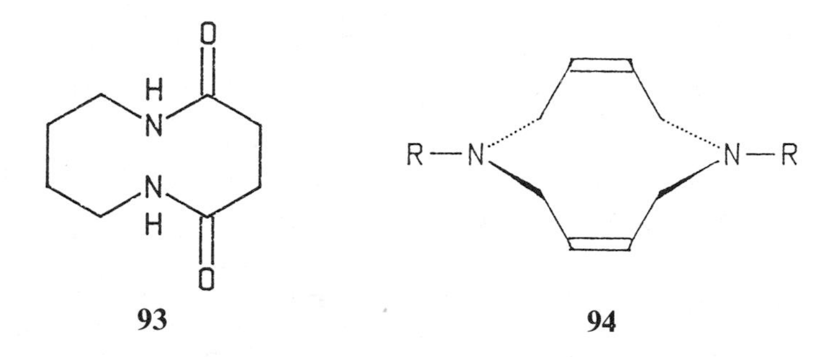

A number of compounds with two *cis* double bonds or benzo fusions in 1,6-relationship have been found to adopt "chair" structures, for example, **94**.[145] Ring inversion of **94** occurs through a boat form, with barriers of 53 (R = H), 42 (R = COOMe), and 57 kJ/mol (R = Me). Compound **94** and a number of related heterocycles were derived from the corresponding diacetylenes and similar monoaza compounds are known.[146] These compounds might be expected to adopt almost strain-free "stretched" chair or boat conformations but there appear to be no data on this. Compounds such as **95**, R = Me and CH_2Ph, have structures like **94**, and similar ring inversion barriers, for example, 45 kJ/mol for R = Me.[114]

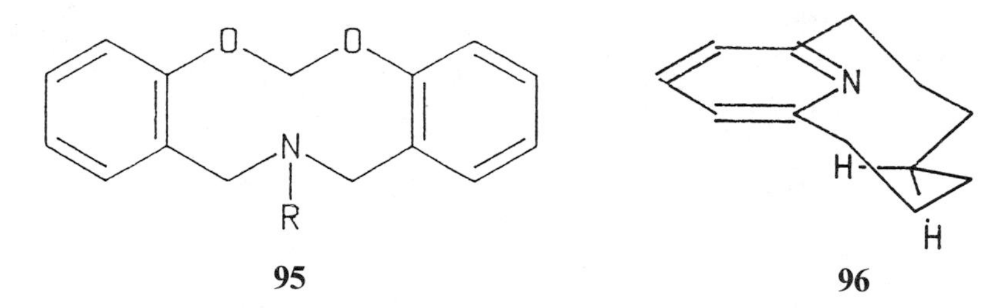

A number of pyridinophanes contain 10-membered rings. [7](2,6)pyridinophane itself appears to adopt a conformation (**96**) that places one proton (chemical shift δ–1.4) under the aromatic ring; the barrier to ring inversion is 38 kJ/mol.[147] Derivatives with gem-dimethyl and dimethylmethylidene substituents have also been studied.[148,149] The barrier to ring inversion in [2.2](2,6)pyridinophane (**97**) is 62 kJ/mol, much lower than that for [2.2]paracyclophane itself (>113 kJ/mol), due to the smaller nonbonded interactions between two lone pairs than between two CH groups.[150] The structure of 1a,9b-dichloro-2,2,10,10-tetramethoxy[2.2](2,6)pyridinophane shows the expected "stepped" conformation with a short N . . . N distance of 2.475 Å.[151] The structure of a 4,16-diaza[2_4](1,2,4,5)cyclophane has also been reported.[152]

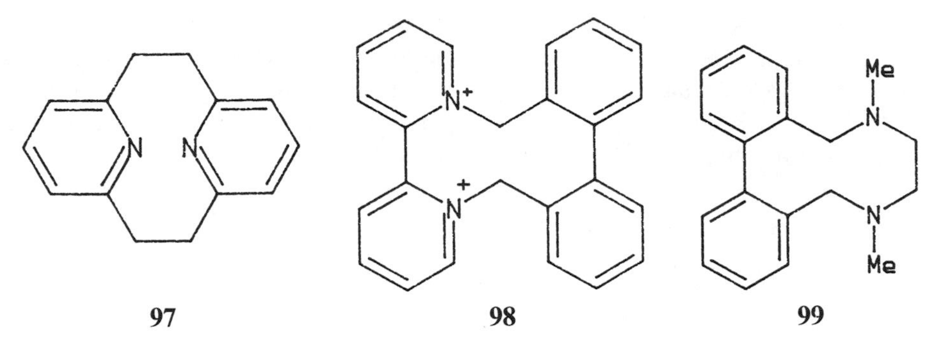

97 **98** **99**

Ten-membered ring derivatives of 2,2-bipyridyl such as **98** are conformationally rigid,[153] while **99** has a C_2 structure with a strongly twisted biphenyl unit.[154]

The N-phenyltriazolinedione adduct of *cis,trans*-cyclodecadiene has been reported.[136] The compound has an *in,out* [6.2.2] structure. 6-Methyl-1,6-diazabicyclo[6.2.2]dodecane, an *in,out* six-atom bridged piperidine, probably adopts a conformation with inside, facing lone pairs.[155] Several complex alkaloid structures containing 10-membered rings have been reported.[156]

3.6. Eleven- and 12-membered Rings

In this section, we first discuss the very small amount of conformational information on 11-membered rings before turning to the much better understood 12-membered rings.

Saturated nitrogen-containing 11-membered rings are still completely undefined. The dinaphtho fused compound (**100**) undergoes more rapid transannular cyclization than a corresponding dibenzo nine-membered analogue and may exist in the conformation shown.[119]

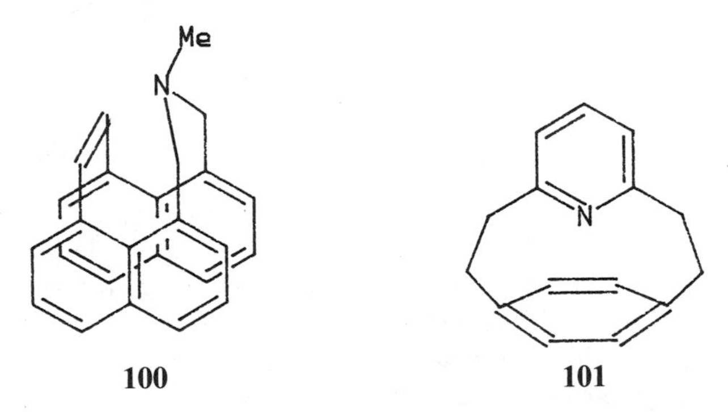

100 **101**

Several pyridinophanes contain 11-membered rings. [2.2](2,6)Pyridinoparacyclophane exists in a conformation (**101**) in which the two aromatic rings are perpendicular to one another,[157] whereas in [2.2](2,5)furano(2,5)pyridinophane, the angle between the two rings is 23°.[158] In [2](1,5)naphthalino[2][2,6]pyridinophane-1,11-diene and its 1,4-naphthalino isomer, the pyridine lone pair is pushed into the naphthalene ring, but when the double bonds are hydrogenated, the interaction between the rings becomes face to face.[159]

Unlike 11-membered rings, saturated 12-membered rings appear to have one strongly preferred conformation, the square [3333] structure. Azacyclodecane hydrochloride adopts this structure with the nitrogen at a corner position.[160] 2R,5R,8R,11R-2,5,8,11-tetraethyl-1,4,7,10-tetraazacyclododecane has a similar structure with C_4 symmetry and a methylene group at the corner.[161] The same [3333] conformation is adopted by several tetraazacyclododecanes and aza-crown ethers; thus tetrapod and 1,4,7,10-tetrakis(2-hydroxyethyl)-1,4,7,10-tetraazacyclododecane have this structure in complexes with Li^+, Na^+, K^+, and H_2O with varying numbers of the oxygen atoms in the arms coordinated; for example, (**102**) shows the K^+ complex.[162] The benzylammonium thiocyanate complex of N,N′-dimethyl-1,7-diaza-4,10-dioxacyclododecane was studied by NMR and by X-ray; the [3333] conformation observed was found to be mobile, with a barrier of about 35 kJ/mol.[163] A complex of the same compound with the dibenzylammonium ion also has the [3333] conformation.[164] Complexes of N-substituted 1-aza-4,7,10-trioxacyclododecane with mono-alkylammonium ions have also been studied.[165] In 4,7,13,16-tetraoxa-1,10-diazabicyclo[8.8.2]icosane, the 12-membered rings are in a [39] conformation in the free ligand but change to [3333] in Li^+, Na^+, and K^+ complexes.[166] In 1,5,9,13-tetraazatricyclo[11.3.1.1^{5,9}]octadecane (**103**), the 12-membered ring is in a [3333] conformation and this is neatly joined with two chair hexahydropyrimidine rings having the anomerically preferred *eq,ax* conformations.[167]

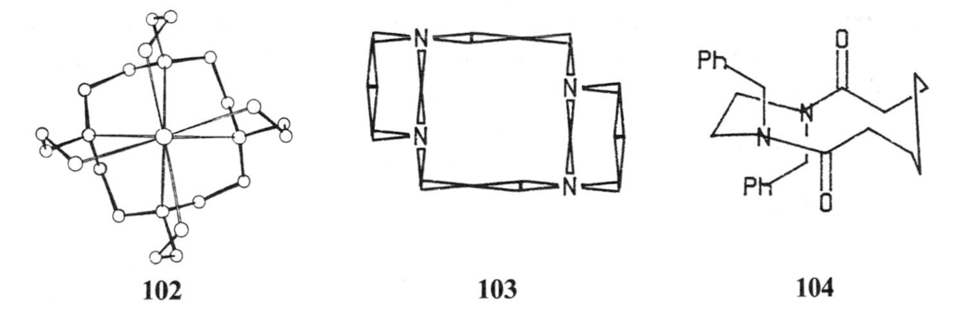

102 **103** **104**

In a dilactam with mirror symmetry, 1,4-dibenzyl-1,4-diazacyclododeca-5,12-dione, both lactams are *trans* and the structure has C_2 symmetry (**104**).[144] A number of tetrapeptide structures have been determined, and several show the interesting regularity of having their NHCO amide links *trans* but their NMeCO links *cis*; for example the sarcosyl (N-methyl glycyl) derivatives cyclo(Glyc-Sar$_3$) monohydrate, cyclo(Gly-Sar-Gly-Sar), and dl-cyclo(Ala-Sar$_3$).[169] However, in dihydrotentoxin (**105**),[168] the situation is reversed. In the structure of cyclo(L-Pro-L-Val-L-Pro-L-Val) dimethylsulfoxide solvate, the Val-NH-CO-Pro amide link is *trans* and there is C_2 symmetry.[170]

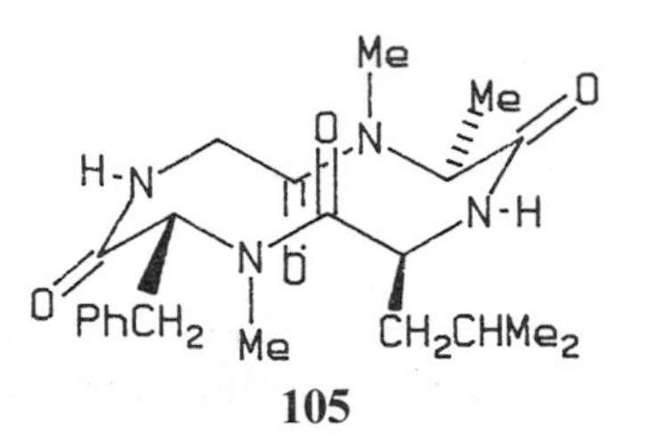

105

Next we turn to the extensive and elegant work by Ollis and co-workers and collaborators on the trianthranilides. The early work has been reviewed.[118] There are two principal conformations to be considered, the propeller with threefold symmetry (**106**) and the helical conformation with, at best, a twofold axis (**107**). Thus, the simple N,N′,N″-trimethyl trianthranilide is largely helical with 5% propeller conformation.[171] N,N′-dibenzyl-trianthranilide toluene solvate and N,N′-dimethyl-trianthranilide are also helical.[172] 5,18-dimethyl-5,11,12,18-tetrahydrotribenzo[b,f,j][1,4]-diazacyclododecine-6,17-dione (**108**), a compound in which one amide link is replaced by a $CH_2—CH_2$ unit, adopts a propeller-like conformation in an *o*-xylene clathrate.[173,174] The N,N′-dibenzyl derivative of (**108**) shows ring inversion between enantiomeric propeller-like conformations with a barrier of 89 kJ/mol. In compounds with methyl groups on the rings *ortho* to the N atoms, enantiomeric helical conformations are adopted. The N,N′-dimethyl-N″-benzyl derivative undergoes spontaneous resolution when it crystallizes as a 1:1 adduct with toluene.[175,176] In a real *tour de force*,[177] 16 trianthranilides with two and three N-substituents were studied by NMR. Both propeller and helical conformations were present in symmetrical derivatives whereas unsymmetrical compounds exist in three helical conformations as well as in a propeller. A reduced compound without carbonyl groups exists as a mixture of enantiomeric helical conformations; barriers to ring inversion are much lower in this case.

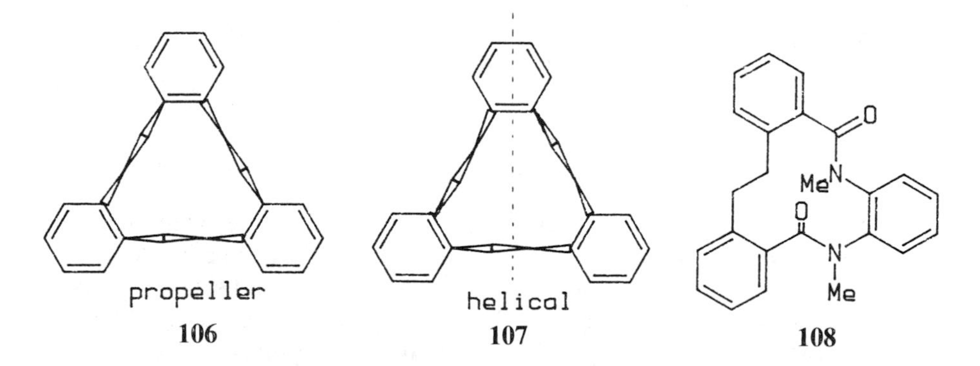

106 **107** **108**

1,3,5-Tri[2,6]pyridacyclohexaphane-2,4,6-trione exists in a nonplanar conformation (**109**) with approximate C_3 symmetry; the nonplanarity was ascribed to N . . . N interactions.[178,179] In the synthetic precursor (**110**), the NH participates in a bifurcated hydrogen bond.

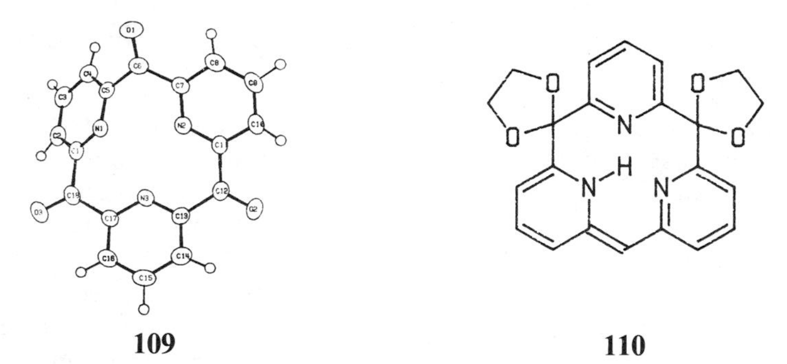

109 **110**

The tribenzohexaaza[12]annulene (**111**) has C_2 symmetry in the crystal, presumably a helical conformation, but in solution, there is rapid rotation about the C—N bonds.[180]

2,11,20-Tritosyl-2,11,20-triaza[3,3,3](1,3,5)cyclophane (**112**) adopts the C_{3h} conformation shown, undergoing a ring inversion process similar to that observed with manxane compounds (see later); the barrier is 53 kJ/mol.[181]

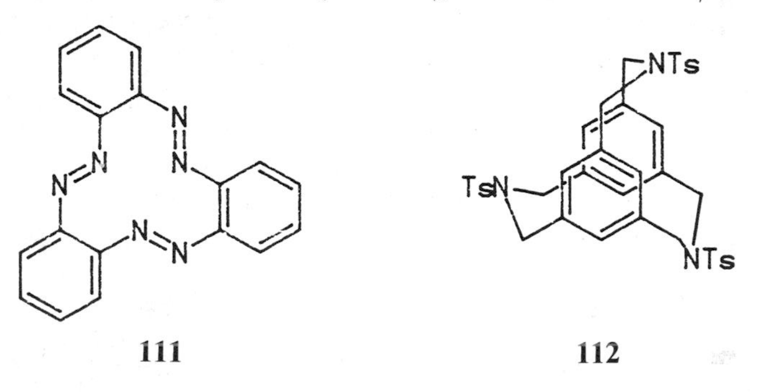

111 **112**

3.7. The Effects of Ring Size on Nitrogen Properties

It might be expected that the expanded bond angles within medium rings would have detectable effects on the properties of nitrogen atoms within the ring. Thus, basicities might be reduced due to the greater ease of flattening of the tricoordinate nitrogen in an amine compared with the tetracoordinate nitrogen in an ammonium ion. For similar reasons, ionization and oxidation potentials might be lowered, since amine radicals prefer a flat geometry. These effects are certainly observed with bridgehead nitrogen atoms in bicyclic compounds (see later), but there is precious little evidence for measurable effects in monocyclic compounds. There are no data on the gas-phase basicities or ionization potentials or the aqueous pK_a values of the simple azacycloalkanes with medium rings, but the pK_a values in 80% methyl cellosolve have been measured.[182] The values are (ring size, pK_a): 6, 9.99; 7, 10.00; 8, 9.77; 9, 9.39; 10, 9.14; 11, 9.04; 12, 9.14. The expected decline and subsequent rise in pK_a occur but appear to be superimposed on a general trend to lower values, since azacyclohexadecane, which is presumably strain-free, has a pK_a of only 9.26.

3.8. Transannular Interactions Involving Nitrogen

Transannular interactions are a major feature of medium-ring chemistry, and some of the earliest to be discovered concerned nitrogen. Leonard's early work has been reviewed more than once.[7] Leonard showed that transannular interaction between a ring amino nitrogen and a suitably placed carbonyl group resulted in a reduced CO stretching frequency in the infrared spectrum; this transannular interaction could be converted into a transannular reaction when the compound was protonated, since protonation occurred on oxygen with the concomitant formation of an N—C bond (no C=O stretch). Evidence from several other spectroscopic techniques followed.[8] From the lack of ring inversion in dynamic NMR studies,[82] it was concluded that the attraction was worth 8–12 kJ/mol. The early spectroscopic work was followed by structural confirmation from X-ray studies. The work of several groups[183] was brought together by Dunitz and his co-workers to form the first thorough study of a reaction coordinate by structural analysis.[9,10,184,185] In neutral N . . . CO compounds, the N . . . C length is still long (over 2 Å), and the relatively weak interaction results in some lengthening of the C=O bond (up to 1.26 Å) and slight pyramidalization of the carbon atom, so that we are still near the beginning of the reaction coordinate. A typical structure is shown in **113**. When the oxygen is protonated, the "reaction" is almost complete, with a short N—C distance, an almost tetrahedral carbon, and a C—O distance typical of a single bond. A typical structure of this type is shown in **114**.

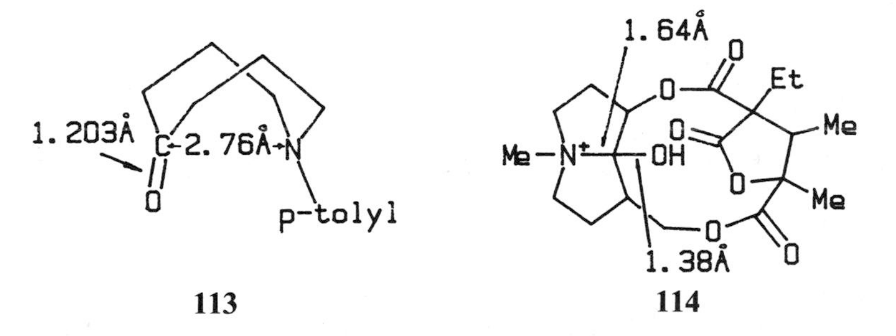

113 **114**

Whereas the N . . . CO interaction is probably attractive at all distances, interaction of an amino nitrogen with a C=C may well be repulsive at most distances. In terms of the molecular orbitals involved, the N lone pair/π* mixing is important in the N . . . C=O case but not in the N . . . C=C; in the latter case, the repulsive (four-electron) N lone pair/π mixing will be dominant. These effects are seen in the photoelectron spectra[11] of aminoketones and aminoalkenes.[186] Thus, in aminoketones (**115a** and **116a**), the nitrogen lone pair ionization energy is lowered and the oxygen lone pair ionization energy is raised. Rademacher considers the latter to be the better evidence for a transannular interaction. In aminoalkene (**115b**), the N lone pair ionization energy is raised by 0.2 eV, and the C=C π ionization is lowered by a similar amount. This must mean that this compound adopts a conformation, as shown, in which the two

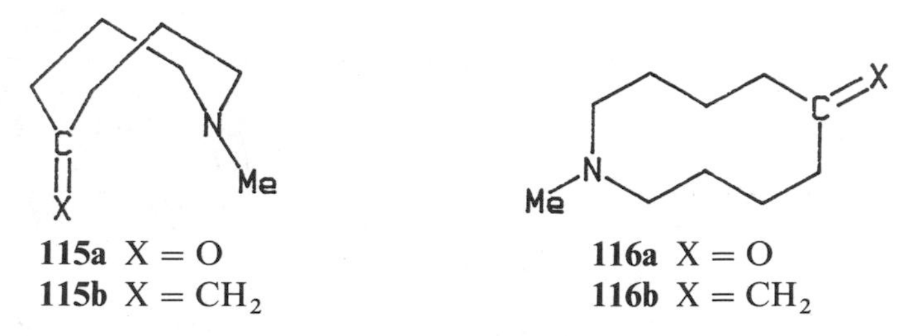

115a X = O
115b X = CH_2

116a X = O
116b X = CH_2

groups interact. Perhaps the N . . . C=C interaction is sufficiently less repulsive than a CH_2 . . . CH_2 interaction that a BB conformation is adopted. In the ten-membered ring analogue (**116b**), there is no detectable interaction, so presumably a conformation is adopted in which the two functional groups are far apart. A more dramatic example of the N . . . C=C interaction is discussed in section 3.11 on intrabridgehead interactions. Protonation can result in transannular reaction in systems capable of N . . . C=C interaction. Thus, compound **117** is converted to **118** on protonation,[187] and other cases are discussed elsewhere in this chapter.[119] The most thorough mechanistic study was on N-methyltetrahydroberberine.[188]

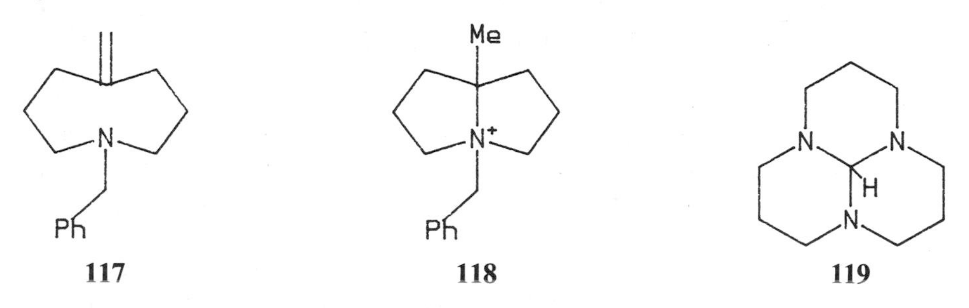

117 **118** **119**

Another interesting transannular interaction arises when tricyclic orthoamides such as **119** are treated with electrophiles; bicyclic amidinium ions with a weak transannular interaction (**120**) can result.[189,190] On the other hand, **121** gives the closed species (**122**).[191] The outcome of these reactions is apparently controlled by ring strain.

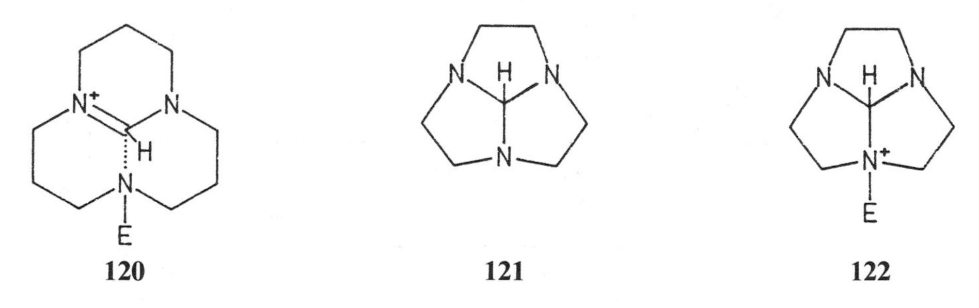

120 **121** **122**

Surprisingly, there has been very little evidence for transannular hydrogen bonding until recently. The case of 1,6-dimethyl-1,6-cyclodecane was discussed in the section on 10-membered rings, but this simply represents the best case in a series, as can be seen from the data in Table 3-6.[192] In general terms, the

TABLE 3-6

Diamine	p	q	pK_{a1} in H_2O	pK_{a2} in H_2O	pK_{a2} w.r. to 2,7-Dimethoxyl-1,8-bis(dimethylamino)-naphthalene[a] in d_6–DMSO
Me-N $(CH_2)_p$ $(CH_2)_q$ N-Me	3	3	5.54	11.9	−1.64
	4	3	3.12	>11	−1.11
	5	3	1.5	>11	−0.89
	4	4	0.4	>12	+0.48
	5	4	1.5	>11	

[a] pK_a 16.1 in 35% aqueous DMSO: Alder, R. W.; Goode, N. C.; Miller, N.; Hibbert, F.; Hunte, K. P. P.; Robbins, H. J. *J. Chem. Soc., Chem. Commun.* **1978**, 89.

basicity enhancement is greatest when the hydrogen bond can be nearly linear, but ring strain relief probably also is important.

3.9. Bicyclic Nitrogen Heterocycles

3.9.1. Medium-ring Bicyclic Compounds

As stated in the introduction, only bicyclic structures that are entirely constructed from medium rings will be considered in this section. There are 10 bicyclic ring systems containing only eight- to 11-membered rings, and 27 systems if seven- and 12-membered rings are included (Table 3-7). The chemistry of these

TABLE 3-7

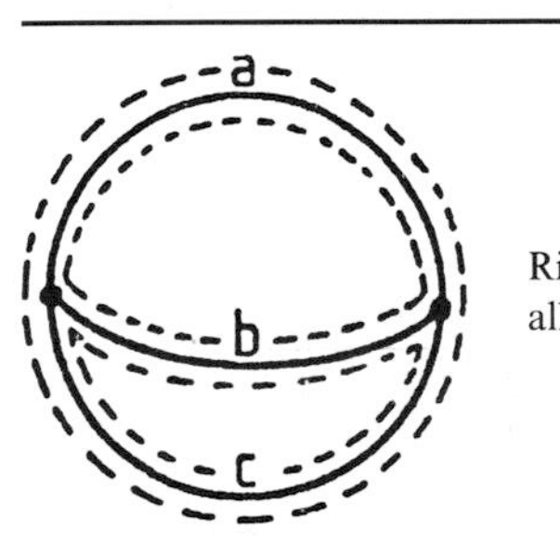

Rings *a*, *b*, and *c* all seven to 12 membered

Number of Skeletal Atoms							
10	11	12	13	14	15	16	17
3.3.2	3.3.3	4.3.3	4.4.3	4.4.4	5.4.4	5.5.4	5.5.5
	4.3.2	4.4.2	5.3.3	5.4.3	5.5.3	6.4.4	
	4.4.1	5.3.2	5.4.2	6.3.3	6.4.3		
		5.4.1	6.3.2	5.5.2	7.3.3		
			5.5.1	6.4.2			
			6.4.1	7.3.2			

ring systems was investigated very little until about 10 years ago, and much of the recent work has been done in our research group and concerns bridgehead amines. The Bristol group has prepared[12,193–195] 21 of the possible 27 systems as saturated bridgehead diamines (**123**), and nitrogen derivatives of all the ring systems, except [5.5.1], [7.3.2], [7.3.3], and [6.4.4], are known. Although the conformational properties of these ring systems are far from solved, a discussion of the problems involved, and of the understanding gained so far, seems appropriate to this chapter, since far less is known about carbocyclic or other heteroatom versions of these systems.

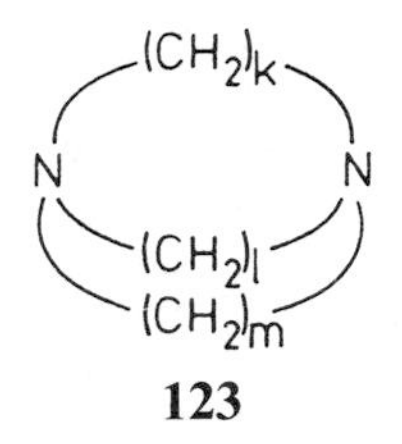

123

It is worth noting that any bicyclic ring system can have one or three even-membered rings, but systems with one or three odd-membered rings are impossible. Interest has naturally centered on the symmetrical systems bicyclo[3.3.3]undecane, bicyclo[4.4.4]tetradecane, and bicyclo[5.5.5]heptadecane, built from eight-, 10-, and 12-membered rings, and the conformations of these systems are now reasonably well understood. Before passing on to the consideration of individual ring systems, however, the general problems of conformational analysis of these bicycles need discussion.

The introduction of a bridge across a medium ring so as to create two new medium-sized rings is about the most strain-enhancing trick one could play on the structure, and these bicyclic structures are considerably more strained than their moncyclic counterparts. This is illustrated in Fig. 3-3, which compares the strain energies and average bond angles of cyclodecane (experimental data[196,197]) and *out,out*-bicyclo[4.4.4]decane (from MM2 calculations[1]). The latter compound is unknown but the corresponding diprotonated diamine doubtless has similar strain energy, and this is responsible for its very low pK_a of -3.2.[198] In this case, the 10-membered rings in the bicycle cannot adopt the favored BCB conformation of the monocycle, and, in fact, most of the conformations calculated to be of low energy[199] for 10-membered rings are out of the question for this bicyclic structure. This is not always the case, of course, and in the C_3 manxane structure adopted by essentially all derivatives (see later) of bicyclo[3.3.3]undecane, all the eight-membered rings are in BC-like conformations. Even here, however, the opposite sides of the BC conformation have to be prised apart in order to insert the third $(CH_2)_3$ bridge, so that bond angles are widened still further and torsion angles made less favorable.

One consequence of the increased strain in the bicyclic systems is that transannular interactions are more strongly driven by strain relief than in monocyclic systems. More significantly, the transannular interactions become selective as illustrated for the *out,out*-bicyclo[4.4.4]tetradecane structure in

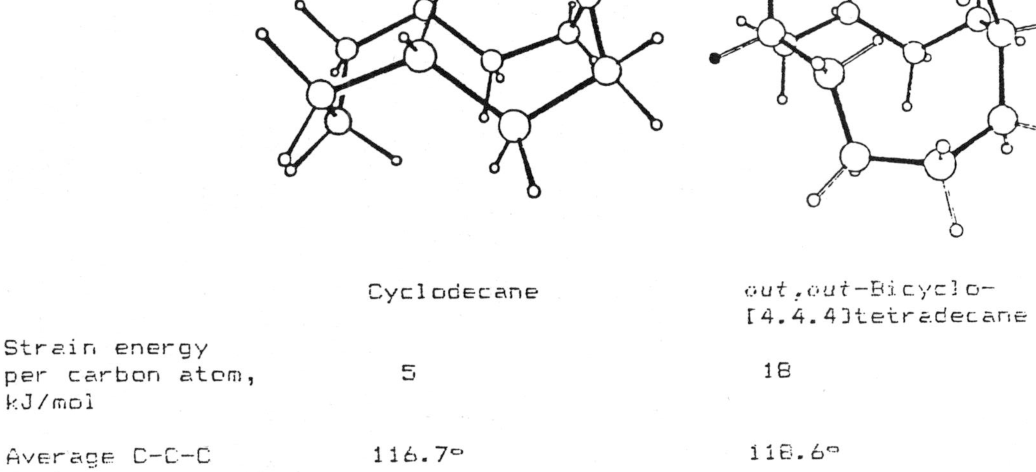

	Cyclodecane	*out,out*-Bicyclo-[4.4.4]tetradecane
Strain energy per carbon atom, kJ/mol	5	18
Average C-C-C angle	116.7°	118.6°

Figure 3-3. Strain-energy comparison of cyclodecane and bicyclo[4.4.4]tetradecane.

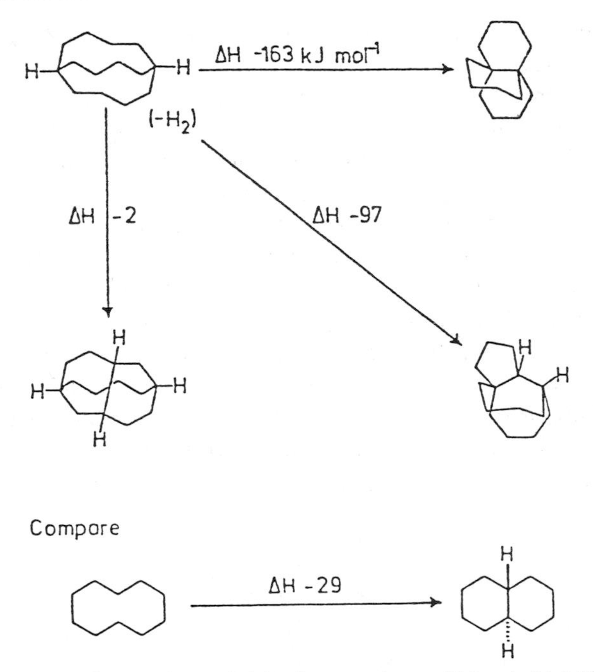

Figure 3-4. Heats of some formal dehydrogenations of bicyclo[4.4.4]tetradecane.

Fig. 3-4. While some transannular bonding possibilities hardly relieve strain at all, intrabridgehead bond formation is particularly favored, and, in fact, in this example the strain relief on bond *formation* becomes comparable to that found for bond *breaking* in small-ring bicyclics and propellanes (Table 3-8). In fact, any chemical process that results in flattening or inward pyramidalization of the bridgehead atoms will give substantial relief of strain.

In,out and even *in,in* structures become viable alternatives to *out,out* isomers in surprisingly small rings. Thus, in the same bicyclo[4.4.4]tetradecane structure, the *in,out* isomer ($\Delta H_f - 91$ kJ/mol) is calculated to be much more stable

TABLE 3-8. Heats of Some Formal Dehydrogenations[a]

Hydrocarbon	Dehydrogenation Product	ΔH, kJ/mol
2 × propane	2,3-dimethylbutane	+29
Cyclopentane	Bicyclo[2.1.0]pentane	+234
Cyclodecane	*Trans*-decalin	−29
2 × 2-methyl-butane	2,2,3,3-tetramethyl-butane	+46
Bicyclo[1.1.1]pentane	[1.1.1]propellane	+163
Bicyclo[2.2.2]octane	[2.2.2]propellane	+280
Bicyclo[3.3.3]undecane	[3.3.3]propellane	−25
Out,out-bicyclo-[4.4.4]tetradecane	[4.4.4]propellane	−163

[a] See Ref. 12.

than the *out,out* isomer (ΔH_f −41), and even the *in,in* isomer (ΔH_f −32) is only a little less stable than the *out,out*, in spite of a severe transannular interaction between the two inside hydrogens. As we shall see, bridgehead nitrogens frequently have inside lone pairs. The *in/out* nomenclature[200] for these isomers has become well established, although it is not entirely unambiguous or consistent,[201] since *out,out* and *in,in* isomers are in reality different conformers of the same configuration, whereas *out,out* and *out,in* isomers are genuinely different configurations. It is worth noting that the Cahn-Ingold-Prelog R/S nomenclature, as recently extended,[202] can handle all possibilities. The problem is trivial, of course, if the three bridges are all different; see **124**. When two bridges are identical, the R/S system is simple to apply, see **125**, and even when all the bridges are the same, this system gives an unambiguous answer.[203] Thus, the ion (**126**) is the *s,s* isomer. Although the availability of an unambiguous system is reassuring, the *in/out* nomenclature is so graphic that it will undoubtedly be retained, and we shall use it in the remainder of this chapter.

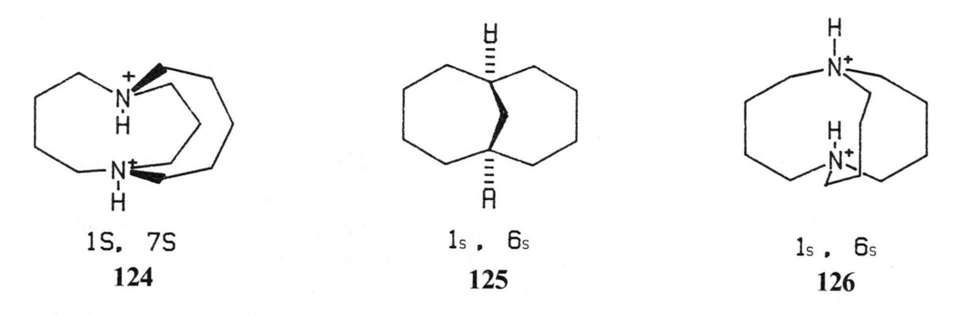

124 **125** **126**

Turning to the conformational possibilities for the rings in bicyclic medium-ring compounds, the introduction of the extra bridge must restrict the conformations open to the medium ring. However, it is far from obvious how these restrictions are to be defined, and all the compounds will have many conformational minima. The recent attack on the problem of hunting for the global minimum by Saunders[204] promises to be vitally important for molecular mechanics calculations on these compounds, but it is also necessary to develop an understanding of the conformational possibilities. For this purpose, there are two possible approaches to analyzing the situation. We can consider individual rings or we can consider individual bridges; we believe that there are major advantages to the latter analysis. In the first place, the conformation of the constituent rings in bicyclic medium-ring compounds may bear little resemblance to those in monocyclic counterparts. Second, the bridgeheads are anchor points that have restricted conformational possibilities. We suggest that the conformations open to individual bridges are determined by the four parameters defining the geometry of the bridgehead atoms and the atoms attached to them; see **127** and **128**. The parameters α, β, γ, and d are set by the *in,out* nature of the bicycle under consideration. The torsional parameter γ is perhaps most crucial because it appears only to be able to adopt a very restricted range of values, about $\pm 30°$. It should be noted that in bicyclic systems built from small and common-sized rings, γ is essentially 0°. The bridge conformation is then set by the torsion angles of the bonds not involving the bridgehead atoms.

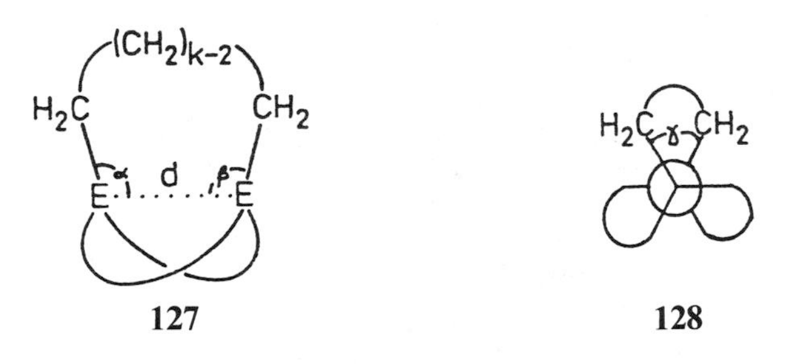

127 **128**

The conformational possibilities for $(CH_2)_n$ bridges can now be defined. Monomethylene bridges are clearly not a problem, and, in fact, neither are $(CH_2)_2$ bridges, since the single torsion angle within the bridge must have the same sign as γ, which can be chosen arbitrarily. These $(CH_2)_2$ bridges have an approximate local axis of symmetry. In principle, trimethylene bridges can have a local axis or plane of symmetry related to the twist and envelope conformations of cyclopentane. In practice, as two carbons of cyclopentane are pulled apart to become the bridge atoms of a trimethylene bridge (see Fig. 3-5), γ increases rapidly in the twist (C_2) conformation, and this is always likely to be a high-energy conformation for such bridges. Thus, $(CH_2)_3$ bridges are always found to have an approximate local plane of symmetry in medium-ring bicyclic compounds. Of course, the bridge may be flipped in either direction in an unsymmetrical compound, and this will be decided by van der Waals interactions with neighboring bridges. This analysis nicely explains the exclusive occurrence of manxane conformations for bicyclo[3.3.3]undecanes (see later).

The possible conformations for $(CH_2)_4$ bridges can be derived in a similar way from the chair and twist conformations of cyclohexane; see Fig. 3-6. The

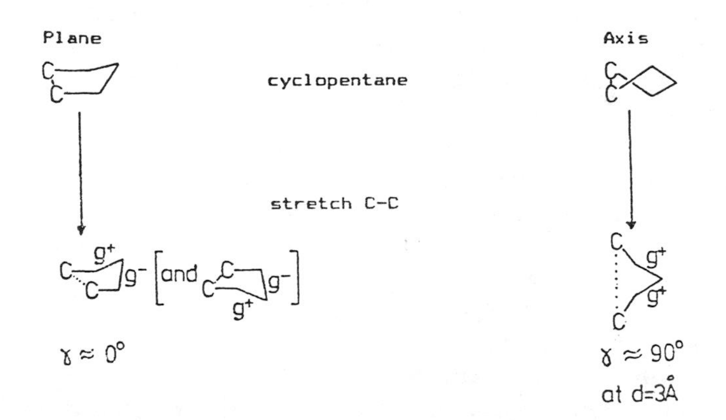

Figure 3-5. Conformations of trimethylene bridges.

Conformations corresponding to the chair and to three twist forms of cyclohexane are possible when γ is non-zero and when d is not too large.

Assume γ is positive

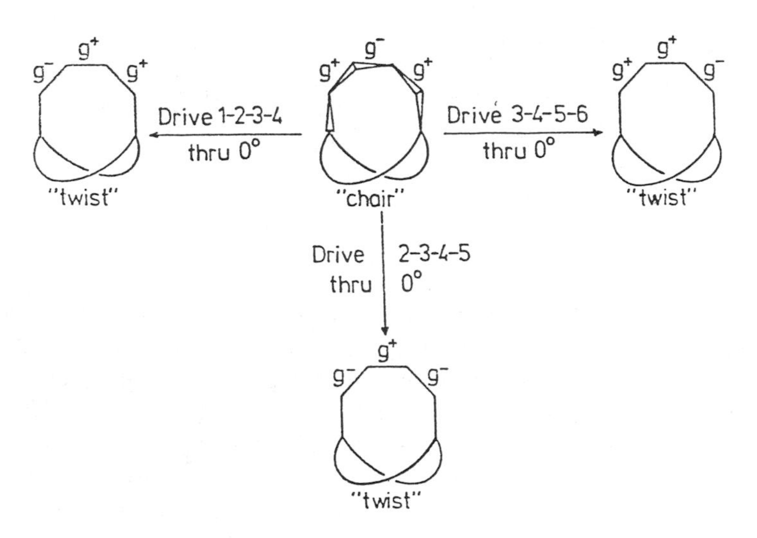

When γ is zero, the only energy minima may be the $g^+g^-g^+$ and the $g^-g^+g^-$ forms, corresponding to the two half-chair forms of cyclohexene

When d is very large (>3Å), conformations with one torsion angle over 120° are possible.

Figure 3-6. Conformations of tetramethylene bridges.

analysis predicts four possible energy minima, so long as no torsion angles exceed 120° (this becomes increasingly likely as *d* increases). When *d* is small, the strong initial preference for the chair conformation will ensure that a "stretched chair" is preferred, but this preference will diminish as *d* increases. In practice, the "stretched chair" conformation with a local axis of symmetry has been found for 22 out of 23 $(CH_2)_4$ bridges examined so far by X-ray structure determinations.[205–208]

In a similar way, the possible conformations for a $(CH_2)_5$ bridge can be derived from cycloheptane; see Fig. 3-7. Unlike the trimethylene bridge case, there is no clear choice between conformations with a local axis or plane of symmetry; in both cases, γ can be kept to the small values required. "Stretched TC" (axis of symmetry) and "stretched chair" (plane of symmetry) conformations are both perfectly viable, and indeed the initial preference for the TC conformation will disappear as the eclipsed bond is stretched. In practice, all but one of the pentamethylene bridges we have determined by X-ray analysis have an approximate plane of symmetry.[208] At present, our information on $(CH_2)_6$ bridges is too fragmentary for any conclusions to be drawn.

We will now examine individual bicyclic nitrogen heterocycles, beginning with the symmetrical [3.3.3], [4.4.4], and [5.5.5] systems.

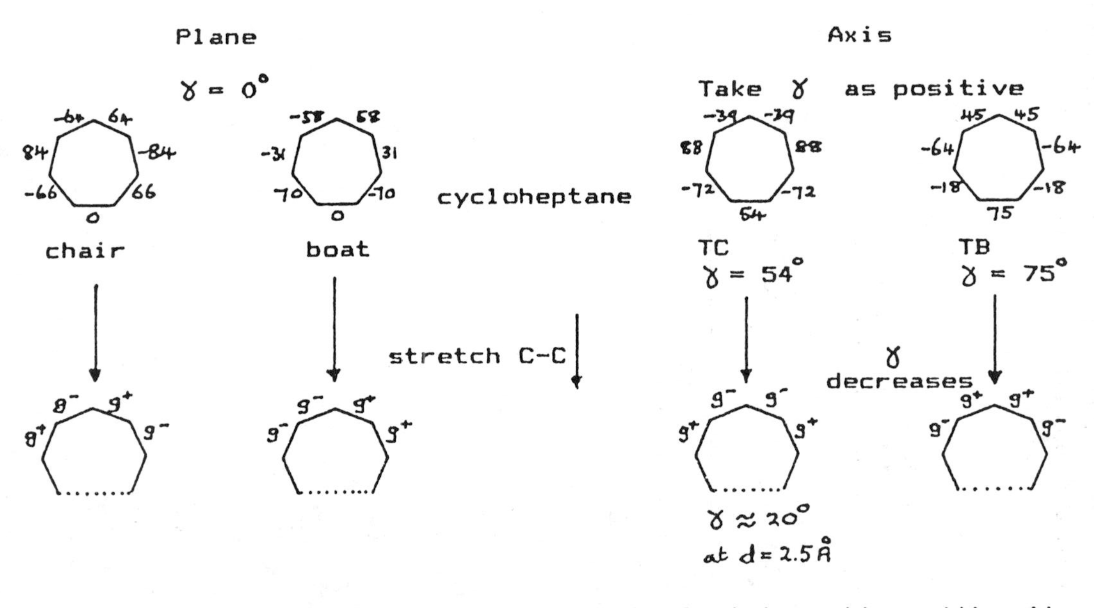

The boat and TB have greater van der Waals interaction with adjacent bridges than the chair and TC; since they are already less stable for cycloheptane, they are unlikely to be favoured.

Unsymmetrical forms with sequences of the type g+g+g-g+ are possible.

Figure 3-7. Conformations of pentamethylene bridges.

3.9.2. *Azabicyclo[3.3.3]undecanes*

The following structures have been determined by X-ray analysis: 1-azabicyclo[3.3.3]undecane hydrochloride[209] (**129**), a naphtho-fused 1,5-diazabicyclo-[3.3.3]undecane[210] (**130**), and a bridged bisimino[14]annulene[211] (**131**) and its derived radical cation.[212] In all cases, the bridgehead atoms are strongly flattened or, in the case of the radical cation, inwardly pyramidalized. All the $(CH_2)_3$

TABLE 3-9. Barriers ($\Delta G^{\dagger}$) in kJ mol^{-1}; N,N Distances (d) in Å

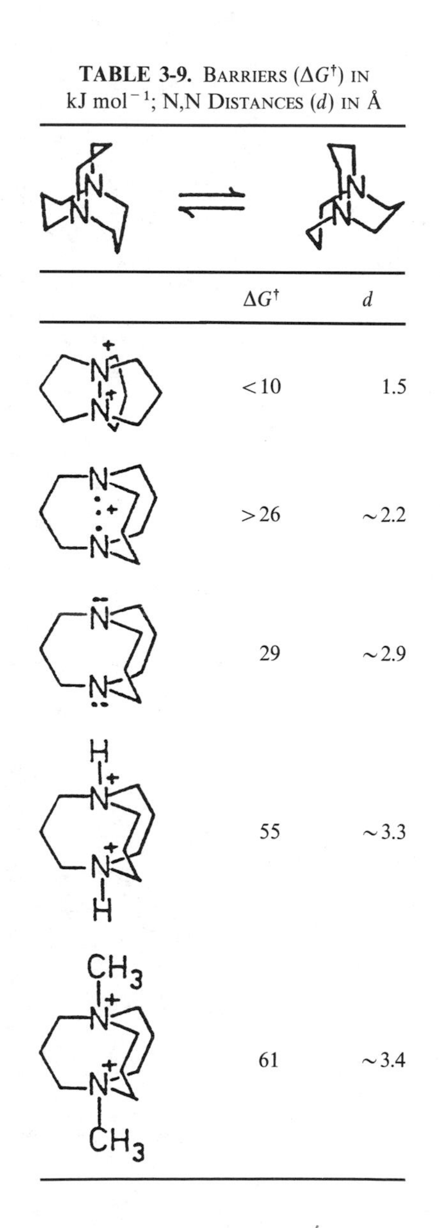

	$\Delta G^{\dagger}$	d
	<10	1.5
	>26	~2.2
	29	~2.9
	55	~3.3
	61	~3.4

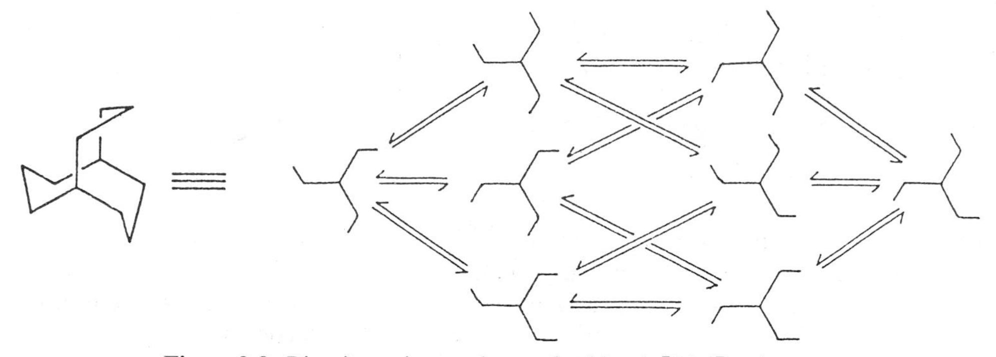

Figure 3-8. Ring-inversion pathways for bicyclo[3.3.3]undecanes.

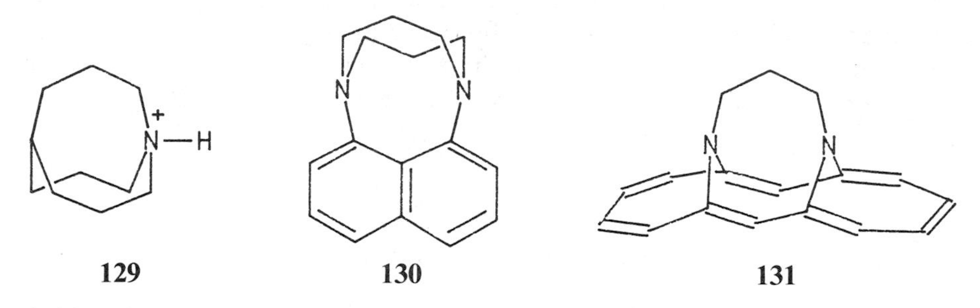

bridges have an approximate local plane of symmetry, bearing out the analysis presented above.

The manxane structure (C_3 or C_{3v} symmetry) undergoes bridge flipping at rates that are convenient for NMR study. The barriers have been determined for a number of cases; see Table 3-9.[12,211–213] A general trend toward higher barriers can be discerned as the bridgehead atoms are pushed further apart. The mechanism of bridge flipping is certainly stepwise and presumably proceeds as shown in Fig. 3-8, with the rate-limiting step being the first. Whether the transition state retains a plane of symmetry perpendicular to the intrabridgehead axis is not known, but the torsion angles in the flipping bridge must be nearly zero. The strain in such an arrangement increases as the intrabridgehead distance d increases, accounting for the trend in the barriers. ^{13}C NMR spectra show that the naphtho-fused compound (**130**) is a mixture of a C_{2v} and a C_2 conformation below $-100°$.

3.9.3. Azabicyclo[4.4.4]tetradecanes

Solid-state structures have been determined for 1,6-diazabicyclo[4.4.4]tetradecane (the [4.4.4]diamine) (**132**), its inside protonated ion,[205] its derived radical cation, and its dication (a [4.4.4]propellane).[207] The remarkable trisperoxide (**133**), first reported in 1885, has a related structure,[206] but with completely flat nitrogens (possibly due to lone-pair donation into the antibonding C—O orbitals) and larger torsion angles about the O—O bonds, as might be expected.[214] The decrease in strain as the two nitrogens are brought together is very obvious

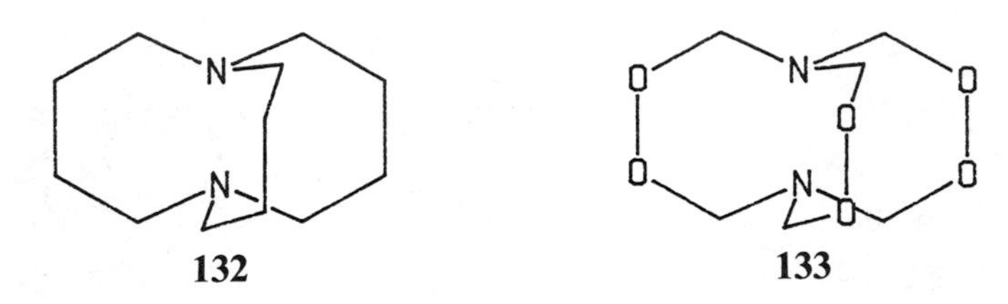

in these structures when the trends in bond angles and torsion angles are examined, but in all cases the bridge conformations are of the "stretched chair" type (see Fig. 3-6). It is remarkable that the preference for a chair structure is preserved out to an intrabridgehead distance of 3.1 Å. Barriers to bridge flipping can be measured for most of these structures; see Table 3-10. The trend in the barriers is the reverse of that in the [3.3.3] series; increasing the intrabridgehead distance d leads to a decrease in the barrier. Although this may seem rather surprising at first, it presumably arises from a decrease in the energetic advantage enjoyed by the ground state "stretched chair" as d increases. The mecha-

TABLE 3-10. Barriers ($\Delta G^{\dagger}$) in kJ mol^{-1}; N,N Distances (d) in Å

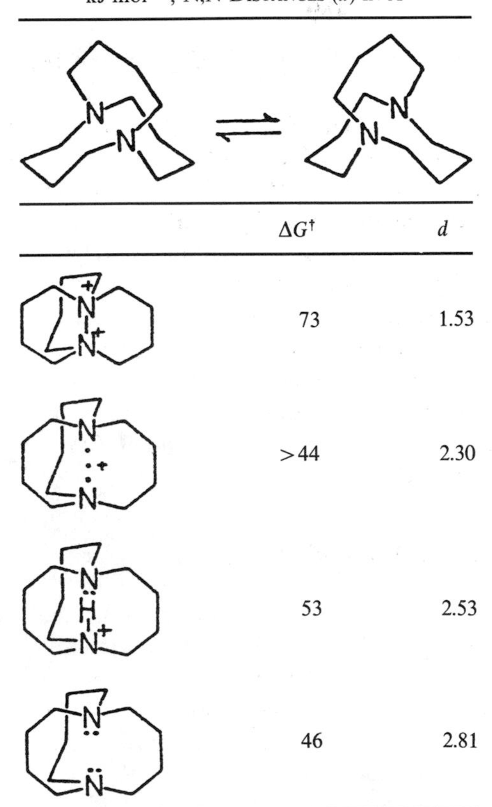

	$\Delta G^{\dagger}$	d
	73	1.53
	>44	2.30
	53	2.53
	46	2.81

nism for bridge flipping is not known in detail, but may be like that of the corresponding propellane, which is stepwise.[215]

When one of the bridgeheads is *out*, as in the outside protonated ion from the [4.4.4]diamine, the conformational situation is apparently finely balanced; we have determined structure of the bromide and the camphorsulfonate salt of this cation.[208] The former has a C_3 structure, with all the bridges in the same basic conformation as the diamine itself, but in the latter salt, one bridge is flipped to a $g^+g^-g^-$ conformation. Calculations using MM2 on monoamine **134**[216] in corresponding conformations show a difference in heats of formation of 5 kJ/mol, with the C_3 conformation preferred, but this might easily be canceled out by packing forces. In both structures, the nonprotonated nitrogen is quite strongly inwardly pyramidalized (C—N—C angles average 112°).

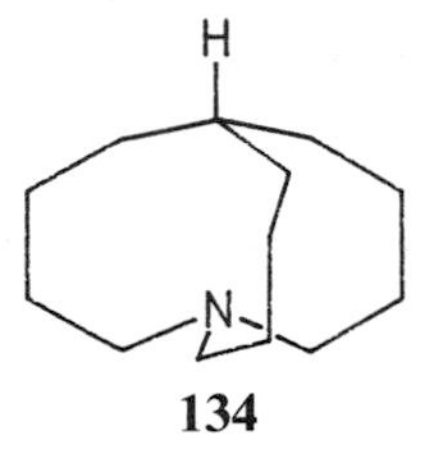

134

Nothing is known experimentally about *out,out*-[4.4.4] structures; in calculations on the hydrocarbon, a D_3 structure has usually been assumed,[217] but we found a totally unsymmetrical structure that is 4 kJ/mol more stable by MM2. Even if this is not correct, it serves as a warning that symmetry must not be assumed in these highly strained structures.

3.9.4. Azabicyclo[5.5.5]heptadecanes

The only compound with this ring system that has been studied is Lehn's [1.1.1]cryptand.[218] Structures have recently been reported for the diamine (**135**) and its mono- and di-inside protonated ion.[219] The structures have C_{3v} or C_3 symmetry, with all bridges in the same conformation, essentially obtained by stretching chair cycloheptane. Unlike the [4.4.4]diamine, there is not a strong hydrogen bond in the monoprotonated ion and the N . . . N distance increases from the cryptand to the mono- and the di-protonated ion. Outside-protonated ions can be observed in solution,[218] but little is known about their structure. However, bridge-flipping barriers have been determined.

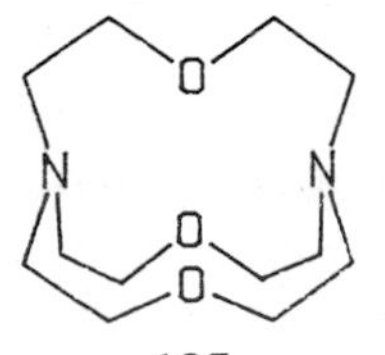

135

3.9.5. *Other Mono- and Diaza Medium-ring Bicyclics*

The only other bicyclic system built entirely from medium rings for which there are significant amounts of conformational data is the [4.4.1] system; even here only fairly heavily constrained systems have been studied to any degree. The structure of the bridgehead alkene 2-phenyl-3-aza-4,5-benzobicyclo[4.4.1] undec-1,10-en-11-one has been reported.[220] In 8,16-imino-*cis*-[2,2]metacyclophane, the dihedral angle betwen the aromatic rings is 93°.[221] Quite a number of aza-substituted versions of Vogel's 1,6-methano-bridged [10]annulenes and syn-1,6;8,13-dimethano-bridged [14]annulenes have been made and their structures determined. The standard conformation of the former is exemplified by 1,6-methano-3-aza[10]annulene (**136**).[222] The bond lengths are, in general, typical for a heteroarene, but the relatively short 1,6 distance suggests that there may be some contribution from the homoisoquinoline structure. A similar conformation is observed for 10-bromo-1,6-methano-2-aza[10]annulene.[223]

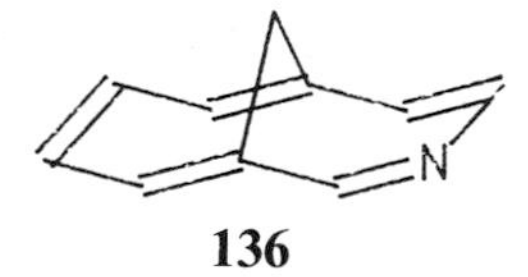

136

In the [14]annulene series, compounds are known with nitrogens in the 2-position[224] and also ones with one[225] or two imino bridges.[226] In the latter case, a series of compounds have been made with a further bridge linking the imino bridges.[217] When this bridge consists of three methylenes, the compound behaves like a 1,5-diazabicyclo[3.3.3]undecane and forms a stable radical cation, whose structure has already been commented upon. Although they are not [4.4.1] systems, tricyclo[6.4.1.1^{2,7}]tetradecanes often have conformations related to these annulenes. Structures are known for an N-methylazepin dimer with *anti* nitrogens,[227] for a 1,2,7,8-tetraaza-4,5,10,11-tetraoxa derivative[214] (this shows torsion angles about O—O bonds of about 100°, like hydrogen peroxide), a dimer of a benzodiazepin,[228] and the adduct of ethoxycarbonylazepin with 2,5-dimethoxycarbonyl-3,4-diphenylcyclopentadienone.[227]

The structure of a complex [6.4.1] fused to an indole[230] and a bridged [3,3]metapyridinophane[231] with an intriguing 7–12–7 ring system were reported recently.

3.10. Bridgehead Nitrogen: Geometry and Properties

As was discussed earlier, there is little evidence for major effects on nitrogen geometries and properties in saturated monocyclic medium-ring heterocycles. At least the effects are small compared with those encountered with bridgehead nitrogens in bicyclic compounds. The reasons for this are connected with the greater strain in the bicyclic compounds and the selective relief of strain by geometry changes at bridgehead atoms, as discussed earlier. The classic series is quinuclidine (**137**), manxine (**138**),[213] and *out*-6H-1-azabicyclo[4.4.4]tetradecane[213] (**134**), which we named hiddenamine. These show properties characteristic of an outside lone pair, a flat nitrogen, and an inside lone pair.[12] Some properties depend on the actual geometry of nitrogen, for example, the ionization energy,[232,233] and the UVspectrum,[234] while others depend on the steric accessibility of the lone pair and the strain energy changes associated with reaction from outside (basicity and nucleophilicity). Table 3-11 summarizes our present knowledge.

The remarkably low basicity of the hiddenamine is the result of the large strain increase when the nitrogen pyramidalizes outward. Molecular mechanics calculations (MM2) indicate a strain difference of 89 kJ/mol between the *in* and *out* structures. Even for the (unknown) *in*-6H isomer, an inside lone pair is preferred by 31 kJ/mol. These strain considerations obviously play their part with diamines, too, where the compounds can choose *out,out*, *in,out*, and *in,in* structures at will. The changeover will obviously occur as ring sizes increase. The borderline is still ill defined but probably occurs with bicyclododecanes and bicyclotridecanes. As discussed earlier, [3.3.3] structures are *out,out* while [4.4.4] are *in,in*. Just how common *in,out* are also remains to be determined; models indicate that these have considerable angle strain in many cases.

TABLE 3-11. Properties of Bridgehead Monoamines

	Quinuclidine (**137**)	Manxine (**138**)	Hiddenamine (**134**)
pK_a	10.95	9.9	0.6
Proton Affinity, kJ/mol	977	977	908
PA of acyclic model	979 (Et_3N)	988 ($n-Pr_3N$)	994 ($n-Bu_3N$)
Ionization Energy	8.06	7.13	7.84
UV spectrum			220 nm (ϵ 4800) in isooctane *or* $EtOH/H_2O$
Hydrogen bonding with PhOH in CCl_4	Yes	Yes	No
Solubility in water			$<10^{-6}$ M
Reaction with iodomethane	Yes	Yes	No
Reaction with $MeSO_3F$	Violent	Violent	Half-life 94 hr in CH_2Cl_2 at 25°

3.11. Intrabridgehead Interactions

These are the bicyclic counterpart of transannular interactions, and may be seen as the logical extension of the series intermolecular, intramolecular, transannular, intrabridgehead.[12] Since the strain relief from bond formation may be substantially greater in the intrabridgehead case, and the control of orbital alignment is much more precise, more weakly bonding, and even antibonding, interactions can be readily observed in the intrabridgehead situation. Thus, hydrogen bonds can be readily observed in both transannular and intrabridgehead cases, but stable N . . . N three-electron bonded species are only observed in the intrabridgehead situation, and then only with some systems, even though the strengths of the two types of bonds are very similar.[235] The explanation seems to be that the three-electron bonded species are kinetically unstable with respect to hydrogen atom transfer reactions and stable species are seen only when attainment of the necessary geometry for this reaction is conformationally difficult. Once again, more detailed understanding awaits a better knowledge of conformations in these molecules. The same applies to the photoelectron spectra of bridgehead diamines.[236] In all cases, there is evidence for intrabridgehead interaction, but detailed analysis is impossible at present.

1-Azabicyclo[4.4.4]tetradec-5-ene (**139**) shows both spectroscopic and chemical evidence of interaction between the amino nitrogen and the bridgehead double bond.[237] In the photoelectron spectrum, the lone-pair ionization energy is lowered by 0.5 eV and the alkene ionization energy is raised by a similar amount. Compound **139** reacts rapidly with acids to give the propellane (**140**), protonation being accompanied by transannular attack by the amino group.

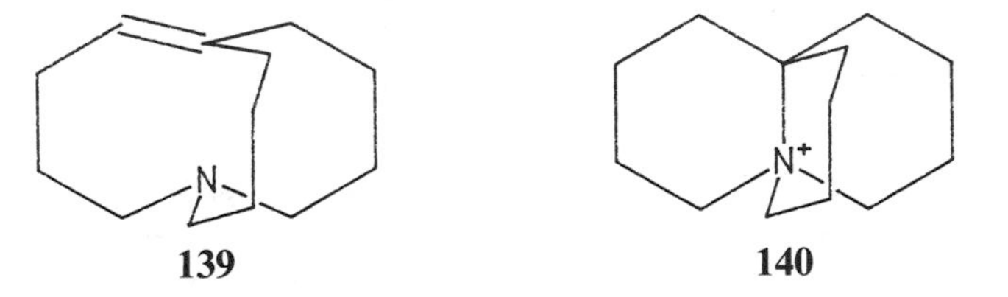

There are, of course, many examples of intrabridgehead interaction of nitrogen with other heteroatoms, such as boron and phosphorous, but these are outside the scope of this chapter.

3.12. Polycyclic Medium-Ring Compounds

The number of polycyclic compounds built entirely of medium rings (i.e., not containing other ring sizes) is still extremely small; they will be discussed here. Certain polycyclic compounds that also contain common-sized rings (five or six members) were discussed earlier under the appropriate mono- or bicyclic system. Based on the much higher ring strain in bicyclic as compared with monocyclic compounds (see above), polycyclic systems should be extremely strained. Thus, taking hexamethylenetetramine with an adamantane skeleton and replacing each CH_2 bridge with a CH_2CH_2 bridge gives a molecule containing only nine-membered rings, which is very strained according to molecular models. The nitrogen lone pairs would probably be forced inside and into close proximity.

Almost the only truly polycyclic medium-ring molecules known are 1,3,6,8-tetraazatricyclo[4.4.1.1^{3,8}]dodecane (**141**) and several derivatives, prepared from formaldehyde and ethylenediamine and its derivatives. These compounds contain only seven-membered rings. Crystal structures have been reported of the parent compound,[238] a dicyclohexano derivative,[239] and a dibenzo version.[240] There is considerable flattening at the nitrogen atoms. The parent compound gives a relatively long-lived radical on oxidation in solution,[241] as well as having a low first ionization potential.[242] It has been suggested[243] on the basis of the splittings observed in the ESR spectrum that whereas the gas-phase radical cation is D_{2d} like the parent diamine, with the odd electron delocalized over all four nitrogens in a b_2 orbital, the solution radical is less symmetrical (C_{2v}) with the spin localized at two nitrogens that are engaged in a relatively short three-electron bond (calculated to be 2.16 Å by MINDO/3); the other pair of nitrogens is calculated to be 2.76 Å apart. Electron transfer between the two pairs of nitrogens is still fast on the ESR time scale, however.

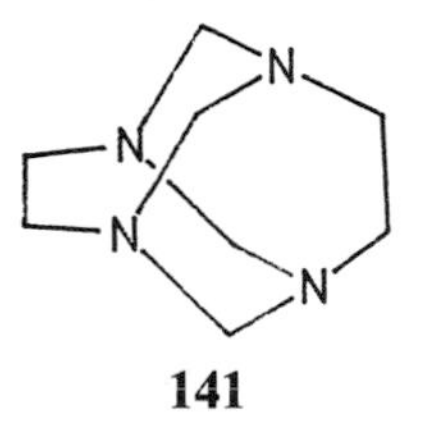

141

References

1. Riddell, F. G. "The Conformational Analysis of Heterocyclic Compounds." Academic Press: London, 1980.
2. Armarego, W. L. F. In "General Heterocyclic Chemistry Series," Vol. 5. Wiley: New York, 1977; Pt. I, p. 433, Pt. II, p. 494.
3. "Comprehensive Heterocyclic Chemistry," Katritzky, A. R.; Rees, C. W., Eds. Vol. 7, "Small and Large Rings," Lwowski, W., Ed. Pergamon: Oxford, 1984.
4. Anet, F. A. L. *Isr. J. Chem.* **1980**, *20*, 72.
5. Dale, J. *Isr. J. Chem.* **1980**, *20*, 3.
6. "Aliphatic, Alicyclic, and Saturated Heterocyclic Chemistry," covering 1970–71, was succeeded by "Saturated Heterocyclic Chemistry," Vols. 2–5, covering up to 1975; this in turn was succeeded by "Heterocyclic Chemistry," Vols. 1–5, covering the period up to 1983. Dynamic studies by NMR are covered in the volumes on NMR, the latest being Vol. 15, Chap. 9, by Berg, U.; Sandstrom, J.
7. Leonard, N. J. *Rec. Chem. Progr.* **1956**, *17*, 243. *Idem. Chimia* **1960**, *14*, 231.
8. Leonard, N. J. *Acc. Chem. Res.* **1979**, *12*, 423.
9. Burgi, H. B.; Dunitz, J. D.; Shefter, E. *J. Am. Chem. Soc.* **1973**, *95*, 5065.
10. Dunitz, J. D. "X-Ray Structure Analysis and the Structure of Organic Molecules." Cornell University Press: Ithaca, N.Y., 1979.

11. Martin, H.-D.; Mayer, B. *Angew. Chem. Int. Ed. Engl.* **1983**, *22*, 283 (transannular effects in photoelectron spectra).
12. Alder, R. W. *Acc. Chem. Res.* **1983**, *16*, 321.
13. "Cyclophanes," Vols. I and II, Keehn, P. M.; Rosenfeld, S. M., Eds. Academic Press: New York, 1983.
14. Haisa, M.; Kashino, S.; Kataoka, S.; Sasahara, N. *Bull. Chem. Soc. Jpn.* **1981**, *54*, 962.
15. Cameron, T. S.; Scheeren, H. W. *J. Chem. Soc., Chem. Commun.* **1977**, 939.
16. Csoregh, I.; Palm, T. B. *Acta Cryst.* **1978**, *B34*, 138.
17. Coetzer, J.; Stern, P. S. *Acta Cryst.* **1973**, *B29*, 685.
18. Noro, T.; Fukushima, S.; Ueno, A.; Miyase, T.; Iitaka, Y. *Chem. Pharm. Bull.* **1979**, *27*, 1495.
19. Alder, R. W.; Brammer, L.; Dunne, B.; Orpen, A. G.; White, J. M. Unpublished data.
20. Elliot, A. J.; Guzik, H.; Puar, M. S. *J. Chem. Soc., Perkin Trans. II* **1983**, 1599.
21. Iizuka, H.; Masaki, M. *Cryst. Struct. Commun.* **1973**, *2*, 375.
22. Kamenar, B.; Matkovic, D.; Nagl, A. *Croat. Chim. Acta* **1983**, *56*, 87.
23. Fukiyama, K.; Shimizu, S.; Kashino, S. *Bull. Chem. Soc. Jpn.* **1974**, *47*, 1117.
24. Ried, W.; Broft, G. W.; Bots, J. W. *Chem. Ber.* **1983**, *116*, 1547.
25. Dionne, P.; St. Jacques, M. *J. Am. Chem. Soc.* **1987**, *109*, 2616, and references cited therein.
26. Pederson, I.; Morokuma, K. *J. Chem. Phys.* **1967**, *46*, 3941.
27. Rice, K. C.; Wasylishen, R. E.; Weiss, U. *Cand. J. Chem.* **1975**, *53*, 414.
28. Borgen, G. *Acta Chem. Scand.* **1972**, *26*, 1738.
29. Dunitz, J. D.; Winkler, F. W. *Acta Cryst.* **1975**, *B31*, 268.
30. Ermer, O.; Lifson, S. *J. Am. Chem. Soc.* **1973**, *95*, 4121.
31. Baenziger, N. C.; Nelsa, A. D.; Tulinsky, A.; Bloor, J. H.; Popov, A. I. *J. Am. Chem. Soc.* **1967**, *87*, 6463.
32. Lattes, A.; Oliveros, E.; Riviere, M.; Belzecki, C.; Mostowicz, D.; Abramskj, W.; Piccinni-Leopardi, C.; Germain, G.; Van Meerssche, M. *J. Am. Chem. Soc.* **1982**, *104*, 3929.
33. Noe, E. A.; Roberts, J. D. *J. Am. Chem. Soc.* **1971**, *93*, 7261.
34. Neuman Jr, R. C.; Jonas, V. *J. Am. Chem. Soc.* **1968**, *90*, 1970.
35. Hunter, P. W. W.; Webb, G. A. *Tetrahedron* **1973**, *29*, 147.
36. St. Amour, R.; St. Jacques, M. *Tetrahedron Lett.* **1985**, *26*, 13.
37. Grouse, K. M.; Cuthbertson, E.; Mackenzie, R. K.; MacNicol, D. D.; Mills, H. H.; Williamson, D. G.; Wilson, F. B. *Spectrochim. Acta* **1980**, *A36*, 333.
38. Isaksson, R.; Liljefors, T. *J. Chem. Soc., Perkin Trans. II* **1981**, 1344.
39. Carter, R. E.; Sandstrom, J. *J. Phys. Chem.* **1972**, *76*, 642.
40. Hall, D. M. *Progr. Stereochem.* **1969**, *4*, 1, and references cited therein.
41. Glass, M. A. W.; Mislow, K.; Hopps, H. B.; Simon, E.; Wahl, G. H. Jr. *J. Am. Chem. Soc.* **1964**, *86*, 1710.
42. Coleman, M. W.; Monro, A. M.; Potter, G. W. H. *J. Hetero. Chem.* **1975**, *12*, 611.
43. Lloyd, D.; Mackie, R. K.; McNab, H.; Marshall, D. R. *J. Chem. Soc., Perkin Trans. II* **1973**, 1729.
44. Lloyd, D.; Marshall, D. R. *J. Chem. Soc.* **1958**, 118.
45. Ferguson, G.; Marsh, W. C.; Lloyd, D.; Marshall, D. R. *J. Chem. Soc., Perkin Trans. II* **1980**, 74.
46. Abraham, R. J.; Kricka, L. J.; Ledwith, A. *J. Chem. Soc., Perkin Trans. II* **1974**, 1648.
47. Ellefson, C. R.; Swenton, L.; Bible Jr, R. H.; Green, P. M. *Tetrahedron* **1976**, *32*, 1081.
48. Post, M. L.; Horn, A. S. *Acta Cryst.* **1977**, *B33*, 2590.
49. Hallberg, A.; Hintermeister, N. M.; Martin, A. R. *Acta Cryst.* **1984**, *C40*, 2110.
50. Post, M. L.; Horn, A. S.; Kennard, O. *Acta Cryst.* **1975**, *B31*, 1008.
51. Atkinson, R. S.; Gawad, N. M. *J. Chem. Soc., Chem. Commun.* **1984**, 557.
52. Paul, I. C.; Johnson, S. M.; Paquette, L. A.; Barnett, J. H.; Haluska, R. J. *J. Am. Chem. Soc.* **1968**, *90*, 5023.
53. Anet, F. A. L. *J. Am. Chem. Soc.* **1964**, *86*, 458.
54. Mannschreck, A.; Rissman, G.; Vogtle, F.; Wild, D. *Chem. Ber.* **1967**, *100*, 335.
55. Buchardt, O.; Pederson, C. L.; Svanholm, U. *Acta Chem. Scand.* **1969**, *23*, 3125.
56. Harris, D. J.; Thomas, M. T.; Sniekus, V. Klingsberg, E. *Can. J. Chem.* **1974**, *52*, 2805.
57. Steigel, A.; Sauer, J.; Kleier, D. A.; Binsch, G. *J. Am. Chem. Soc.* **1972**, *94*, 2770.
58. **X-ray structures of [3.2.2] systems:** Tsuda, Y.; Kaneda, M.; Takagi, S.; Yamaki, M.; Iitaka, Y. *Tetrahedron Lett.* **1978**, 1199. Eberbach, W.; Hadicke, E.; Trostmann, U. *Tetrahedron Lett.* **1981**, *22*, 4953. Quast, H.; Eckert, P.; Seiferling, B.; Peters, E.-M.; Peters, K.; von Schnering, H. G. *Chem. Ber.* **1985**, *118*, 3058.
    X-ray structures of [4.2.1] systems: Murray-Rust, J.; Murray-Rust, P. *Acta Cryst.* **1975**, *B31*, 589. Mercier, R.; Vebrel, J.; Schmidt, M.; Sheldrick, G. M. *Acta Cryst.* **1984**, *C40*, 1626.

Marton-Meresz, M.; Kuszmann, J.; Pelczer, I.; Parkanyi, L.; Koritsanszky, T.; Kalman, A. *Tetrahedron* **1983**, *39*, 275. Sugio, S.; Mizuno, H.; Kitamura, K.; Hamada, K.; Ikehara, M.; Tomita, K. *Acta. Cryst.* **1983**, *C39*, 745. Sugio, S.; Kitamura, K.; Mizuno, H.; Wakahara, A.; Ikehara, M.; Tomita, K. *Acta Cryst.* **1984**, *C40*, 1608. Asahi, K.; Anzai, K.; Suzuki, S. *Chem. Lett.* **1973**, 1197. Janssen, H.-W.; Mohr, S.; Mondon, A. *Chem. Ber.* **1981**, *114*, 2158. **X-ray structures of [4.3.1] systems:** Florencio, F.; Smith-Verdier, P.; Garcia-Blanco, S. *Acta Cryst.* **1982**, *B38*, 2089. Foces-Foces, C.; Cano, F. H.; Garcia-Blanco, S. *Cryst. Struct. Commun.* **1977**, *6*, 537. Kutney, J. P.; Fuji, K.; Treasurywala, A. M.; Fayos, J.; Clardy, J.; Scott, A. I.; Wei, C. C. *J. Am. Chem. Soc.* **1973**, *95*, 5407. Brackman, J.-C.; Hootele, C.; Miller, N.; Declercq, J.-P.; Germain, C.; van Meersche, M.; *Can. J. Chem.* **1979**, *57*, 1691. Inubushi, Y.; Hraryama, T. *Chem. Pharm. Bull.* **1981**, *29*, 3418. Hawley, D. M.; Ferguson, G.; Robertson, J. M. *J. Chem. Soc.*, *B* **1968**, 1255. Ayer, W. A.; Fukazawa, Y.; Singer, P. P.; Altenkirk, B. *Tetrahedron Lett.* **1973**, 5045. Sasaki, K.; Hirata, Y. *Acta Cryst.* **1973**, *B29*, 547.

59. Yavari, I. *J. Mol. Struct.* **1980**, *67*, 293.
60. Anet, F. A. L.; Degan, P. J.; Yavari, I. *J. Org. Chem.* **1978**, *43*, 3021.
61. Lambert, J. B.; Khan, S. A. *J. Org. Chem.* **1975**, *40*, 369.
62. Cady, H. H.; Larson, A. C.; Cromer, D. T. *Acta Cryst.* **1963**, *16*, 617.
63. Cobbledick, R. E.; Small, R. W. H. *Acta Cryst.* **1974**, *B30*, 1918.
64. Cobbledick, R. E.; Main, P.; Small, R. W. H. *Acta Cryst.* **1985**, *C41*, 1351.
65. Abel, J. E.; Choi, C. S.; Dickens, B.; Stewart, J. M. *Acta Cryst.* **1973**, *B29*, 651.
66. Brill, T. B.; Haller, T. M.; Karpowicz, R. J.; Reingold, A. L. *J. Phys. Chem.* **1984**, *88*, 4138.
67. Schenk, H. *Acta Cryst.* **1971**, *B27*, 185.
68. Choi, C. S.; Boutin, H. P. *Acta Cryst.* **1970**, *B26*, 1235; see also Ref. 62.
69. Santoro, A.; Choi, C. S.; Abel, J. E. *Acta Cryst.* **1975**, *B31*, 2126.
70. Ammon, H. L.; Bhattacharjee, S. K.; Gilardi, R. D. *Acta Cryst.* **1983**, *C39*, 1680.
71. Winkler, F. K.; Sieler, P. *Acta Cryst.* **1979**, *B35*, 1920.
72. Dunitz, J. D.; Winkler, F. K. *Acta Cryst.* **1975**, *B31*, 273.
73. Lefebvre-Soubeyran, P. O. *Acta Cryst.* **1976**, *B32*, 1305.
74. Overberger, C. G.; Chi, M. S.; Pucci, D. G.; Barry, J. A. *Tetrahedron Lett.* **1972**, 4564.
75. Vitt, G.; Hadicke, E.; Quinkert, G. *Chem. Ber.* **1976**, *109*, 518.
76. Anet, F. A. L.; Yavari, I. *J. Am. Chem. Soc.* **1978**, *100*, 7814.
77. Aoe, K.; Ban, Y.; Oda, K.; Ohnuma, T. *J. Am. Chem. Soc.* **1984**, *106*, 5378.
78. Browne, E. J.; Englehardt, L. M.; White, A. H. *Aust. J. Chem.* **1982**, *36*, 2555.
79. (a) Hardy, A. D.; Ahmed, F. R. *Acta Cryst.* **1974**, *B30*, 1670. (b) *Idem.*, *ibid.* **1974**, *B30*, 1674.
80. Renaud, R. N.; Layton, R. B.; Fraser, R. R. *Can. J. Chem.* **1973**, *51*, 3380.
81. Fraser, R. R.; Layton, R. B.; Raza, M. A.; Renaud, R. N. *Can. J. Chem.* **1975**, *53*, 167.
82. Gellatly, R. P.; Ollis, W. D.; Sutherland, I. O. *J. Chem. Soc., Perkin Trans. I* **1976**, 913.
83. Dunitz, J. D.; White, D. N. J. *Isr. J. Chem.* **1972**, *10*, 249.
84. White, D. N. J.; Guy, M. P. H. *J. Chem. Soc., Perkin Trans. II* **1975**, 43.
85. Elguero, J.; Katritzky, A. R.; El-Osta, B. S.; Harlow, R. L.; Simonsen, S. H. *J. Chem. Soc., Perkin Trans. I* **1976**, 312.
86. Crossely, R.; Downing, A. P.; Nogradi, M.; Braga de Olivera, A.; Ollis, W. D.; Sutherland, I. O. *J. Chem. Soc., Perkin Trans. I* **1973**, 205.
87. Lindquist, A.; Sandstrom, J. *J. Chem. Scr.* **1974**, *5*, 52.
88. Anet, F. A. L.; Yavari, Y. *Tetrahedron Lett.* **1975**, 4221.
89. Somekawa, K.; Kumamoto, S.; Ueda, I. *Tetrahedron* **1980**, *36*, 81.
90. Anet, F. A. L. *Tetrahedron Lett.* **1980**, 2133.
91. Somekawa, K.; Kumamoto, S. *J. Org. Chem.* **1982**, *47*, 1564.
92. Ueda, I.; Somekawa, K.; Kumamoto, B.; Matsuo, T. *Acta Cryst.* **1979**, *B35*, 778.
93. Ollis, W. D.; Stoddart, J. F. *J. Chem. Soc., Chem. Commun.* **1973**, 571.
94. Breuninger, M.; Schwesinger, R.; Gallenkamp, B.; Muller, K.; Fritz, H.; Hunkler, D.; Prinzbach, H. *Chem. Ber.* **1980**, *113*, 3161.
95. Fletschinger, M.; Zipperer, B.; Fritz, H.; Prinzbach, H. *Tetrahedron Lett.* **1987**, *28*, 2517.
96. Ruxer, J. M.; Solladie, G. *J. Chem. Res.* **1978**, 408.
97. Olliero, D.; Ruxer, J. M.; Solladie-Cavallo, A.; Solladie, G. *J. Chem. Soc., Chem. Commun.* **1976**, 276.
98. Singh, A.; Archer, S.; Hoogsteen, K.; Hirshfield, J. *J. Org. Chem.* **1982**, *47*, 752.
99. Kalaus, G.; Malkieh, N.; Katona, I.; Kajtar-Peredy, M.; Koritsanszky, T.; Kalman, A.; Szabo, L.; Szantary, C. *J. Org. Chem.* **1985**, *50*, 3760. Jones D. S.; Karle, I. L. *Acta Cryst.* **1974**, *B30*, 617.
100. Tokuma, Y.; Koda, S.; Miyoshi, T.; Morimoto, Y. *Bull. Chem. Soc. Jpn.* **1974**, *47*, 18.

101. Meier, H.; Echter, T.; Bleckmann, K.; Winter, W. *Chem. Ber.* **1985**, *118*, 221. Adam, W.; Klug, G.; Peters, E.-M.; Peters, K.; von Schnering, H. G. *Tetrahedron* **1985**, *41*, 2045.
102. Allinger, N. L.; Tribble, M. T.; Miller, M. A.; Wertz, D. H. *J. Am. Chem. Soc.* **1971**, *93*, 1637.
103. Richman, J. E.; Atkins, T. J. *J. Am. Chem. Soc.* **1974**, *96*, 2268. Atkins, T. J.; Richman, J. E.; Oettle, W. F. *Org. Synth.* **1978**, *58*, 86.
104. Weisman, G. R. Personal communication.
105. (a) Zompa, L. J.; Margulis, T. N. *Inorg. Chem. Acta* **1978**, *28*, L157. (b) Weighardt, K.; Schmidt, W.; Endres, H.; Wolfe, C. R. *Chem. Ber.* **1979**, *112*, 2837. (c) Weighardt, K.; Schmidt, W.; Nuber, B.; Prikner, B.; Weiss, J. *Chem. Ber.* **1980**, *113*, 36. (d) Weighardt, K.; Schmidt, W.; van Eldik, R.; Nuber, B.; Weiss, J. *Inorg. Chem.* **1980**, *19*, 2922. (e) Weighardt, K.; Pohl, K.; Gebert, W. *Angew. Chem., Int. Ed. Engl.* **1983**, *22*, 727. (f) van der Merwe, M. J.; Boeyens, J. C. A.; Hancock, R. D. *Inorg. Chem.* **1983**, *22*, 3489. (g) Weighardt, K.; Pohl, K.; Jibril, I.; Huttner, G. *Angew. Chem., Int. Ed. Engl.* **1984**, *23*, 77. (h) Chaudhuri, P.; Weighardt, K.; Jibril, I.; Huttner, G. *Z. Naturforsch. Teil B* **1984**, *39*, 1172. (i) Haight, G. P. Jr.; Hambley, T. W.; Hendry, P.; Lawrence, G. A.; Sargeson, A. M. *J. Chem. Soc., Chem. Commun.* **1985**, 488. (j) Weighardt, K.; Ventur, D.; Tsai, Y. H.; Kruger, C. *Inorg. Chem. Acta*, **1985**, *99*, L25.
106. Huisgen, R.; Brade, H.; Walz, H.; Glogger, I. *Chem. Ber.* **1957**, *90*, 1437.
107. Hallam, H. E.; Jones, C. M. *J. Mol. Struct.* **1967**, *1*, 413.
108. Dunitz, J. D.; Winkler, F. K. *Acta Cryst.* **1975**, *B31*, 251.
109. (a) Winkler, F. K.; Dunitz, J. D. *Acta Cryst.* **1975**, *B31*, 276. (b) See Ref. 29.
110. Hossain, M. B.; Baker, J. R.; van der Helm, D. *Acta Cryst.* **1981**, *B37*, 575.
111. Damrauer, R.; Soucy, D.; Winkler, P.; Eby, S. *J. Org. Chem.* **1980**, *45*, 1315.
112. Anet, F. A. L.; Jochims, J. C.; Bradley, C. H. *J. Am. Chem. Soc.* **1970**, *92*, 2558.
113. Anet, F. A. L.; Yavari, I. *J. Am. Chem. Soc.* **1977**, *99*, 7640.
114. Ollis, W. D.; Stoddart, J. F. *Angew. Chem., Int. Ed. Engl.* **1974**, *13*, 728, 730.
115. Brickwood, D. J.; Ollis, W. D.; Stoddart, J. F. *Angew. Chem., Int. Ed. Engl.* **1974**, *13*, 731.
116. Ollis, W. D.; Stoddart, J. F. *J. Chem. Soc., Perkin Trans. I* **1976**, 926.
117. Brickwood, D. J.; Ollis, W. D.; Stoddart, J. F. *J. Chem. Soc., Perkin Trans. I* **1978**, 1385.
118. Ollis, W. D.; Stoddart, J. F.; Sutherland, I. O. *Tetrahedron* **1974**, *30*, 1903.
119. Brickwood, D. J.; Hassan, A. M.; Ollis, W. D.; Stephanatou, J. S.; Stoddart, J. F. *J. Chem. Soc., Perkin Trans. I* **1978**, 1393.
120. Berson, J. A. *J. Am. Chem. Soc.* **1956**, *78*, 4170.
121. Oberti, R.; Albini, A.; Fasani, E. *J. Heterocyclic Chem.* **1983**, *20*, 1007.
122. Bremner, J. B.; Engelhardt, L. M.; White, A. H.; Winzenberg, K. N. *J. Am. Chem. Soc.* **1985**, *107*, 3910.
123. Rothe, M.; Steffen, K. D.; Rothe, I. *Angew. Chem., Int. Ed. Engl.* **1965**, *4*, 356.
124. Dale, J.; Titlestad, K. *J. Chem. Soc., Chem. Comm.* **1969**, 656.
125. Kessler, H.; Kondor, P.; Krack, G.; Kramer, P. *J. Am. Chem. Soc.* **1978**, *100*, 2548.
126. Kessler, H.; Schuck, R.; Siegmeier, R. *J. Am. Chem. Soc.* **1982**, *104*, 4486. See also Kessler, H.; Friedrich, A.; Krack, G.; Hull, W. E. *Pept. Synth., Struct., Funct., Proc. Am. Pept. Symp., 7th* **1981**, 335; *Chem. Abs. 97*, 56239.
127. Bats, J. W.; Fuess, H. *J. Am. Chem. Soc.* **1980**, *102*, 2065.
128. Anastassiou, A. G.; Kasmai, H. S. *Adv. Heterocyclic Chem.* **1978**, *23*, 55.
129. Anastassiou, A. G. In "Comprehensive Heterocyclic Chemistry," Katritzky, A. R.; Rees, C.W., Eds. Vol. 7, Chap. 5.20.
130. Perlmutter, H. D.; Trattner, R. B. *Adv. Heterocyclic Chem.* **1982**, *31*, 115.
131. Chiang, C. S.; Paul, I.; Anastassiou, A. G.; Eachus, S. W. *J. Am. Chem. Soc.* **1974**, *96*, 1636.
132. Hojo, K.; Masamune, S. *J. Am. Chem. Soc.* **1970**, *92*, 6690. See Anastassiou, A. G.; Cellura, R. P.; Gebrian, H. J. *J. Chem. Soc., Chem. Commun.* **1970**, 375.
133. Anastassiou, A. G.; Reichmanis, E. *Angew. Chem., Int. Ed. Engl.* **1974**, *13*, 404; *idem, J. Chem. Soc., Chem. Commun.* **1975**, 149.
134. Anastassiou, A. G.; Sabahi, M.; Badri, R. *Tetrahedron Lett.* **1978**, 4755.
135. Hemmi, K.; Nakai, H.; Naruto, S.; Yonemitsu, O. *J. Chem. Soc., Perkin Trans. II* **1972**, 2252.
136. Gassman, P.G.; Hoye, R. C. *J. Am. Chem. Soc.* **1981**, *103*, 2498.
137. Langlois, Y.; Gueritte, F.; Andriamialisoa, R. Z.; Langlois, N.; Potier, P.; Chiaroni, A.; Riche, C. *Tetrahedron* **1976**, *32*, 945. Inubushi, Y.; Harayama, T. *Chem. Pharm. Bull.* **1981**, *29*, 3418. Sundberg, R. J.; Luis, J. G.; Parton, R. L.; Schreiber, S.; Srinivasan, P. C.; Lamb, P.; Forcier, P.; Bryan, R. F. *J. Org. Chem.* **1978**, *43*, 4859. Goh, S. H.; Wei, C.; Ali, A. R. M. *Tetrahedron Lett.* **1984**, *25*, 3483. Sakai, S.; Aimi, N.; Yamaguchi, K.; Hitotsuyanagi, Y.; Watanabe, C.; Yokose, K.; Koyama, Y.; Shudo, K.; *Itai*, A. *Chem. Pharm. Bull.* **1984**, *32*, 354, see also p. 377. Mangeney, P.; Langlois, N.; Leroy, C.; Riche, C.; Langlois, Y. *J. Org. Chem.* **1982**, *47*, 4261.

138. Dunitz, J. D. *Perspect. Struct. Chem.* **1968**, *2*, 1.
139. Alder, R. W.; Orpen, A. G.; White, J. M. Unpublished.
140. Moss, R. E. Ph.D. thesis, Bristol, 1984.
141. Gabe, E. J.; Le Page, Y.; Prasad, L.; Weisman, G. R. *Acta Cryst.* **1982**, *B38*, 2752.
142. Alcock, N. W.; Moore, P.; Mok, K. F. *J. Chem. Soc., Perkin Trans. II* **1980**, 1186.
143. Winkler, T.; Leutert, T. *Helv. Chim. Acta* **1982**, *65*, 1760.
144. Schwartz, E.; Gottleib, H. E.; Frolow, F.; Shanzer, A. *J. Org. Chem.* **1985**, *50*, 5469.
145. Feigenbaum A.; Lehn, J.-M. *Bull. Soc. Chim. Fr.* **1973**, 198.
146. Epsztein R.; Le Goff, N. *Tetrahedron Lett.* **1985**, *26*, 3203.
147. Fujita S.; Nozaki, H. *Bull. Chem. Soc. Jpn.* **1971**, *44*, 2827.
148. Fujita, S.; Imamura, K.; Nozaki, H. *Bull. Chem. Soc. Jpn.* **1973**, *46*, 1579.
149. Fujita, S.; Imamura, K.; Nozaki, H. *Bull. Chem. Soc. Jpn.* **1972**, *45*, 1881.
150. Gault, I.; Price, B. J.; Sutherland, I. O. *Chem. Commun.* **1967**, 540.
151. Taylor, D. *Aust. J. Chem.* **1978**, *31*, 1953.
152. Hanson, A. W. *Cryst. Struct. Commun.* **1981**, *10*, 313.
153. Vogtle F.; Brombach, D. *Chem. Ber.* **1975**, *108*, 1682.
154. van Koten, G.; Marsman, J. W.; Breedijk, G. A. K.; Spek, A. L. *J. Chem. Soc., Perkin Trans. II* **1977**, 1942.
155. Nelsen S. F.; Gannett, P. M. *J. Am. Chem. Soc.* **1982**, *104*, 4698.
156. Massiot, G.; Lavaud, C.; Vercanteren, J.; Lemeu Olivier, L.; Levy, J.; Guilheim, J.; Pascard, C. *Helv. Chim. Acta* **1983**, *66*, 2414. Atkinson, R. S.; Fawcett, J.; Gawad, N. A.; Russell, D. R.; Sherry, J. S. *J. Chem. Soc., Chem. Commun.* **1984**, 1072. Chevolot, L.; Chevolot, A.-M.; Gajhede, M.; Larsen, C.; Anthoni, U.; Christophersen, C. *J. Am. Chem. Soc.* **1985**, *107*, 4542.
157. Boekelheide, V.; Galuszko, K.; Szeto, K. S. *J. Am. Chem. Soc.* **1974**, *96*, 1578.; Weaver L. H.; Matthews, B. W. *Ibid.* **1974**, *96*, 1581.
158. Attwood, J. L.; Hunter, W. E.; Wong, C.; Paudler, W. W. *J. Heterocyclic Chem.* **1975**, *12*, 433.
159. Haenel, M. W.; Lintner, B.; Benn, R.; Rufinska, A.; Schroth, G.; Kruger, C.; Hirsch, S.; Irngartinger, H.; Schweitzer, D. *Chem. Ber.* **1985**, *118*, 4884.
160. Dunitz, J. D.; Weber, H. P. *Helv. Chim. Acta* **1964**, *47*, 1138.
161. Tsuboyama, K.; Tsuboyama, S.; Uzawa, J.; Kobayashi, K.; Sakurai, T. *Tetrahedron Lett.* **1977**, 4603.; Sakurai, T.; Kobayashi, K.; Tsuboyama, K.; Tsuboyama, S. *Acta Cryst.* **1978**, *B34*, 1144.
162. Buden, S.; Dale, J.; Groth, P.; Krane, J. *J. Chem. Soc., Chem. Commun.* **1982**, 1172.
163. Metcalfe, J. C.; Stoddart, J. F.; Jones, G.; Hull, W. E.; Atkinson, A.; Kerr, I. S.; Williams, D. J. *J. Chem. Soc., Chem. Commun.* **1980**, 540.
164. Krane, J.; Aune, O. *Acta Chem. Scand.* **1980**, *B34*, 397.
165. Johnson, M. R.; Sutherland, I. O.; Newton, R. F. *J. Chem. Soc., Perkin Trans. I* **1979**, 357.
166. Alfeim, T.; Dale, J.; Groth, P.; Krautwurst, K. O. *J. Chem. Soc., Chem. Commun.* **1984**, 1502.
167. Gabe, E. J.; Le Page, Y.; Prasad, L.; Weisman, G. R. *Acta Cryst.* **1983**, *C39*, 275.
168. Meyer, W. L.; Kuyper, L. F.; Phelps, D. W.; Cordes, A. W. *J. Chem. Soc., Chem. Commun.* **1974**, 339.
169. Declercq, J. P.; Germain, G.; van Meersche, M.; Debaerdemaeker, T.; Dale, J.; Titlestad, E. K. *Bull. Soc. Chim. Belg.* **1975**, *84*, 275.
170. Ueda, I.; Ueda, T.; Sada, I.; Kato, T.; Mikuriya, M.; Kida, S.; Izumiya, N. *Acta Cryst.* **1984**, *C40*, 111.
171. Ollis, W. D.; Price, J. A.; Stephanatou, J. S.; Stoddart, J. F. *Angew. Chem. Int. Ed. Engl.* **1975**, *14*, 169.
172. Williams, D. J. *J. Chem. Soc., Chem. Commun.* **1977**, 170.
173. Ollis, W. D.; Stephanatou, J. S.; Stoddart, J. F.; Unal, T. G. G.; Williams, D. J. *Tetrahedron Lett.* **1981**, 2225.
174. Ollis, W. D., Stephanatou, J. S., Stoddart, J. F. *J. Chem. Soc., Perkin Trans. I* **1982**, 1715.
175. Edge, S. J.; Ollis, W. D.; Stephanatou, J. S.; Williams, D. J.; Woode, K. A. *Tetrahedron Lett.* **1981**, 2229.
176. Edge, S. J.; Ollis, W. D.; Stephanatou, J. S.; Stoddart, J. F. *J. Chem. Soc., Perkin Trans. I* **1982**, 1701.
177. Hoofar, A.; Ollis, W. D.; Price, J. A.; Stoddart, J. F. *J. Chem. Soc., Perkin Trans. I* **1982**, 1649.
178. Newkome, G. R.; Sauer, J. D.; Mattschei, P. K.; Nayak, A. *Heterocycles* **1978**, *9*, 1555
179. Newkome, G. R.; Joo, Y. J.; Theriot, K. J.; Fronzek, F. R. *J. Am. Chem. Soc.* **1986**, *108*, 6074.
180. Hilpert, H.; Hoesch, L.; Drieding, A. S. *Helv. Chim. Acta* **1985**, *68*, 325.
181. Vogtle, F.; Neumann, P. *Chem. Commun.* **1970**, 1464.
182. Ruzicka, L.; Kobelt, M.; Hafliger, O.; Prelog, V. *Helv. Chim. Acta* **1949**, *32*, 544.

183. Wunderlich, J. A. *Acta Cryst.* **1967**, *B23*, 846. Tulinsky, A.; van der Hende, J. H. *J. Am. Chem. Soc.* **1967**, *89*, 2905. Hall, S. R.; Ahmed F. R. *Acta Cryst.* **1968**, *B24*, 337. *Idem, ibid.* **1968**, *B24*, 346. Birnbaum, K. B. *Ibid.* **1972**, *B28*, 2825. Burgi, H. B.; Dunitz, J. D.; Shefter, E. *Cryst. Struct. Commun.* **1973**, *2*, 667. Bye, E. *Acta Chem. Scand.* **1974**, *B28*, 5. Birnbaum, G. I. *J. Am. Chem. Soc.* **1974**, *96*, 6165. Kaftory, M.; Dunitz, J. D. *Acta Cryst.* **1975**, *B31*, 2912. *Idem, ibid.* **1975**, *B31*, 2914. *Idem, ibid.* **1976**, *B32*, 1. Bye, E.; Dunitz, J. D. *Ibid.* **1978**, *B34*, 3245. Mostad, A. *Acta Chem. Scand.* **1984**, *B38*, 381.
184. Burgi, H. B.; Dunitz, J. D.; Lehn, J.-M.; Wipff, G. *Tetrahedron* **1974**, *30*, 1563.
185. Burgi, H. B. *Angew. Chem. Int. Ed. Engl.* **1975**, *14*, 460.
186. Spanka, G.; Rademacher, P. *J. Org. Chem.* **1986**, *51*, 592.
187. Johnson, R. A. *J. Org. Chem.* **1972**, *37*, 312.
188. Kirby, A. J.; Logan, C. J. *J. Chem. Soc., Perkin Trans. II* **1978**, 642.
189. Erhardt, J. M.; Wuest, J. D. *J. Am. Chem. Soc.* **1980**, *102*, 6363.
190. In Ref. 120, R=Me, the N . . . C distance is 2.56A; Weisman, G. R. Personal communication.
191. Atkins, T. J. *J. Am. Chem. Soc.* **1980**, *102*, 6364.
192. Alder, R. W.; Eastment, P.; Moss, R. E. Unpublished.
193. Alder, R. W.; Sessions, R. B.; Bennet, A. J.; Moss, R. E. *J. Chem. Soc., Perkin I* **1982**, 603.
194. Alder, R. W.; Sessions, R. B. *Tetrahedron Lett.* **1982**, 1121.
195. Alder, R. W.; Eastment, P.; Moss, R. E.; Sessions, R. B.; Stringfellow, M. A. *Tetrahedron Lett.* **1982**, 4181.
196. Cox, J. D.; Pilcher, G. "Thermochemistry of Organic and Organometallic Compounds." Academic Press: London, 1970.
197. Hilderbrandt, R. L.; Wieser, J. D.; Montgomery, L. K. *J. Am. Chem. Soc.* **1973**, *95*, 8598.
198. Alder, R. W.; Casson, A.; Sessions, R. B. *J. Am. Chem. Soc.* **1979**, *101*, 3652.
199. Hendrickson, J. B. *J. Am. Chem. Soc.* **1967**, *89*, 7047; see also Ref. 197.
200. Simmons, H. E.; Park, C. H. *J. Am. Chem. Soc.* **1968**, *90*, 2429.
201. Haines, A. H.; Karntiang, P. *J. Chem. Soc., Perkin Trans. I* **1979**, 2577.
202. Prelog, V.; Helmchen, G. *Angew. Chem. Int. Ed. Engl.* **1982**, *21*, 567.
203. Prelog, V. Personal communication to R. W. Alder.
204. Saunders, M. *J. Am. Chem. Soc.* **1987**, *109*, 3150.
205. Alder, R. W.; Orpen, A. G.; Sessions, R. B. *J. Chem. Soc., Chem. Commun.* **1983**, 999.
206. Schafer, W. P.; Fourkas, J. T.; Tiemann, B. G. *J. Am. Chem. Soc.* **1985**, *107*, 2461.
207. Alder, R. W.; Orpen, A. G.; White, J. M. *J. Chem. Soc., Chem. Commun.* **1985**, 949.
208. Alder, R. W.; Orpen, A. G.; White, J. M. Unpublished work.
209. Wang, A.H.-J.; Missavage, R. J.; Bryn, S. R.; Paul, I. C. *J. Am. Chem. Soc.* **1972**, *94*, 7100.
210. Alder, R. W.; Goode, N. C.; King, T. J.; Mellor, J. M.; Miller, B. W. *J. Chem. Soc., Chem. Commun.* **1976**, 173.
211. Destro, R.; Pilati, T.; Simonetta, M.; Vogel, E. *J. Am. Chem. Soc.* **1985**, *107*, 3192.
212. Gerson, F.; Knobel, J.; Buser, U.; Vogel, E.; Zehnder, M. *J. Am. Chem. Soc.* **1986**, *108*, 3781.
213. Coll, J.C.; Crist, D. R.; Barrio, M. d. C. G.; Leonard, N. J. *J. Am. Chem. Soc.* **1972**, *94*, 7092.
214. See Whittleton, S. N.; Seiler, P.; Dunitz, J. D. *Helv. Chim. Acta* **1981**, *64*, 2614.
215. Gilboa, H.; Altman, J.; Loewenstein, A. *J. Am. Chem. Soc.* **1969**, *91*, 6062. Conversion of the last chair to a twist is probably concerned with rotation about the intrabridgehead bond, as with *cis*-decalin. See Baas, J. M. A.; van der Graaf, B.; Tavernier, D.; Vanhee, P. *J. Am. Chem. Soc.* **1981**, *103*, 5014.
216. Alder, R. W.; Arrowsmith, R. J. *J. Chem. Res. (S)* **1980**, 163; *J. Chem. Res. (M)* **1980**, 2301.
217. Maier, W. F.; Schleyer, P. v. R. *J. Am. Chem. Soc.* **1981**, *103*, 1891. McEwan, A. B.; Schleyer, P. v. R. *Ibid.* **1986**, *108*, 3951.
218. Cheney, J.; Lehn, J.-M. *J. Chem. Soc., Chem. Commun.* **1972**, 487. Cheney, J.; Kintzinger, J. P.; Lehn, J.-M. *Nouv. J. Chem.* **1978**, *2*, 411. Smith, P. B.; Dye, J. L.; Cheney, J.; Lehn, J.-M. *J. Am. Chem. Soc.* **1981**, *103*, 6037.
219. Brugge, H.-J.; Carboo, D.; von Deuten, K.; Knochel, A.; Kopf, J.; Dressig, W. *J. Am. Chem. Soc.* **1986**, *108*, 107.
220. Zoch, H.-G.; Sziemes, G.; Romer, R.; Germain, G.; Declercq, J.-P. *Chem. Ber.* **1983**, *116*, 2285.
221. Hanson, A. W.; Huml, K. *Acta Cryst.* **1971**, *B27*, 459.
222. Tuckmantel, W.; Andree, G.; Seidel, A.; Schmickler, H.; Lex, J.; Krafa, E.; Haug, M.; Cremer, D.; Vogel, E. *Angew. Chem., Int. Ed. Engl.* **1985**, *24*, 592.
223. Destro, R.; Simonetta, M.; Vogel, E. *J. Am. Chem. Soc.* **1981**, *103*, 2863.
224. Lex, J.; Vogel, E.; Simonetta, M. *Acta Cryst.* **1983**, *B39*, 770.
225. Destro, R.; Gavezotti, A.; Simonetta, M. *Acta Cryst.* **1982**, *B38*, 1352. Destro, R.; Ortoleva, A.; Simonetta, M. *Acta Cryst.* **1983**, *C39*, 1282.

226. Destro, R.; Pilati, T.; Simonetta, M.; Vogel, E. *J. Am. Chem. Soc.* **1985**, *107*, 3185.
227. Gottlicher, S.; Habermehl, G. *Chem. Ber.* **1971**, *104*, 524.
228. Motion, K. R.; Munro, D. P.; Sharp, J. T.; Walkinshaw, M. D. *J. Chem. Soc., Perkin I* **1984**, 2027.
229. Harano, K.; Ban, T.; Yasuda, M.; Kanematsu, K. *Tetrahedron Lett.* **1979**, 1599.
230. Leitcher, R. M.; Lai, T.-F.; Mak, T. C. W. *J. Chem. Soc., Perkin I* **1985**, 1921.
231. Cabral, J. de O.; Cabral, M. F.; Drew, M. G. B.; Escho, F. S.; Haas, D.; Nelson, S. M. *J. Chem. Soc., Chem. Commun.* **1982**, 1066.
232. Aue, D. H.; Webb, H. M.; Bowers, M. T. *J. Am. Chem. Soc.* **1975**, *97*, 4136.
233. Alder, R. W.; Arrowsmith, R. J.; Casson, A.; Sessions, R. B.; Heilbronner, E.; Kovac, B.; Huber, H.; Taagepera, M. *J. Am. Chem. Soc.* **1981**, *103*, 6137.
234. Halpern, A. M. *J. Am. Chem. Soc.* **1974**, *96*, 7655.
235. Nelsen, S. F.; Alder, R. W.; Sessions, R. B.; Asmus, K.-D.; Hillier, K.-O.; Gobl, M. *J. Am. Chem. Soc.* **1980**, *102*, 1429.
236. Honegger, E.; Yang, Z.-Z.; Heilbronner, E.; Alder, R. W.; Moss, R. E.; Sessions, R. B. *J. Electron Spectros. Related Phenom.* **1985**, *36*, 297.
237. Alder, R. W.; Arrowsmith, R. J.; Boothby, C. St. J.; Heilbronner, E.; Zhong-Zhi, Y. *J. Chem. Soc., Chem. Commun.* **1982**, 940.
238. Riddell, F. G. *J. Chem. Soc., Perkin Trans. II* **1974**, 1136.
239. Murray-Rust, P.; Riddell, F. G. *Can. J. Chem.* **1975**, *53*, 1933.
240. Murray-Rust, P.; Smith, I. *Acta Cryst.* **1975**, *B31*, 583.
241. Nelsen, S. F.; Hintz, P. *J. Am. Chem. Soc.* **1972**, *94*, 7114.
242. Nelsen, S. F.; Buschek, J. M. *J. Am. Chem. Soc.* **1974**, *96*, 6424.
243. Nelsen, S. F.; Haselbach, E.; Gschwind, R.; Klemm, U.; Lanyova, S. *J. Am. Chem. Soc.* **1978**, *100*, 4367.

4

Conformational Analysis of Medium-sized Sulfur-containing Heterocycles

William N. Setzer
THE UNIVERSITY OF ALABAMA IN HUNTSVILLE,
HUNTSVILLE, AL

Richard S. Glass
THE UNIVERSITY OF ARIZONA,
TUCSON, AZ

4.1. INTRODUCTION

This review will present a discussion of the conformational analysis of medium-sized ring polythioethers[1] and will deal primarily with seven-, eight-, nine-, and 10-membered rings. A discussion of six-membered and smaller rings is not included.[2] Conformational analyses of 11-membered and larger rings are also not included. The macrocyclic polythioethers have been reviewed[3] and the interested reader is referred to this review and recent related work that includes their metal complexes.[4–11] Conformational analysis of sulfur-containing cyclophanes with large rings (see section 4.6 for those containing medium-sized rings) has also been reviewed[12] and is of continued current interest.[13] The purpose of this chapter is to describe the effect of sulfur substitution in these rings as well as to illustrate the various techniques useful in the analyses. Only compounds with divalent sulfur (thioethers, disulfides, and polysulfides) are discussed.

Replacement of methylene groups in cycloalkanes with sulfur may have important consequences in the preferred conformations of these compounds as well as the conformational equilibria involved.[14] The geometrical parameters involving sulfur are notably different from those for carbon, and these differences can affect the degree of ring pucker, transannular steric interactions, and barriers to interconversion. If the heterocycle contains more than one heteroatom or other functional groups, the geometrical disposition of the sulfur atom may have important effects on the chemistry of these compounds (e.g., redox chemistry, transition metal complexation).

In simple dialkyl sulfides, C—S bond lengths are around 1.82 Å[15–18] and C—S—C bond angles are about 100–105°.[18,19] Dimethylsulfide has a C—S bond length and C—S—C bond angle of 1.802 Å and 98.9° respectively.[20] Owing to these differences in molecular parameters, the repulsive interaction in the twist forms of sulfur-substituted six-membered rings is less than in cyclohexane. Sulfur–sulfur bond lengths of organic di- and trisulfides range about 2.03–2.08 Å and the range of C—S—S bond angles is 101–106°.[21] X-ray crystal structures of a number of acyclic disulfides[22–25] show the gauche conformation about the disulfide bond to be preferred, with an average C—S—S—C dihedral angle of about 85°. Such a conformational preference for disulfides is consistent with the "gauche effect,"[26] and has also been found using molecular orbital[27–31] and force-field[32,33] calculations. In addition, the barrier to rotation about an S—S bond is greater than that about a CH_2–CH_2 bond.[34] The van der Waals radius for sulfur is 1.80[35]–1.85[36] Å, but this is probably not spherically uniform about the sulfur atom. Based on studies of close contacts to divalent sulfur found by X-ray determinations, there are directional preferences for nonbonded contacts.[37] Attention has been called to two other special features associated with substituting sulfur atoms for CH_2 groups that have conformational consequences. There is a slight preference for the gauche conformer about a C—S bond (as in ethyl methyl sulfide) in contrast to the preference for the *anti* conformer about a C—C bond (as in *n*-butane).[38] This factor contributes to the gauche conformation about all of the C—S bonds in the 18-membered ring of 1,4,7,10,13,16-hexathiacyclooctadecane (hexathia-18-crown-6 or 18-ane-S_6) in the solid state.[7] In systems with an SCCS moiety, the gauche conformation about the C—C bond is destabilized by about 1.5 kcal/mol by the so-called "gauche-repulsive effect," which is neither steric nor electrostatic in origin.[39]

4.2. Methodology

The principal physical methods used in conformational analysis of medium-sized sulfur-containing heterocycles are single-crystal X-ray structural analysis in the solid state, ^1H and ^{13}C NMR spectroscopic analysis in liquid solution, and photoelectron spectroscopy in combination with computational analysis in the gas phase. If the heterocycle contains an S—S bond, then several methods are useful for providing information on the dihedral angle about this linkage. The absorption maximum of the first band in the ultraviolet (UV) region due to an electronic transition of disulfides correlates with the CSSC dihedral angle.[40] For disulfides, there is also a correlation between the S—S stretching frequency, which gives rise to a strong band in Raman spectra, and the CSSC dihedral angle.[41] For cyclic sulfur allotropes (S_n),[42] their oxides,[43,44] and cyclic selenium sulfides,[45] analysis of their Raman spectra has resulted in elucidating their molecular structures. Photoelectron spectroscopy has also been used to adduce evidence on the torsion angle about S—S bonds as presented in more detail in the following. Variable-temperature NMR spectroscopic studies are the most valuable for experimentally determining energy differences between conformations and energy barriers for their interconversion. Variable-temperature photoelectron spectroscopy[46] also can be used to determine energy differences between conformations but has been little utilized as yet. Most of the efforts on conformational analysis of medium-sized sulfur-containing heterocycles have focused on discerning local energy minima, and very little work has been done on the potential energy surfaces. It is important to determine these surfaces more adequately because of the preconceptions about their nature.

A brief overview of the use of photoelectron spectroscopy in conformational analysis of sulfur-containing heterocycles is provided here. Detailed results are presented throughout the rest of this chapter. For a more extensive and general treatment of conformational analysis by this method, the reader is referred to two recent reviews.[47,48] A review on the use of photoelectron spectroscopy to discern nonbonding and transannular interactions, which is relevant to conformational analysis as explained below, has been published.[49] In addition, the photoelectron spectroscopy of organic sulfur compounds,[50] and heterocyclic compounds in general,[51] has also been reviewed.

To obtain a photoelectron spectrum,[52] the sulfur-containing heterocycle is irradiated in the gas phase with high-energy monochromatic UV light. Typically, He(I) irradiation, whose photons have an energy of 21.2 eV, is used. This irradiation results in the ejection of photoelectrons, whose kinetic energy is then measured. The difference between the energy of the He(I) photon and the kinetic energy of the photoelectron is the ionization potential for removal of an electron from the valence shell of the heterocycle. If monochromatic X-ray irradiation is used, inner-shell electrons may be ejected, but this method is rarely used for conformational analysis.[53] Since ionization by He(I) photons is a Franck-Condon process, vertical ionization potentials are generally obtained. However, for ionization of localized, nonbonding electrons on sulfur atoms, vertical and adiabatic ionization potentials are close because there is little change in geometry on ionization. The value of measuring these ionization potentials is that they correspond to the negative of the energy of the molecular orbital from which the electron is removed, provided that Koopmans' theorem[54] is valid. Fortunately, Koopmans' theorem is valid for removal of electrons from the molecular orbitals of highest energy. The energies of these molecular orbitals provide the structural information we seek. Ionization of electrons in molecular orbitals of lower energy may not obey Koopmans' theorem.[50] Structural information is obtained most easily when the sulfur-containing heterocycle contains π-systems or other atoms with unshared electron pairs. This results because the highest-energy molecular orbitals in such compounds generally reflect the interaction of the *p*-type nonbonding orbital on sulfur with the π-system or other nonbonding orbitals. Such interaction can be easily conceptualized in terms of through-bond and through-space effects.[55] The crucial point for conformational analysis is that such effects are dependent on geometry in a predictable way.

To understand the basis for such effects, consider first a dialkyl thioether.[50] The highest occupied molecular orbital in such a compound is the *p*-type nonbonding orbital localized on the sulfur atom.[50,56] If there are two divalent sulfur moieties in a molecule, then the *p*-type lone-pair orbital on each of the two sulfur atoms may interact by through-bond, through-space, or both effects.[47–50,57–62]

Through-bond effects are well illustrated in the splitting of the highest occupied molecular orbitals in the chair conformation of 1,4-dithiane.[57] The interaction between the *p*-type lone-pair orbital on each of the two sulfur atoms is substantial. This interaction results in a splitting in the otherwise degenerate orbitals such that two ionization potentials differing by almost 0.5 eV[57,58,63] are observed in the photoelectron spectrum for this compound. These two ionization potentials are those of lowest energy measured in the photoelectron spectrum and are confidently assigned to ionization of *p*-type lone-pair electrons on sulfur for two reasons. They occur in the energy region characteristic for such ionizations and are relatively sharp. Ionizations corresponding to the removal of localized, nonbonding electrons are generally sharp. Line-shape analysis of the ionization peak and comparison of its relative intensity with He(I) versus He(II) irradiation are important methods for assigning peaks. The basis for the substantial interaction between the *p*-type lone-pair orbitals in 1,4-dithiane is a through-bond effect.[57] The *p*-orbitals on sulfur and the C(2)–C(3) and C(5)–C(6) sigma orbitals in the chair conformation of 1,4-dithiane are well aligned for overlap and of appropriate symmetry to result in extensive interaction and consequent orbital splitting. Interestingly, the antisymmetric combination of *p*-orbitals that overlap with σ^* (see Fig. 4-1*a*) is of lower energy than the symmetric combination that interacts with the σ-MOs (see Fig. 4-1*b*). These orbitals are shown schematically in Fig. 4-1.

Through-space effects are well illustrated in medium-sized ring dithioethers[62] such as 1,5-dithiacyclooctane, in which transannular effects thrust two sulfur atoms close to one another despite the substantial number of bonds separating them. These cases in which through-space effects dominate the lone-pair/lone-pair interaction are discussed in detail below. It is important to note that the extent of the interaction—that is, the extent of orbital overlap—depends not only on the S · · · S nonbonded distance but also on the dihedral angle between the two CSC planes. This latter geometric parameter determines whether the highly directional *p*-type orbitals are pointed directly at each other, are some-

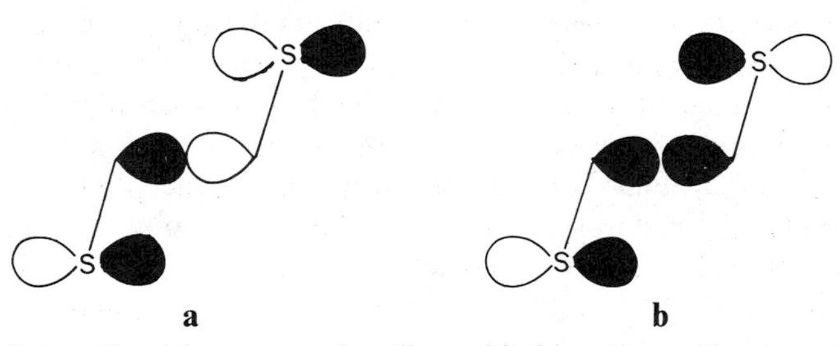

Figure 4-1. Interaction of *p*-type nonbonding orbitals on two sulfur atoms in an SCCS fragment through the central σ-orbital.

what askew, or are parallel. Clearly, for a given S · · · S separation, if the *p*-type orbitals are pointed directly at each other, the overlap, and hence orbital splitting, will be greater than if the orbitals are somewhat askew or parallel to each other. The observed splittings in the photoelectron spectrum of these compounds thereby provide insight into possible S · · · S distances and CSC dihedral angles. A conformation is deduced by fitting the spectrum with the molecular orbital energies obtained by computations using the geometries of conformational minima obtained by molecular mechanics methods or semiempirical MO calculations such as MINDO/3, MNDO, or AM-1. This procedure works remarkably well considering its inherent flaws. More satisfactory results may be obtained by using better parameters for lone-pair/lone-pair interactions (either those obtained experimentally or by *ab initio* calculations) or by complete *ab initio* calculations pending the availability of faster computers. For a more complete discussion of these problems and future possibilities, see Chapter 1 of this volume. If the sulfur atoms are directly attached to each other, as in disulfides, then the lone-pair splitting depends on the CSSC torsion angle.[48,50,58–61,64] Specifically, the splitting is proportional to the cosine of the dihedral angle. An interesting consequence of this relationship is that there are two values for the torsion angle that give comparable splittings (one between 0° and 90°, where the antisymmetric combination of lone-pair orbitals is higher in energy than the symmetric combination, and the other between 90° and 180°, where the antisymmetric combination is lower in energy). Specific examples illustrating the use of this geometry-dependent through-space effect for the conformational analysis of medium-sized ring disulfides are given later in this chapter.

4.3. Seven-membered Rings

The twist–chair (**1**) conformation has been found to be the lowest-energy conformation, by vibrational spectral analysis, for cycloheptane, oxepane, and 1,3-dioxepane.[65] This is in agreement with several strain-energy calculations on the conformation of cycloheptane.[66–68] A conformational analysis from temperature-dependent circular dichroism measurements also indicates a twist–chair to be the most stable conformation for cycloheptanone.[69] Empirical force-field calculations for cycloheptasulfur predict the twist–chair to be the most stable conformation,[70] in apparent conflict with X-ray crystal structure analysis,[71,72] as well as vibrational spectral analysis,[73] which indicate a chair (**2**) conformation. However, the twist–chair conformation for cycloheptasulfur is calculated by Kao and Allinger[70] to be only 0.3 kcal/mol more stable than the chair.

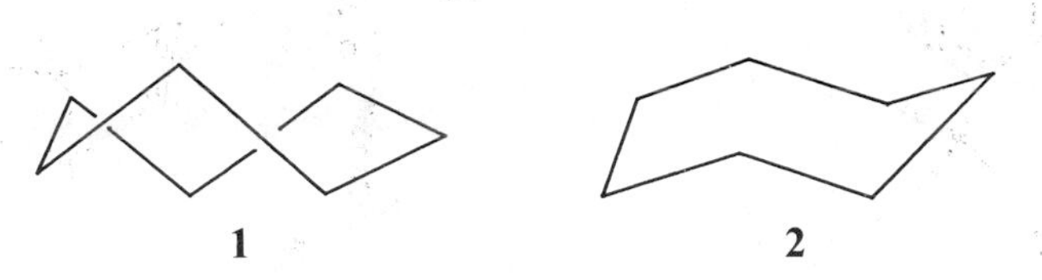

Conformational analysis of 1,4-dithiacycloheptane indicates the preferred conformation to be a twist–chair with C_2 symmetry (**3**). This twist–chair is calculated, using molecular mechanics, to be the lowest-energy conformation,[62] and is found in the solid state by X-ray crystallography[74] and in the gas phase by photoelectron spectroscopy.[62] The antimony trichloride complex of 1,4-dithiacycloheptane has been analyzed by X-ray crystallography and the heterocycle is found to adopt the twist–chair conformation.[75] Proton NMR spectra of a number of 1,4-dithiacycloheptan-6-ones (**4** and **5**) are consistent with equilibrating twist–chair conformations.[76] Similary, carbon-13 NMR analysis of a series of bicyclic 1,4-dithiacycloheptane spiranes (**6**) indicates the preferred conformation of the seven-membered ring to be the symmetrical twist–chair conformation in solution.[77]

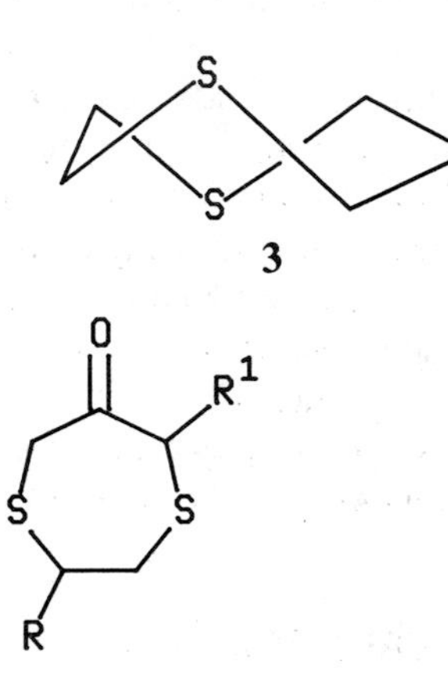

3

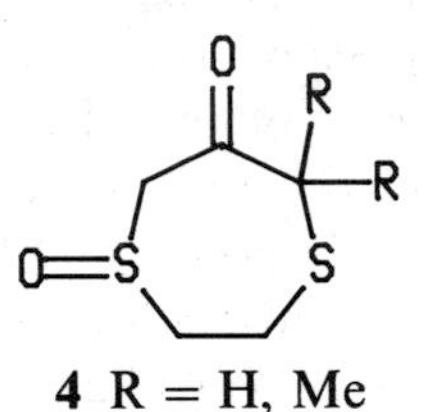

4 R = H, Me

5 R, R^1 = H, H; H, Me; Me, H

6 X = CH_2, O, $(CH_2)_2$, $(CH_2)_3$

The X-ray crystal structure of 3,3,3′,3′,6,6,6′,6′-octamethyl-4,4′-bi(1-thiacycloheptylidene)-5-5′-dione (**7**) shows the thiacycloheptane ring to adopt a distorted twist–chair conformation.[78] The reaction of $Fe(CO)_5$ with 3,3,6,6-tetramethyl-1-thiacycloheptyne gives a dimeric complex (**8**) in which the thiacycloheptyne ring adopts a chair conformation, constrained, evidently, by the unsaturation.[79]

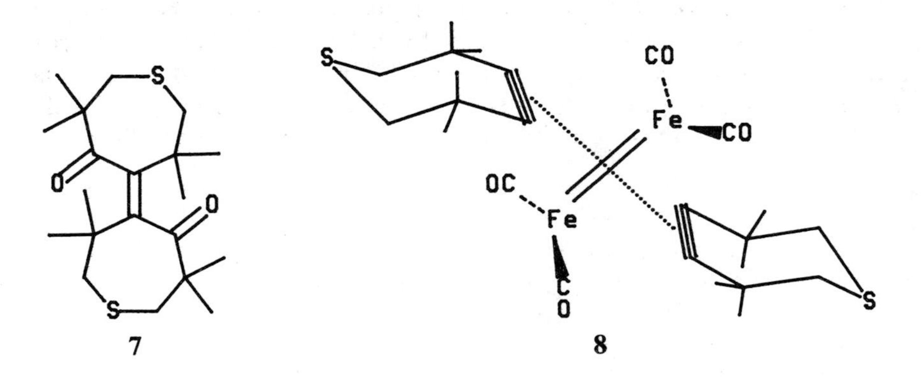

7 **8**

Disulfide and trisulfide groups in seven-membered rings serve to increase the barrier to ring inversion, rendering these compounds amenable to variable-temperature NMR analysis. The temperature-dependent ^{1}H NMR spectrum of the natural product lenthionine (1,2,3,5,6-pentathiacycloheptane, **9**) shows, at -90°C, two singlets with an intensity ratio of 1:1. This spectrum is consistent with two rapidly converting chair conformations (**10**) at this temperature.[80]

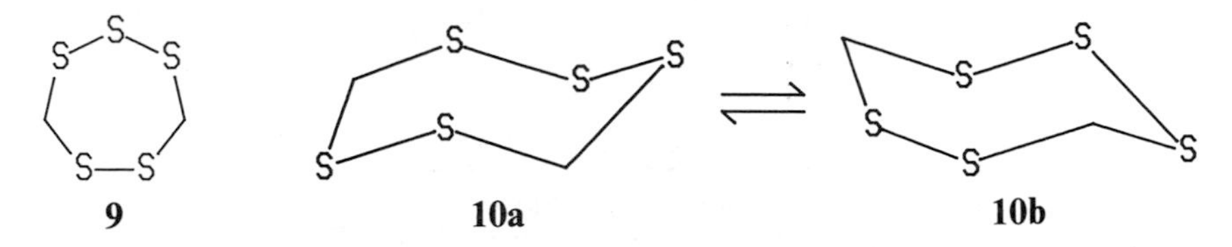

Lenthionine adopts this chair conformation in the solid state, as shown by X-ray crystallography.[81] Apparently, such a high proportion of sulfur atoms in the ring causes the compound to adopt the conformation of cycloheptasulfur rather than cycloheptane.

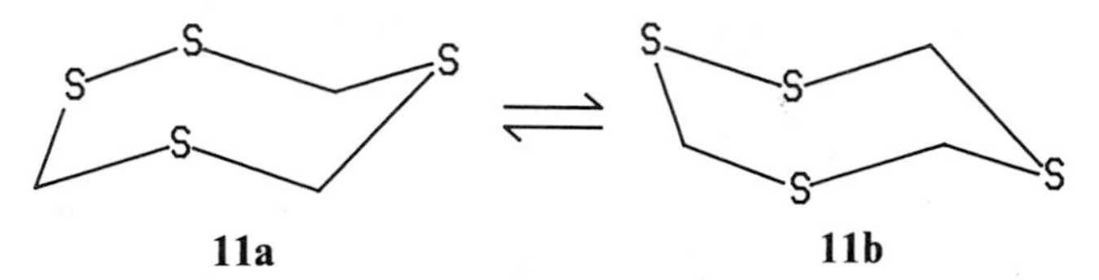

The low-temperature ^{1}H NMR spectrum (-90°C) of 1,2,4,6-tetrathiacycloheptane shows two singlets of intensity ratio 2:1. This was interpreted in terms of two interconverting chairs (**11**) analogous to that seen for lenthionine.[80] Unless two of the signals are overlapped, however, the above equilibrium should give three singlets (1:1:1). An equilibrium between interconverting twist–chair conformations (**12**) was not considered, but is consistent with the observed NMR spectrum.

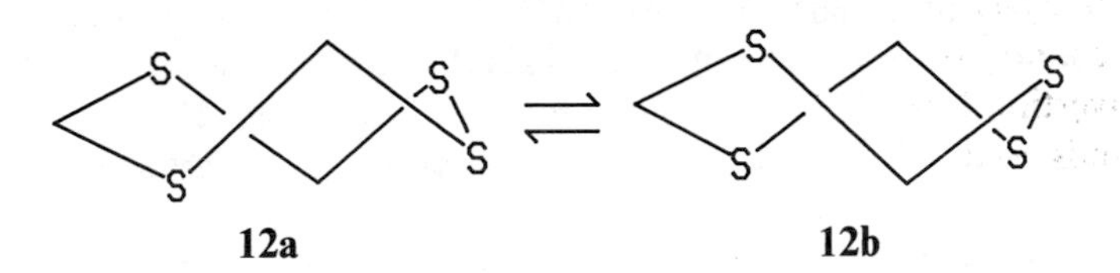

1,2,3-Trithiacycloheptane has been studied by ^{1}H NMR spectroscopy.[82] The compound shows two signals that coalesce at about -130°C. Individual conformations could not be frozen out, however, and the data cannot distinguish between interconverting chairs, twist–chairs, or other forms.

Unsaturation in seven-membered rings serves both to increase the barrier to ring inversion and to favor the chair conformation over the twist–chair. Low-temperature ^{1}H NMR analysis of a number of 1,3-dithiacyclohept-5-enes (**13–16**) is consistent with the chair conformation as the lowest energy. The free energy differences for ring interconversion (from the coalescence temperatures) are shown in Table 4-1.[83]

TABLE 4-1. Free-energy Differences for Ring Inversion of Seven-membered-ring Chair Conformations

Compound	$\Delta G^{\ddagger}$ (kcal/mol)[a]
13	8.5
14	8.2
15	10.9
16	12.1

[a] Free-energy differences for ring interconversion from the coalescence temperatures taken from Ref. 83.

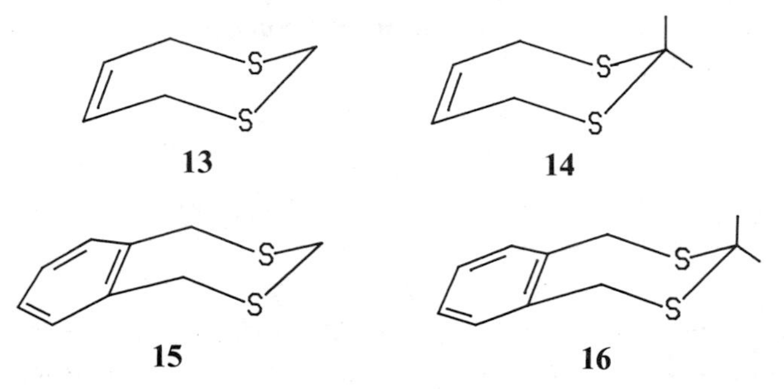

A series of 1,2,3-trithiacyclohept-5-enes has also been analyzed by variable-temperature ^{1}H NMR spectroscopy.[82] At $-90°C$, **17** was shown to adopt only the chair form ($\Delta G^{\ddagger} = 8.9$ kcal/mol). At room temperature, the benzo analogue (**18**) was interpreted as consisting of two different conformers in a 85:15 ratio: the chair form (**19**) and a flexible or interconverting twist/boat form (**20**).[82] A similar analysis of **21** shows, however, that at room temperature the frozen chair accounts for only 45% of the material while 55% adopt the flexible or interconverting forms.[84] The variable-temperature ^{1}H NMR analysis of these compounds and 1,2-dithia-4,5-benzocycloheptane (**22**) indicates interconver-

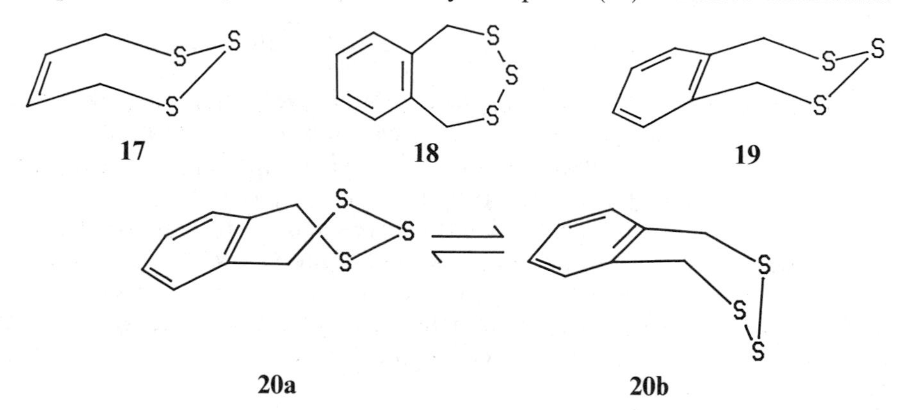

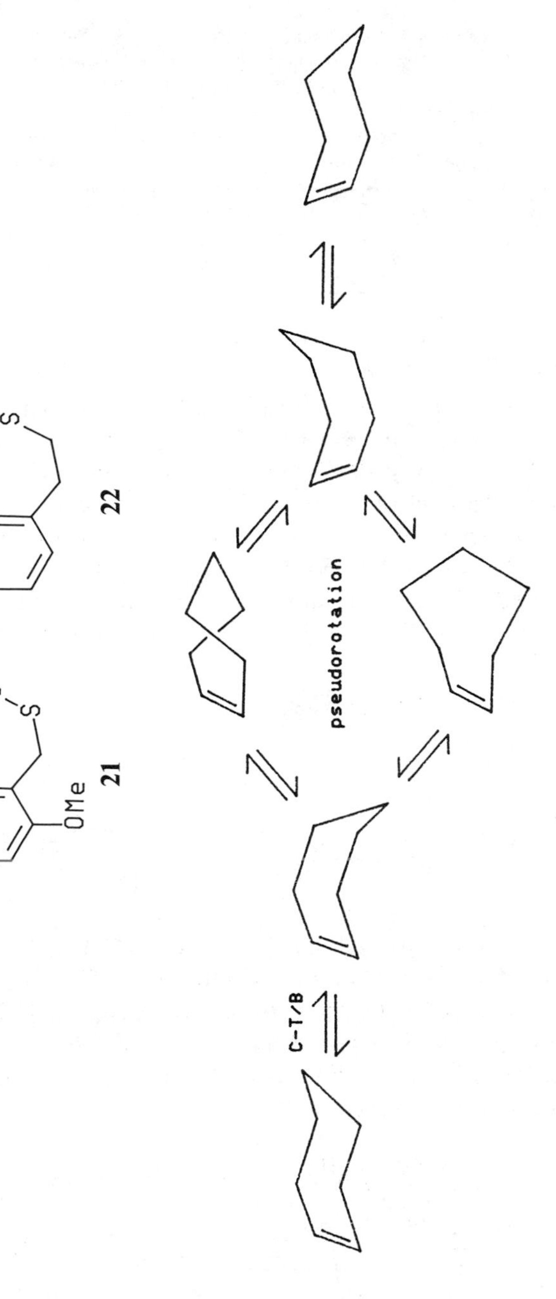
OMe
OMe
S
S
S
21
S
S
22
pseudorotation
C-T/B

TABLE 4-2. Free-Energy Differences and Free Energies of Activation for Benzo Seven-Membered-Ring Interconversions[a]

Compound	$\Delta G°$C → T/B (kcal/mol)	$\Delta G^{\ddagger}$C → T/B (kcal/mol)	$\Delta G^{\ddagger}$pseudorotation (kcal/mol)
18	−0.1	19.8	11.5
15	1.0	17.4	10
19	0.5	13.5	10.4

[a] Data from Ref. 84.

sion of chair forms to involve first a ring flip to the boat form, then pseudorotation to a twist and to the other boat form. Another ring flip converts the boat to the alternative chair form.[84] The free-energy differences and free energies of activation for these processes are shown in Table 4-2.

The X-ray structure of diltiazem, **23**, (+)-3-acetoxy-5-(2-dimethylaminoethyl)-2-(4-methoxyphenyl)-2,3-dihydro-1,5-benzothiazepin-4(5H)-one hydrochloride, which is a calcium antagonist (calcium channel blocker), shows the seven-membered ring to be in the slightly twisted-boat conformation **24**.[85]

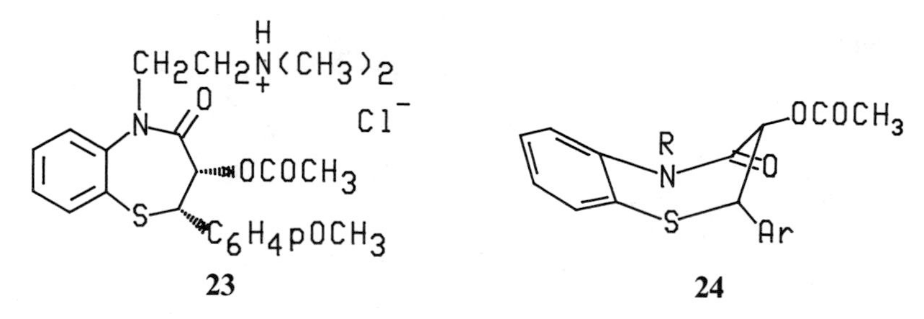

The better to understand the pharmacological activity of the neuropsychiatric drugs, the conformation of dibenzo[b.f]heteroepin drugs has been studied and reviewed.[86] Of these, those with a dibenzo[b.f]thiepin moiety are relevant to this review. The X-ray structure of 3-methoxy-10-methyl-11-phenyldibenzo[b.f]thiepin (**25**) shows that the seven-membered ring is in the boat conformation as shown in **26**.[87]

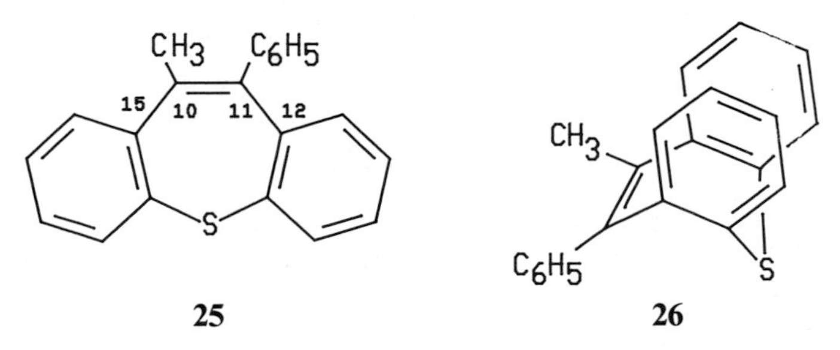

The corresponding 10,11-dihydro compounds **27** adopt twist–boat conformation **28** as shown by X-ray studies on the potent neuroleptic oxy-

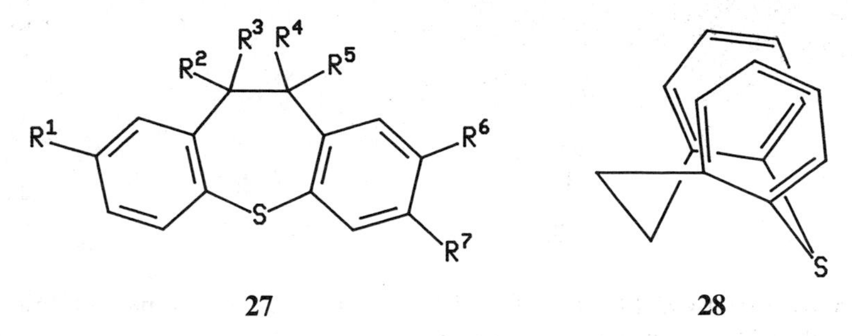

27 **28**

prothepine (**27**),[88] $R^1 = SCH_3$; $R^2 = R^4 = R^5 = R^6 = R^7 = H$; $R^3 =$ —$N(CH_2CH_2)_2N(CH_2)_3OH$, neuroleptic **27**; $R^1 = F$; $R^2 = R^4 = R^5 = R^7 = H$; $R^3 =$ —$N(CH_2CH_2)_2N(CH_2)_2$—$N(CH_2CH_2OCO)$,[89] the neuroleptic drug octoclothepin[90] (**27**); $R^1 = Cl$; $R^2 = R^4 = R^5 = R^6 = H$; $R^7 =$ —$N(CH_2CH_2)_2NCH_3$ and sedative **27**; $R^1 = R^2 = R^6 = H$; $R^3 = CH_3$; $R^4 = OH$; $R^5 = C_6H_5$; $R^7 = OCH_3$.[91]

Variable-temperature NMR studies on 3H,7H-naphtho[1,8-de]-1,2-dithiepin (**29**) revealed an AB spectrum for the benzylic hydrogen atoms, which coalesces at $-73°C$ with $\Delta G^{\ddagger} = 9.5$ kcal/mol for ring inversion of a nonplanar form.[92]

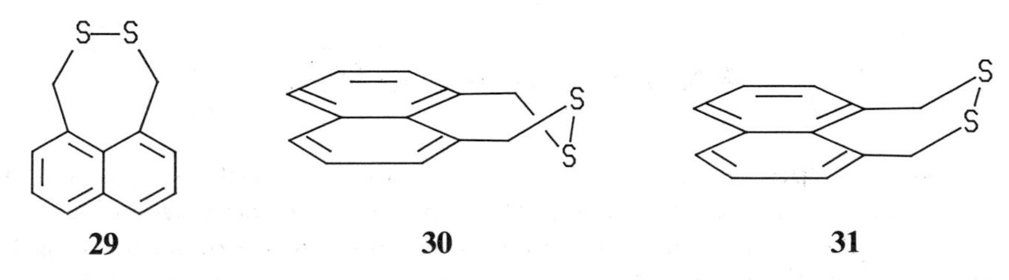

29 **30** **31**

This suggests equilibrating the C_2 twist or C_s half-boat conformer (**30** and **31**, respectively) at low temperature. However, molecular mechanics calculations suggested that the C_1 twist–boat conformer **32** is of lowest energy. Since these conformers differ in their CSSC dihedral angles, photoelectron spectroscopy provided an effective means for distinguishing them. The C_1 twist–boat form was confirmed by the lone-pair splitting. Equilibration of C_1 twist–boat forms results in the equivalence of the CH_2 groups, and this process could not be frozen out in the NMR experiments.

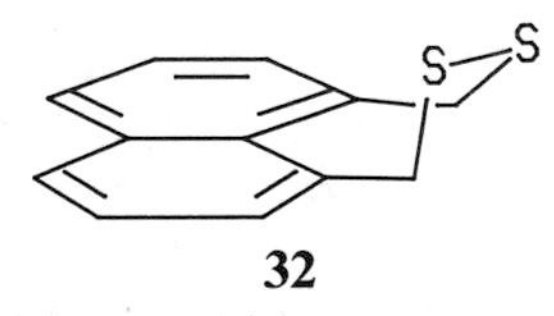

32

Low-temperature NMR studies of 4 4 6,6-tetramethyl-1,2-dithiacycloheptane (**33**) and 2,2,7,7-d_4-5,5-dimethyl-1,2-dithiacycloheptane (**34**) have been reported.[82,93] The methylene protons at positions 3 and 7 of **33** are nonequivalent at $-110°C$ and appear as an AB quartet. Analogously, at $-110°C$, **34** shows two singlets for the 5-*gem*-dimethyl group and an AB quartet for the two

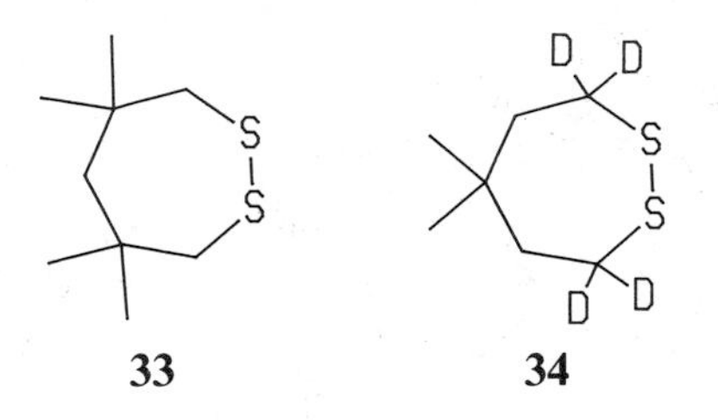

methylene groups at positions 4 and 6. Specific conformations have not been assigned to these compounds, however.

X-ray crystallographic studies[94] on dimethyl 1,3,5,2,4-trithiadiazepine-6,7-dicarboxylate,[95] methyl 1,3,5,2,4,6-trithiatriazepine-7-carboxylate,[95] 1,3,5,2,4-benzotrithiadiazepine,[96] and the parent heterocycle 1,3,5,2,4-trithiadiazepine[96] reveal the seven-membered heterocyclic rings to be planar with bond lengths intermediate between single and double bonds, as expected for a delocalized 10π aromatic system. The dihydro derivative, 6,7-dihydro-1,3,5,2,4-trithiadiazepine, adopts a conformation in which the five heteroatoms are nearly coplanar with the carbon atoms symmetrically displaced above and below this plane, as revealed by X-ray structural studies.[94,96]

4.4. Eight-membered Rings

From the experimental data available, it appears as if the cyclooctane ring system prefers the boat–chair conformation (**35**) except when there is extensive replacement of the ring carbon atoms or extensive substitution of the ring hydrogens.[97] Cyclooctane itself has been found by ^{13}C and ^{1}H NMR analysis to exist predominantly in the boat–chair conformation[98–100] in agreement with theoretical calculations.[66,67,101,102] This conformation has also been observed in many crystalline derivatives of cyclooctane: *cis*- and *trans*-1,2-cyclooctane-dicarboxylic acid,[103] dimeric cyclooctanone peroxide,[104] 1-aminocyclooctane-carboxylic acid hydrobromide,[105] *trans*-1,4-dichlorocyclooctane,[106] cyclooctane-1,5-diol,[107] and cyclooctane-1,5-dione.[107]

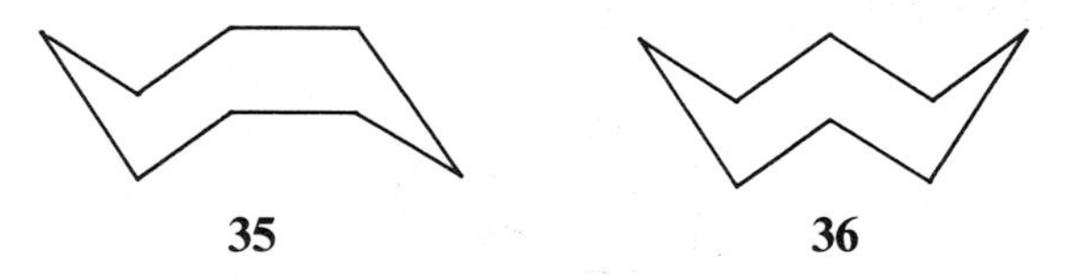

35 **36**

Analysis of the NMR spectra of the oxocanes indicates that oxocane itself adopts predominantly a boat–chair conformation,[108] as does 1,3-dioxocane.[108–110] 1,3,6-Trioxocane and 1,3,5,7-tetroxocane, however, exist as mixtures of boat–chair and chair–chair (crown-type, **36**) conformations.[108–110] Cyclooctasulfur adopts a crown conformation in both the solid state, as determined by X-ray[111–115] and neutron [115] diffraction, and in the gas phase, as determined by photoelectron spectroscopy.[116] Force-field calculations also indicate the crown conformation for cyclooctasulfur to be the most stable.[70]

X-ray crystal structures of a number of derivatives of 1,5-dithiacyclooctane have been determined. In 1-acetonyl-1-thionia-5-thiacyclooctane perchlorate **37**,[117,118] *trans*-bis(1,5-dithiacyclooctane)tetrachlorotin(IV) (**38**),[119] and *trans*-3-methoxy-1,5-dithiacyclooctane-1-oxide (**39**),[120] the dithiacyclooctane ring is seen to adopt a boat–chair conformation (**40**).

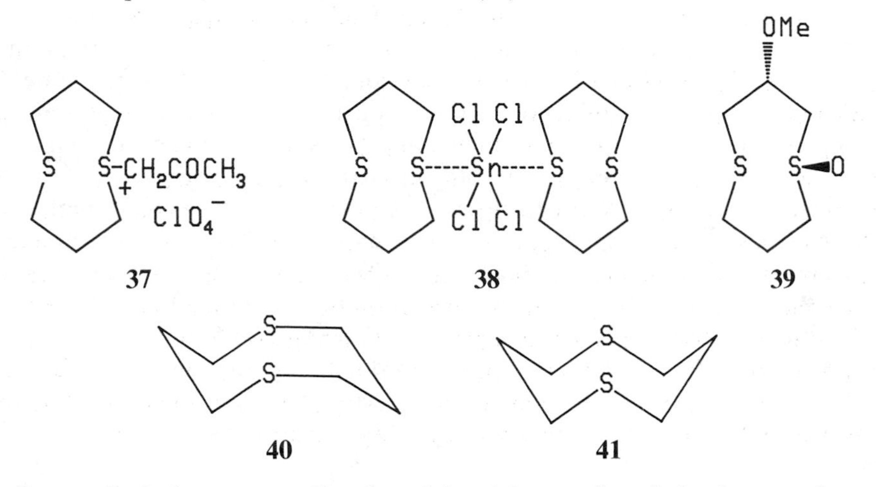

Interestingly, however, a disorder of the eight-membered ring between boat–chair **40** and chair–chair **41** is found in the X-ray structure of bis(1,5-dithiacyclooctane)nickel(II) chloride (**42**).[121] The crystal structure of the iodine complex of 1,5-dithiacyclooctane (**43**) shows both the boat–chair and the chair–chair conformations to be present in the asymmetric unit.[122,123] 1-Methyl-1-thionia-5-thiacyclooctane iodide (**44**) adopts the chair–chair conformation in the crystalline state.[124] Apparently, the energy difference between the boat–chair and chair–chair conformations for 1,5-dithiacyclooctane is very small. Indeed, molecular mechanics calculations indicate the boat–chair conformation to be only 0.98 kcal/mol more stable than the chair–chair.[62]

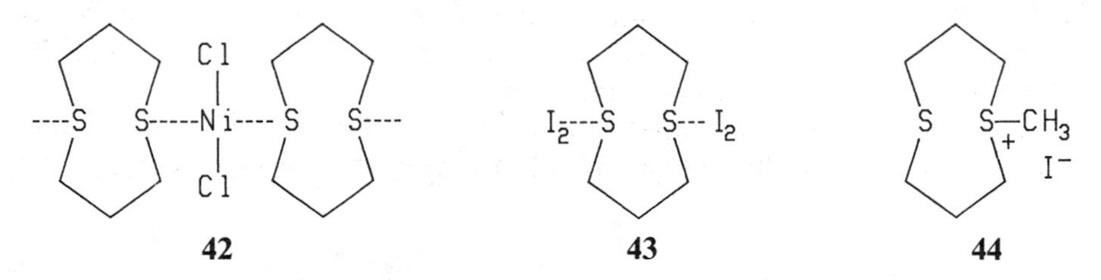

An important feature in both of these conformations is that the sulfur atoms are on the same side of the ring, in close proximity to each other. Photoelectron spectroscopic analysis of 1,5-dithiacyclooctane reflects this geometrical arrangement of the sulfur atoms.[62] The sulfur lone pairs show a splitting of 0.43 eV, indicating a large degree of mixing. 1,4-Dithiacycloheptane and 1,5-dithiacyclononane, on the other hand, show little or no splitting of the sulfur lone pairs.

An apparent consequence of the large transannular interaction of the sulfur lone pairs is both the relatively low ionization potential (8.30 eV compared with 8.68 eV and 8.36 eV for 1,4-dithiacycloheptane and 1,5-dithiacyclononane, respectively),[62] and electrochemical oxidation potential (0.34 volt compared with 0.84 volt and 0.42 volt for 1,4-dithiacycloheptane and 1,5-dithiacyclononane, respectively) of 1,5-dithiacyclooctane.[63]

A second consequence of the preferred conformation of 1,5-dithiacyclooctane with the two thioether moieties on the same side of the ring is that the molecule can serve as a bidentate ligand without undergoing a conformational change. Thus, bis(3-hydroxy-1,5-dithiacyclooctane)copper(II) perchlorate[125] and bis(3-methoxy-1,5-dithiacyclooctane)copper(II) perchlorate[126] (**45**) are in complexes in which both sulfur atoms bind the same metal and the ligand retains a boat–chair conformation. Similarly, the ligand in bis(3-hydroxy-1,5-dithiacyclooctane)nickel(II) perchlorate adopts a boat–chair conformation in this complex.[127] In the analogous iron(II) complex, bis(3-hydroxy-1,5-dithiacyclooctane)iron(II) perchlorate (**46**), however, X-ray crystal structure analysis reveals the alternative chair–chair conformation to be adopted by the ligand.[127] Interestingly, the copper(I) complex, bis(3-methoxy-1,5-dithiacyclooctane)-copper(I) perchlorate (**47**) has a pentacoordinate copper. One of the ligands serves as a tridentate ligand and adopts a chair–chair conformation while the other ligand is only bidentate, adopting a boat–chair conformation.[126]

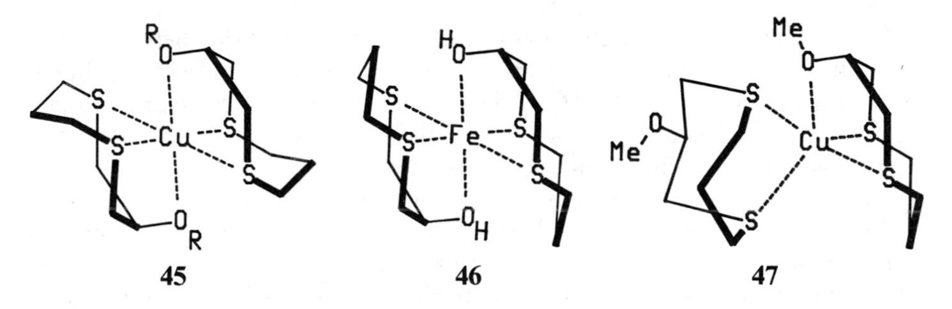

45 **46** **47**

The conformation adopted by 1,5-dithiacyclooctane-3,7-dione bis(ethylene ketal) (**48**) in the solid state is neither a boat–chair nor a chair–chair but rather a conformation such that the six carbon atoms are nearly coplanar and the sulfur atoms lie above and below the plane.[128]

Analysis of the variable-temperature ^1H and ^{13}C NMR of 1-thiacyclooctane-5-one indicates that this compound adopts preferentially an unsymmetrical

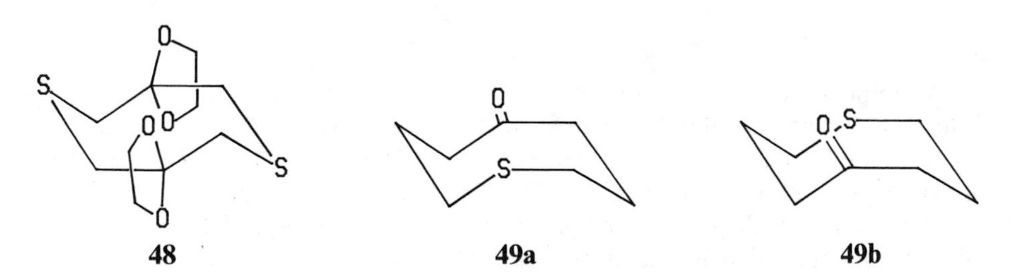

48 **49a** **49b**

boat–chair conformation (**49**) at −150°C. Pseudorotational equilibration of the two boat–chair forms has a free-energy barrier of 6.7 kcal/mol at −115°C.[129]

Curiously, 1,3,5,7-tetrathiocane has been found to adopt, in the crystalline state,[130,131] as well as in solution,[130] the boat–chair conformation **50**. The low-temperature ^{1}H NMR data are not consistent with even minor amounts of chair–chair (crown-type) form. An X-ray crystal structure of di-μ-chloro-μ-(1,3,5,7-tetrathiocane-S^1,S^3)-bis[trimethylplatinum (IV)] shows the tetrathiocane ring to retain the boat–chair conformation upon complexation.[132] A variable-temperature NMR study of pentacarbonyl(1,3,5,7-tetrathiocane)-molybdenum(0) has been done[133] and the data were interpreted in terms of rapidly equilibrating boat–chair conformations of the tetrathiocane ligand, along with fluxional processes involving the molybdenum–sulfur interactions (pyramidal atomic inversion about the sulfur atoms and 1,3-metal shifts).

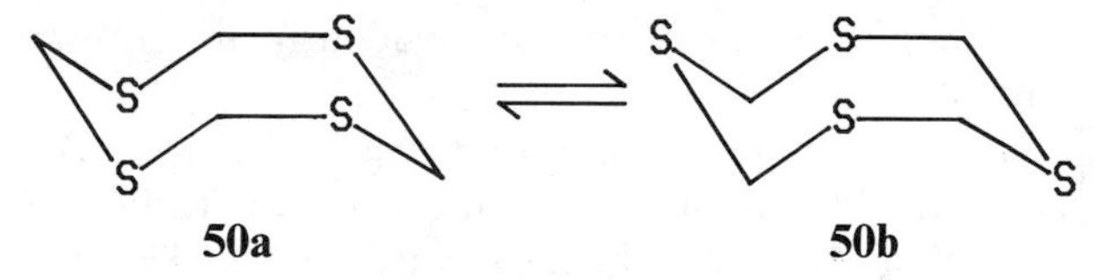

N,N′-Diphenyl-3,7-diaza-1,5-dithiacyclooctane apparently does not adopt a boat–chair conformation but was concluded to prefer a chair–chair conformation (**51**) in solution based upon dipole moment studies.[134] An X-ray crystal structure of the N,N′dimethyl analogue shows this compound to adopt a chair–chair conformation, but with the N-substituents axial (**52**) rather than equatorial.[135]

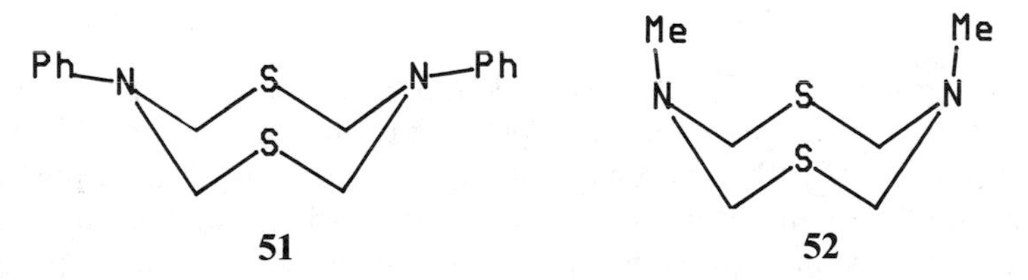

Attractive 1,5-transannular Sn ··· S interaction is well illustrated in 2-chloro-2-phenyl-1,3,6-trithia-2-stannocane, whose X-ray structure has been determined.[136] The ring adopts boat–chair conformation (**53**) in which the tin atom has trigonal bipyramidal geometry and the 1,5-transannular Sn ··· S distance is 2.806 Å. The eight-membered ring in 2,2-diphenyl-1,3,6-trithia-2-stannocane is also in a boat–chair conformation as determined by X-ray studies but the transannular Sn ··· S distance is 3.246 Å and the tin atom is intermediate in geometry between a tetrahedron and a trigonal bipyramid.[137] The structurally

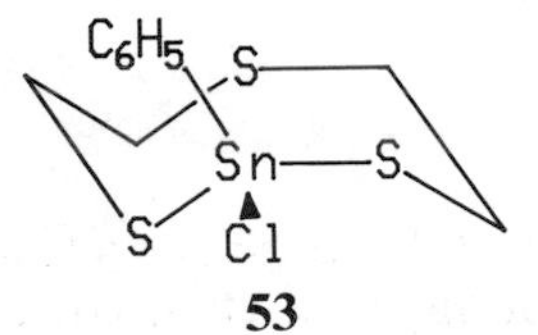

similar bismuth-containing heterocycle, 5-phenyl-1,4,6,5-oxadithiabismocane (**54**), on the other hand, adopts a chair–chair conformation in the solid state.[138]

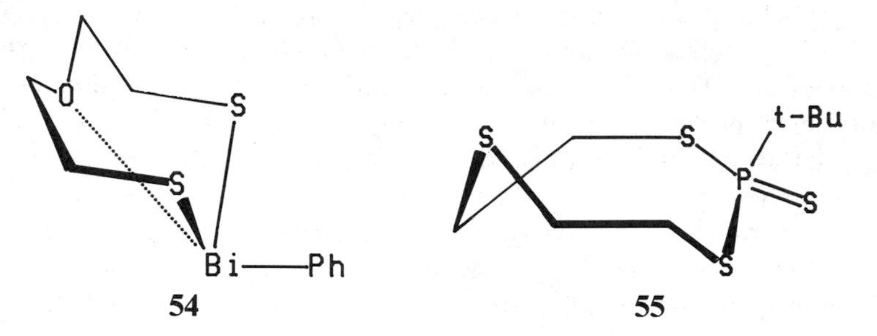

The eight-membered ring phosphorus heterocycle, 2-thio-2-*tert*-butyl-1,3,6,2-trithiaphosphocane (**55**) adopts, in the crystalline state, the conformation shown.[139] In this conformation, there are evidently no transannular phosphorus–sulfur interactions.

The two diastereomeric *trans*-2-methylthiacyclooct-4-enes (**56** and **57**) have been tentatively assigned the twist conformations shown, rather than chair conformations, based upon ^{13}C NMR chemical shift data.[140] These conformations were later confirmed using force-field calculations.[141] The chair conformations are calculated to be about 5 kcal/mol higher in energy than the twist.

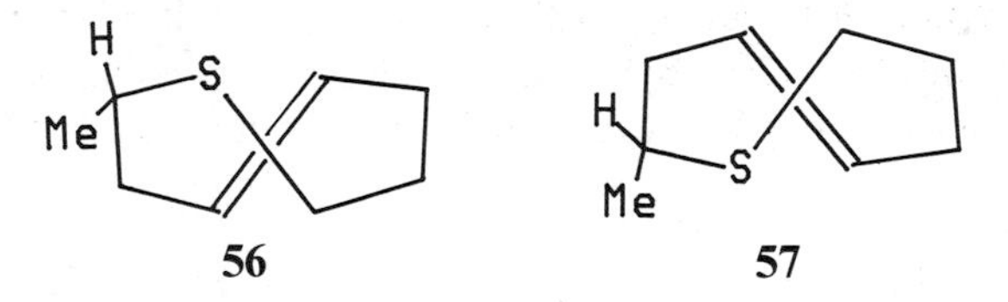

Variable-temperature ^{1}H NMR studies of 3,4,6,7-dibenzothiacyclooctane **58** show the compound to interconvert rapidly at room temperature.[142] At lower temperatures, the ^{1}H NMR is consistent with a rigid boat–chair conformation (**59**) and rapidly equilibrating boat–boat or twist–boat–boat forms. The observed free energy of activation for the boat–chair ⇄ boat–boat process is $\Delta G^{\ddagger} = 16.2$ kcal/mol. Strain-energy calculations indicate the boat–chair conformation to be the lowest-energy conformation for this compound.[142]

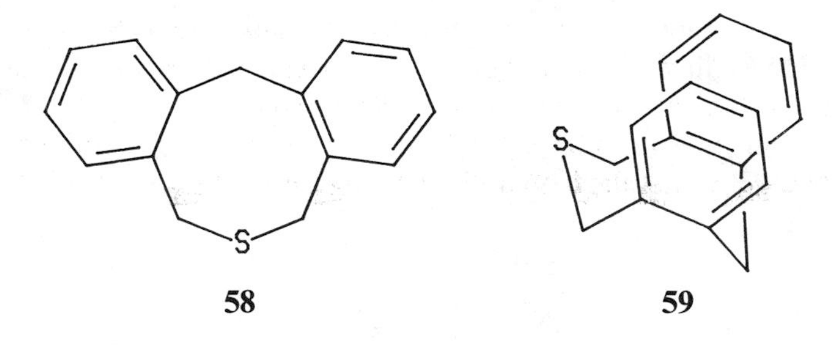

Solution NMR studies of 2,8-dihydrodibenzo[b,f][1,5]-dithiocin (**60**) reveal a 1:4 ratio at room temperature of the chair conformer and a boat-type con-

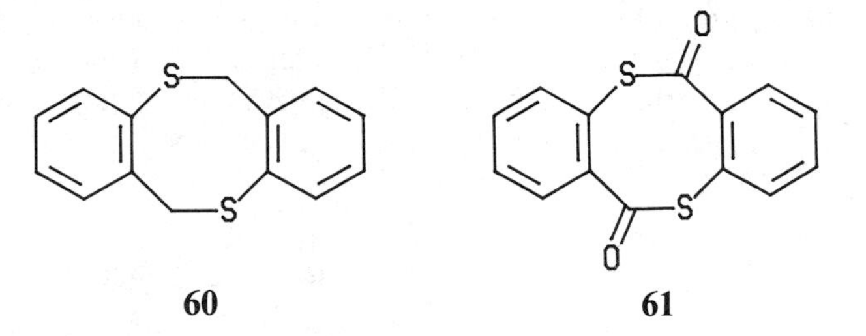

60 61

former.[143] In the solid state, the X-ray structure of dithiosalicylide **61** shows a boatlike conformation for the eight-membered ring.[144] In solution, variable-temperature ^{1}H NMR studies on di-o-thiothymolide revealed a $\Delta G^{\ddagger}$ of 24.6 kcal/mol for the ring inversion between the enantiomeric boat conformations **62**.[145]

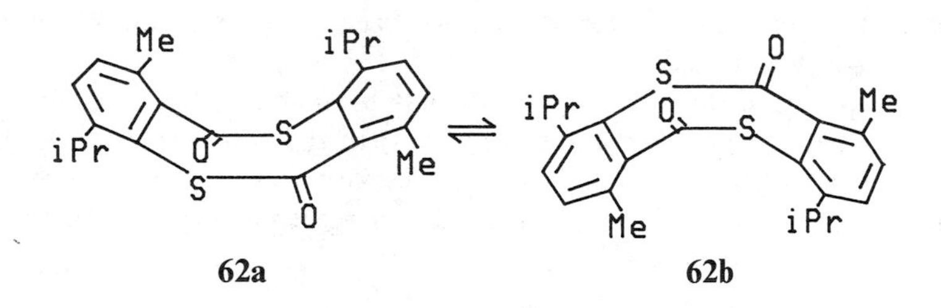

62a 62b

A twist–boat–chair conformation (**63**) has been found in the X-ray crystal structures of both the alkaloid cassipourine (**64**)[146] and the parent heterocycle, 1,2,5,6-tetrathiacyclooctane (**65**).[147] The dibenzo analogue of 1,2,5,6-tetrathiacyclooctane (**66**) adopts a chair conformation in the crystalline state.[148] Note that gauche orientations of the disulfide groups are still present in these conformations.

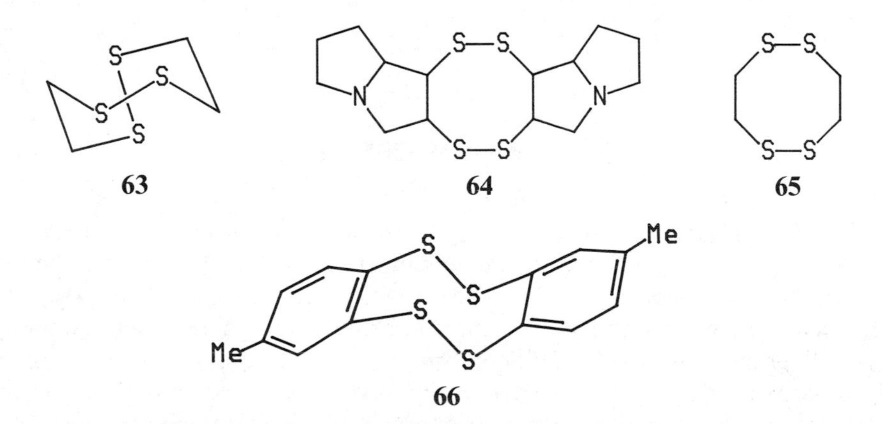

63 64 65

66

The X-ray crystal structure of L-cysteinyl-L-cysteine (**67**), an eight-membered ring cyclic disulfide, shows the solid-state conformation of this compound to be a distorted boat–chair with a *cis* peptide unit and a right-handed helical sense of the disulfide group (**68**).[149] This conformation is also adopted by the compound in solution as evidenced by ^{1}H NMR analysis.[150] Strain-energy calculations, however, had earlier predicted the alternative conformation (**69**) differing in the dihedral angle of the disulfide groups (left-handed helical sense) to be more stable.[151,152]

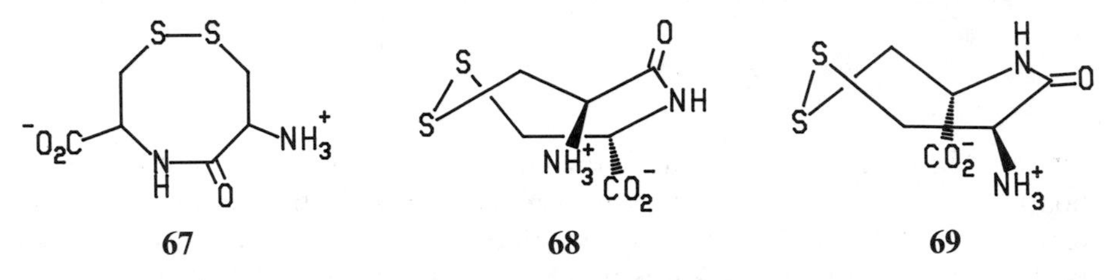

X-ray diffraction analysis of 3,7-diphenyl- and 3,7-bis(p-methoxyphenyl)-1,5-dithia-2,4,6,8-tetrazocine shows the eight-membered heterocyclic ring to be planar with bond lengths intermediate between single and double bonds as expected for a delocalized 10π aromatic ring.[153]

The dibenzo-1,2-dithiacyclooctane (**70**) has been shown by X-ray crystallography to adopt a chairlike conformation (**71**) in the solid state.[154] Molecular mechanics calculations indicate this chair conformation (with a C—S—S—C dihedral angle of about 60°) to be lower in energy than the boat or tub form, **72** (C—S—S—C dihedral angle about 90°).[32]

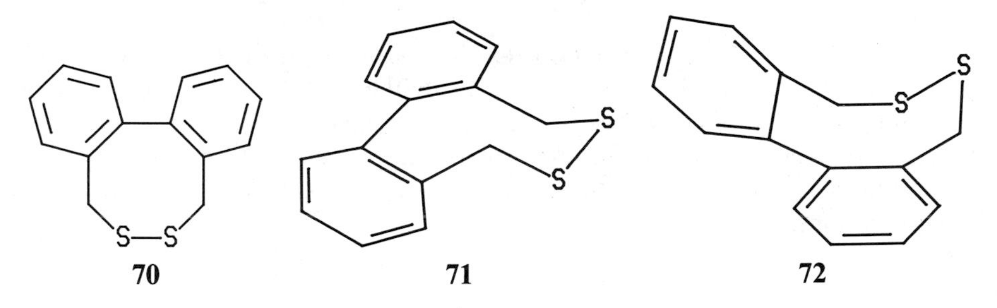

4.5. Nine-membered Rings

Solution NMR spectroscopy,[155,156] as well as strain-energy calculations,[66–68,157] indicate the most stable conformation of cyclononane to be the [333][158] or twist–boat–chair (**73**). The nine-membered ring of 1-hydroxycyclononanyl dimethyl phosphate has this conformation, as determined by X-ray crystallographic analysis.[159] This conformation is also adopted by trimeric acetone peroxide,[160] as well as 1,1,4,4-tetramethylcyclononane and related derivatives.[161] In contrast, 1,4,7-trioxacyclononane has been found, by ^{13}C NMR and IR spectroscopic analysis,[162] to exist both in solution and in the solid state in an unsymmetrical [234][158] conformation (C_1 symmetry, **74**). Force-field calculations indicate the most stable conformation of cyclononasulfur, also, to be the [234] conformation.[70] Some derivatives of cyclononane, cyclononylamine hydrobromide[163] and cyclononanone-mercuric chloride,[164] on the other hand, crystallize in the [12222][158] conformation (C_2 symmetry, **75**).

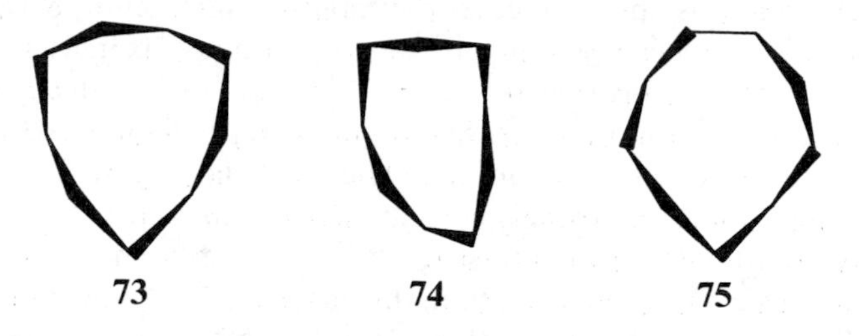

Substitution of two or three carbon atoms of cyclononane with sulfur produces no gross conformational changes in the nine-membered ring. X-ray crystallographic analyses of 1,5-dithiacyclononane[74] and 1,4,7-trithiacyclononane[165] have shown these compounds to adopt [333] or twist–boat–chair conformations in the solid state (**76** and **77** respectively).

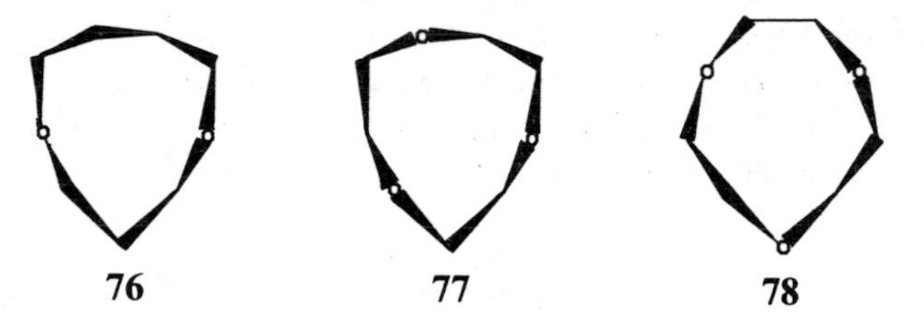

Gas-phase photoelectron spectroscopic analyses are in agreement with the X-ray crystal structures.[62] The photoelectron spectra of these two nine-membered rings reflect the transannular geometrical dispositions of the sulfur atoms with respect to each other. In **76**, the sulfur atoms are on opposite sides of the ring, the lone pairs do not interact, and the PES shows very little splitting of the sulfur lone pairs. In **77**, on the other hand, the sulfur lone pairs do interact and the PES shows a splitting of 0.6 eV.[62] Molecular mechanics calculations[62] show the [333] conformation for 1,5-dithiacyclononane to be the lowest-energy conformation, but erroneously indicate a [12222] conformation for 1,4,7-trithiacyclononane **78** to be more stable than the observed [333].

The conformation adopted by 1,4,7-trithiacyclononane, with the sulfur atoms all on the same side of the ring and the lone pairs pointing toward the cavity of the ring (endodentate), make it a potentially useful ligand for transition metal complexation. This compound may serve as a facially coordinating tridentate ligand without undergoing an intervening conformational change.

Complexes of 1,4,7-trithiacyclononane with iron (II),[166] cobalt (II),[167] cobalt (III),[168] nickel (II),[167] and copper (II),[167] have been prepared and characterized by X-ray diffraction methods. The ligand retains the [333] conformation and induces nearly regular octahedral coordination of these metal centers. Remarkably, there is a relatively modest change in geometry between the free and complexed ligand. The principal change involves decreasing the S—C—C—S torsion angle, which is 58.5, 46, 51, 51.1, 55, and 52° for the free ligand, and its Fe(II), Co(II), Co(III), Ni(II), and Cu(II) complexes respectively. The S—C—C—S torsion angle for the complex ligand is substantially decreased in tricarbonyl(1,4,7-trithiacyclononane)molybdenum(0) where it is 48°.[169] Decreasing this angle increases the repulsive interaction between sulfur atoms but this is partly compensated by the formation of stronger metal–sulfur bonds.[169] Very recent X-ray structure studies[170] of bis(1,4,7-trithiacyclononane) palladium(II) and dibromo(1,4,7-trithiacyclononane) palladium(II) also reveal the [333] conformation for the coordinated heterocyclic ring. In these two cases, two of the sulfur atoms are strongly coordinated and the third, disposed axially, only weakly coordinated. In contrast to these complexes in which the ligand retains its preferred [333] conformation, this heterocycle assumed [12222] conformation **78** in its bis complex with Pt(II).[171] In this complex, the ligand is bidentate and the coordination geometry about platinum is planar. The energy difference between the [333] and [12222] conformations is probably not large, and complexation with a metal involving square planar coordination evidently can force a conformational change in this ligand.

Complexes in which 1-thia-4,7-diazacyclononane is coordinated and their X-ray structures have been reported. In its bis complex with nickel(II),[172] the nine-membered ring adopts the [333] conformation analogous to 1,4,7-trithiacyclononane. However, in its bis complex with copper(II),[173] the heterocycle adopts [234] conformation **79**.

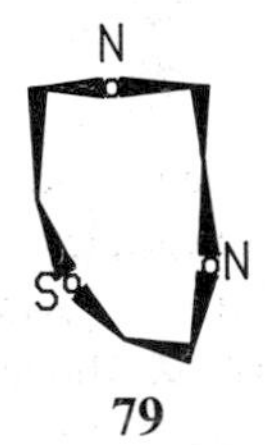

79

Unsaturation, especially benzo substitution in nine-membered rings, serves to restrict the conformational mobility, and these systems are amenable to ^{1}H NMR analysis. 3,4,7,8-Dibenzothiacyclononane has been studied by ^{1}H NMR and molecular mechanics and is concluded preferentially to adopt a C_2 chairlike conformation (**80**).[142] The analogous thiacyclononene (**81**) has been shown by low-temperature ^{1}H NMR analysis, in agreement with molecular mechanics calculations, to prefer a rigid boat–chair conformation in solution.[142] The ob-

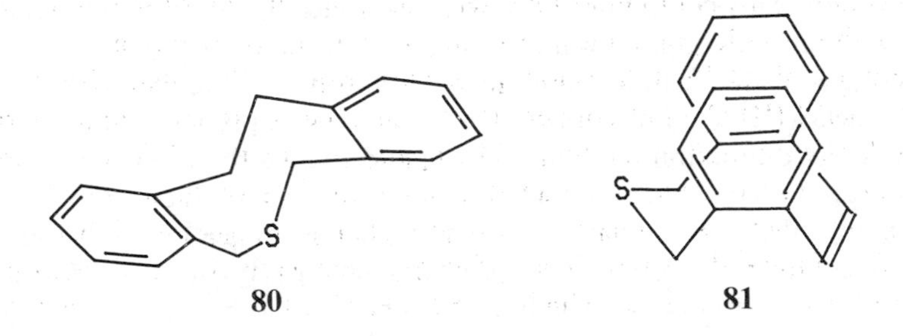

80 **81**

served free energy of activation ($\Delta G^{\ddagger}$ = 15.5 kcal/mol) for boat–chair ⇄ boat–chair interconversion of this compound is in good agreement with the calculated strain-energy difference (16.1 kcal/mol) between the boat–chair and a boatlike transition state geometry.[142]

Increasing aromatic substitution restricts conformational mobility and can lock the nine-membered ring of thiacyclononanes into single conformations. The tribenzo thiacyclononane **82** has been studied by variable-temperature ^{1}H and ^{13}C NMR and has been found to adopt a rigid "crown" conformation at room temperature.[174] The crown conformations do interconvert at higher temperatures, with a coalescence temperature of >150°C. The binaphthyl compound **83** adopts the rigid conformation shown as evidenced by ^{1}H NMR chemical shift data.[175]

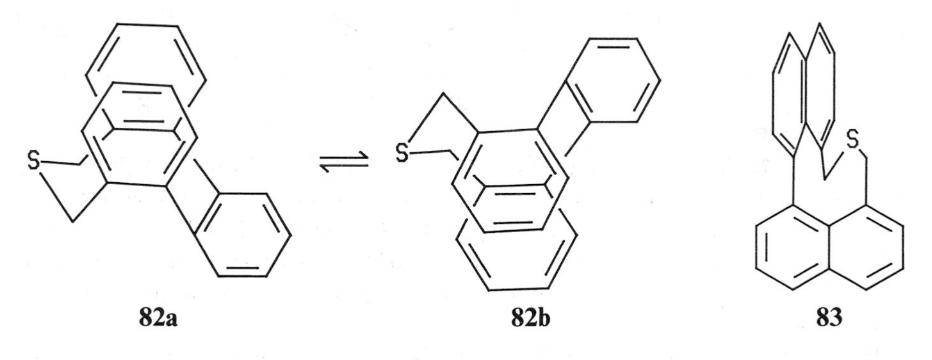

4.6 Ten-membered Rings

Strain-energy conformational analysis of cyclodecane predicts the [2323][158] or boat–chair–boat conformation (**84**) to be the most stable.[66–68] This prediction is in agreement with X-ray crystal structure analyses of various cyclodecane derivatives: *cis*-1,6-diaminocyclodecane dihydrochloride,[176] *trans*-1,6-diaminocyclodecane dihydrochloride,[177] *trans*-1,6-dibromocyclodecane,[178] and cyclodecylamine hydrochloride,[179] which all crystallize in [2323] conformations. Cyclodecasulfur, however, was calculated to prefer a "crown" conformation (D_{5d} symmetry, **85**)[70] but crystallizes in a conformation having D_2 symmetry, **86**.[180] Conformation **86** was not considered in the molecular mechanics calculations of cyclodecasulfur.

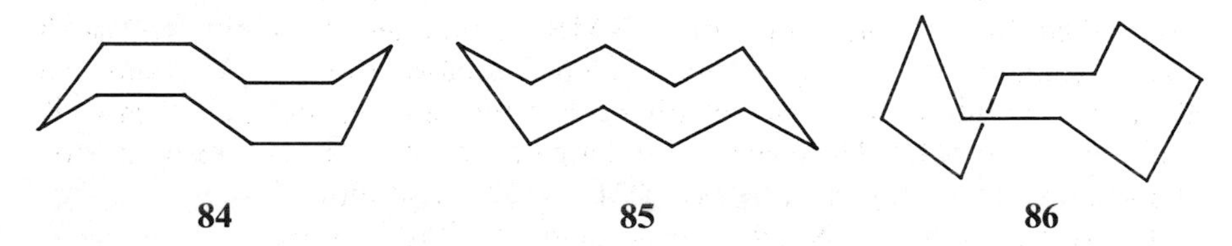

X-ray crystal structure analysis of 1,6-dithiacyclodecane shows this compound to adopt, in the solid state, a [2323] or boat–chair–boat conformation with C_{2h} symmetry, **87**.[74] Interestingly, molecular mechanics and molecular orbital calculations indicate an alternative [2323] conformation (**88**) to be more stable.[62] This alternative conformation has been shown by photoelectron spectroscopic analysis to be the preferred conformation in the gas phase.[62] The MINDO/3 calculated sulfur lone-pair splittings for **88** correlate with the observed PES better than do those calculated for **87**.[62] Both solution and gas-phase infrared (IR) spectra are markedly different from the solid-state spectrum, indicating the gas-phase and solution conformations to be different from the solid-state conformation.[181] The gas-phase and solution IR spectra for 1,4-dithiacycloheptane and 1,5-dithiacyclononane are similar to their respective solid-state IR spectra, however.

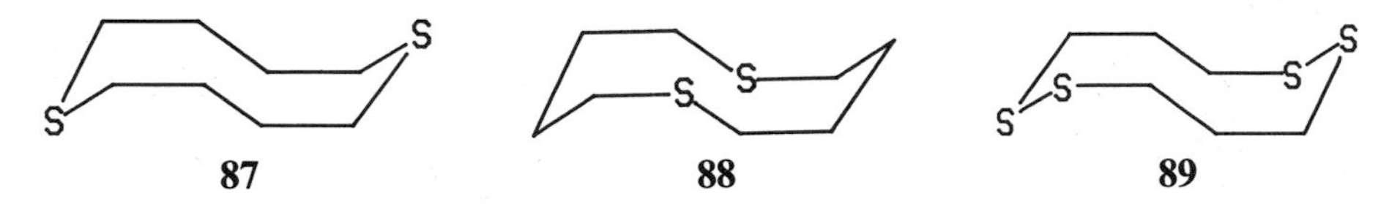

87 **88** **89**

1,2,6,7-Tetrathiacyclodecane also adopts a [2323] conformation (**89**) in the crystalline state.[182] The disulfide group in this molecule replaces an ethylene group of cyclodecane such that a gauche C—S—S—C dihedral angle is adopted. In addition, the long S—S bonds of the disulfide groups (compared with C—C) in these positions may serve to minimize transannular H · · · H interactions.

Increasing the number of sulfur atoms in the cyclodecane ring leads to the D_2 conformation **86**, observed in cyclodecasulfur, to be preferred over the [2323] conformation. Thus, X-ray crystal structures of both the monoclinic[183] and orthorhombic[184] forms of 1,3,5,7,9-pentathiacyclodecane show this compound to adopt the conformation **90** with pseudo D_2 symmetry. 1-Oxa-3,5,7,9-tetrathiacyclodecane also adopts this conformation (**91**) in the crystalline state.[185]

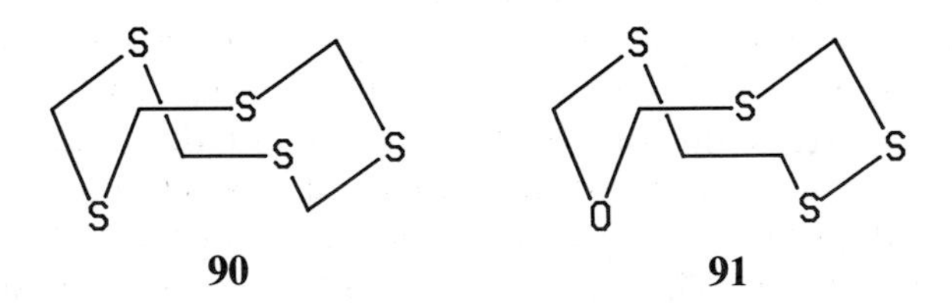

90 **91**

As seen in other medium-sized rings, unsaturation raises the barrier to conformational interconversion in 10-membered rings. These compounds, then, can be studied by variable-temperature NMR techniques. 1,6-Dithia-*cis,cis*-3,8-cyclodecadiene, at $-57°C$, adopts a “chair” conformation.[186] At room temperature, equilibration occurs involving the two equivalent chair forms (**92**) and a boat form (**93**). The coalescence temperature of $-24°C$ corresponds to a free-energy barrier for this process of $\Delta G^{\ddagger} = 12.1$ kcal/mol.[186]

Low-temperature ^1H NMR analysis suggests that 1,3-dioxa-7-thia-4,5,9,10-dibenzocyclodecane preferentially adopts the “chair” conformation with C_2 symmetry (**94**).[142]

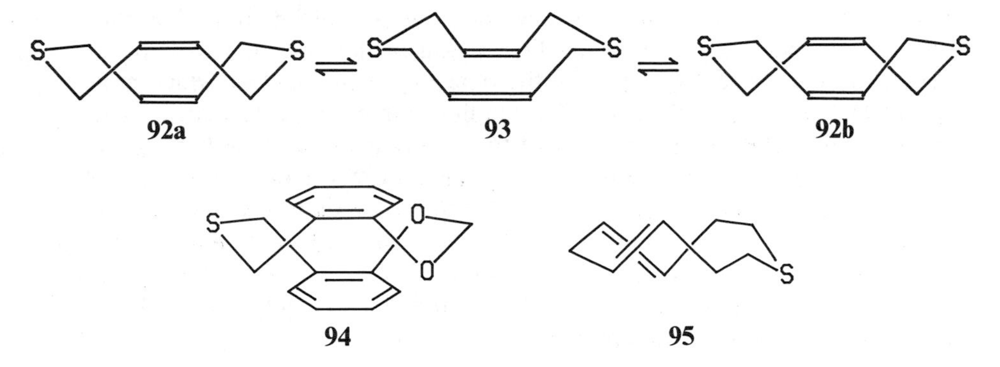

The likely conformation for 1-thia-*trans,trans*-3,6-cyclodecadiene has been suggested to be **95**.[187] Low-temperature ^{1}H NMR spectra of the 6-methyl derivative were interpreted in terms of equilibria between the two equivalent conformers, probably involving a transition conformation with pseudo sigma symmetry.[187]

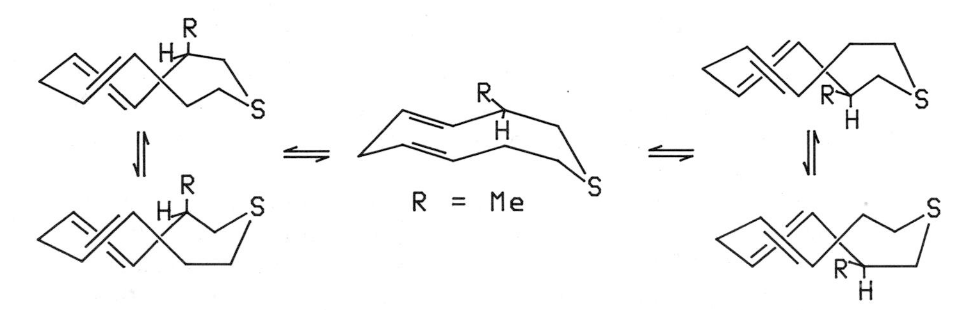

Conformational analysis of cyclophanes is of considerable interest and was reviewed relatively recently.[188] Only a few of these compounds fall within the scope of this chapter and these, with sulfur-substituted 10-membered rings, are discussed below. For a more extensive coverage of cyclophanes containing sulfur heterocycles with greater than 10-membered rings, the reader is referred to the previous review.[188]

Although a detailed conformational analysis of 2,6-dithia[7]metacyclophane (**96**) has not been carried out, variable-temperature ^{1}H NMR spectroscopic analysis revealed important conformational information.[189] At room temperature, both hydrogens at C(4) are equivalent and resonate at an anomalously

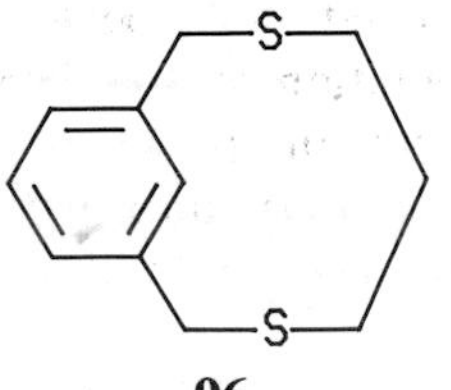

high field (δ 0.45). This suggests that **96** is in a conformation that thrusts H(4) across the ring into the shielding region of the aromatic ring. In addition, there is rapid equilibration of the type shown in eq. (4-1), which renders both H(4) hydrogens equivalent. This equilibration requires the hydrogen atom shown to pass through the interior of the ring. On cooling, the hydrogens attached to C(4) become nonequivalent with a coalescence temperature of $-50°C$ and a calculated energy barrier of 10.2 kcal/mol for the process shown in eq. (3-1).[190] Another process that renders the benzylic hydrogens equivalent can also be frozen out with a coalescence temperature of $-75°C$ and a calculated energy barrier for this process of 10.1 kcal/mol. These results require two different conformers at low temperature.

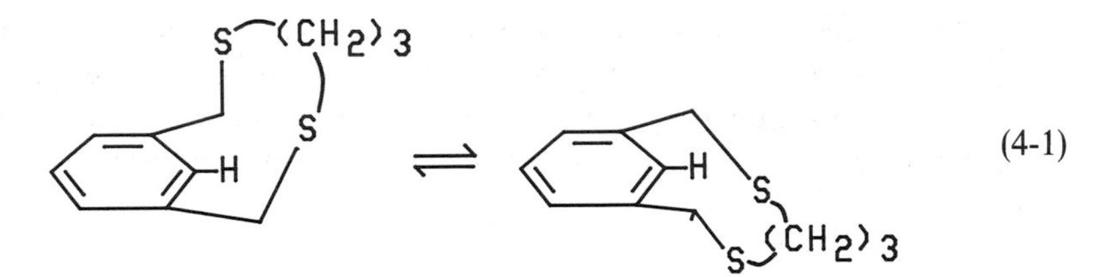

(4-1)

The conformation of 1,10-dithia[2.2]metacyclophane[191] and its 8-nitro derivative[192] have been determined in the solid state by X-ray methods. They adopt an *anti* conformation with respect to the aromatic rings and the 1,5-dithia 10-membered ring moiety is in a crownlike conformation as shown in **97**. Variable-temperature NMR studies on 1,10-dithia[2.2]metacyclophane demonstrated that this is the conformation in solution up to at least 190°C.[193] ^1H NMR spectroscopic studies on 1,2,9,10-tetrathia[2.2]metacyclophane and its derivatives[194] led to the suggestion that it exists in an *anti* conformation analogous to 1,10-dithia[2.2]metacyclophane. However, it was subsequently opined[195] that the ^1H NMR data are more consistent with the *syn* conformer.

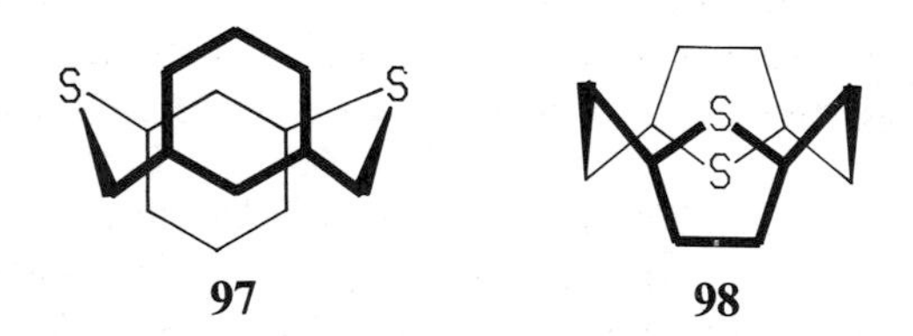

97 **98**

The conformation of [2.2](2,5)-thiophenophane[196] and its 1:1 complex with benzotrifuroxan[197] in the solid state have been determined by X-ray methods. The molecule has the *anti* orientation of thiophene rings and the 10-membered ring is crownlike, as seen in **98**. In the uncomplexed cyclophane, the thiophene rings are in an envelope conformation with the sulfur atom almost 0.2 Å out of the ring plane. ^1H NMR spectroscopic studies in solution have revealed that this cyclophane is immobile.[198]

References and Notes

1. For a general review on conformational analysis of medium-sized rings see Dale, J. "Stereochemistry and Conformational Analysis." Verlag Chemie: New York, 1978, Chap. V.
2. For reviews on saturated six-membered and smaller rings containing sulfur see Zefirov, N. S.; Kuzimirchik, I. V. *Russ. Chem. Rev.* **1974**, *43*, 107. Armarego, W. L. F. "Stereochemistry of Heterocyclic Compounds," Part II. Wiley-Interscience: New York, 1977, Chap. 3. *Heterocyclic Chem., Spec. Per. Rep.* **1980**, *1*, 476. *Sat. Heterocyclic Chem., Spec. Per. Rep.* **1974**, *2*, 182; **1975**, *3*, 154; **1976**, *4*, 148; **1977**, *5*, 122. *Org. Cmpds. Sulphur, Selenium Tellurium, Spec. Per. Rep.* **1970**, *1*, 134; **1973**, *2*, 135; **1975**, *3*, 139; **1977**, *4*, 217; **1979**, *5*, 213; **1981**, *6*, 233. Riddell, F. G. "The Conformational Analysis of Heterocyclic Compounds." Academic Press: London, 1980.
3. For reviews on macrocyclic polythioethers see Izatt, R. M.; Christensen, J. J. "Synthetic Multidentate Macrocyclic Compounds." Academic Press: New York, 1978.
4. Dalley, N. K.; Larson, S. B.; Smith, J. S.; Matheson, K. L.; Izatt, R. M.; Christensen, J. J. *J. Heterocyclic Chem.* **1981**, *18*, 463.
5. Campbell, M. L.; Dalley, N. K.; Izatt, R. M.; Lamb, J. D. *Acta Crystallogr., Sect. B* **1981**, *B37*, 1664.
6. Huffman, J. C.; Campbell, M. L.; Dalley, N. K.; Larson, S. B. *Acta Crystallogr., Sect. B* **1981**, *B37*, 1739.
7. Hartman, J. R.; Wolf, R. E.; Foxman, B. M.; Cooper, S. R. *J. Am. Chem. Soc.* **1983**, *105*, 131.
8. Ammon, H. L.; Bhattacharjee, S. K.; Shinkai, S.; Honda, Y. *J. Am. Chem. Soc.* **1984**, *106*, 262.
9. Pett, V. B.; Diaddario, L. L.; Dockal, E. R.; Cornfield, P. W.; Ceccarelli, C.; Glick, M. D.; Ochrymowycz, L. A.; Rorabacher, D. B. *Inorg. Chem.* **1983**, *22*, 3661.
10. Hintsa, E. J.; Hartman, J. R.; Cooper, S. R. *J. Am. Chem. Soc.* **1983**, *105*, 3738.
11. Dalley, N. K.; Larson, S. B. *Acta Crystallogr., Sect. B* **1981**, *B37*, 2225.
12. Keehn, P. M. In "Cyclophanes," Keehn, P. M.; Rosenfeld, S. M., Eds. Academic Press: New York, 1983, Chap. 3. Mitchell, R. H. In "Cyclophanes," Keehn, P. M.; Rosenfeld, S. M., Eds. Academic Press: New York, 1983, Chap. 4.
13. Mitchell, R. H.; Lai, Y.-H. *J. Org. Chem.* **1984**, *49*, 254.
14. For a general overview of structural parameters in sulfur compounds see Hargittai, I. "The Structure of Volatile Sulphur Compounds." D. Reidel: Dordrecht, 1985. Hargittai, I. In "Organic Sulfur Chemistry," Bernardi, F.; Csizmadia, I. G.; Mangini, A., Eds. Elsevier: Amsterdam, 1985, pp 68–132.
15. Abrahams, S. C. *J. Chem. Soc., Q. Rev.* **1956**, 407.
16. Sutton, L. E. "Interatomic Distances," Special Publication No. 11.; Chemical Society of London: London, 1958.
17. Kooyman, E. C. In "Organosulfur Chemistry," Janssen, M. J., Ed. Wiley Interscience: New York, 1967, Chap. 1.
18. Tagaki, W. In "Organic Chemistry of Sulfur," Oae, S., Ed. Plenum Press: New York, 1977; Chap. 6.
19. Block, E. "Reactions of Organosulfur Compounds." Academic Press: New York, 1978.
20. Pierce, L.; Hayaski, M. *J. Chem. Phys.* **1963**, *38*, 2753.
21. Field, L. In "Organic Chemistry of Sulfur." Oae, S., Ed. Plenum Press: New York, 1977, Chap. 7.
22. Lee, J. D.; Bryant, M. W. R. *Acta Crystallogr., Sect. B.* **1969**, *B25*, 2094; 2497; **1970**, *B26*, 1729; **1971**, *B27*, 2325.
23. Ricci, J. S.; Bernal, I. *J. Am. Chem. Soc.* **1969**, *91*, 4078; *J. Chem. Soc., Sect. B* **1970**, 806.
24. Ottersen, T.; Warner, L. G.; Seff, K. *Acta Crystallogr., Sect B* **1973**, *B29*, 2954.
25. Woodard, C. M.; Brown, D. S.; Lee, J. D.; Massey, A. G. *J. Organometal. Chem.* **1976**, *121*, 333.
26. Wolfe, S. *Acc. Chem. Res.* **1972**, *5*, 102.
27. Boyd, D. B. *J. Am. Chem. Soc.* **1972**, *94*, 8799; *J. Phys. Chem.* **1974**, *78*, 1554.
28. Van Wart, H. E.; Shipman, L. L.; Sheraga, H. A. *J. Phys. Chem.* **1974**, *78*, 1848.
29. Eslava, L. A.; Putnam, J. B.; Pedersen, L. *Int. J. Peptide Protein Res.* **1978**, *11*, 149.
30. Blustin, P. H. *Theor. Chim. Acta* **1978**, *48*, 1.
31. Snyder, J. P.; Carlsen, L. *J. Am. Chem. Soc.* **1977**, *99*, 2931.
32. Allinger, N. L.; Hickey, M. J.; Kao, J. *J. Am. Chem. Soc.* **1976**, *98*, 2741.
33. Jørgensen, F. S.; Snyder, J. P. *Tetrahedron* **1979**, *35*, 1399.

34. Hubbard, W. N.; Douslin, D. R.; McCullough, J. P.; Scott, D. W.; Todd, S. S.; Messerly, J. F.; Hossenlopp, I. A.; George, A.; Waddington, G. *J. Am. Chem. Soc.* **1958**, *80*, 3547. Fraser, R. R.; Boussard, G.; Saunders, J. K.; Lambert, J. B.; Mixan, C. E. *Ibid.* **1971**, *93*, 3822. Steudel, R. *Angew. Chem. Int. Ed. Engl.* **1975**, *14*, 655.
35. Bondi, A. *J. Phys. Chem.* **1964**, *68*, 441.
36. Pauling, L. "The Nature of the Chemical Bond," 3rd ed. Cornell University Press: Ithaca, N. Y., 1960, p. 260.
37. Rosenfield, Jr., R. E.; Parthasarathy, R.; Dunitz, J. D. *J. Am. Chem. Soc.* **1977**, *99*, 4860. Boyd, D. B. *J. Phys. Chem.* **1978**, *82*, 1407. Guru Row, T. N.; Parthasarathy, R. *Ibid.* **1981**, *103*, 477.
38. Sakakibara, M.; Matsuura, H.; Harada, I.; Shimanouchi, T. *Bull. Chem. Soc. Jpn.* **1977**, *50*, 111. Ohta, M.; Ogawa, Y.; Matsuura, H.; Harada, I.; Shimanouchi, T. *Ibid.* **1977**, *50*, 380. Ogawa, Y.; Ohta, M.; Sakakibara, M.; Matsuura, H.; Harada, I.; Shimanouchi, T. *Ibid.* **1977**, *50*, 650. Oyanagi, K.; Kuchitsu, K. *Ibid.* **1978**, *51*, 2243. Durig, J. R.; Compton, D. A. C.; Jalilian, M.-R. *J. Phys. Chem.* **1979**, *83*, 511.
39. Zefirov, N. S.; Gurvich, L. G.; Shashkov, A. S.; Krimer, M. Z.; Vorobeva, E. A. *Tetrahedron* **1976**, *32*, 1211. Eliel, E. L.; Juaristi, E. *J. Am. Chem. Soc.* **1978**, *100*, 6114.
40. Boyd, D. B. *J. Am. Chem. Soc.* **1972**, *94*, 8799. Webb, J.; Strickland, R. W.; Richardson, F. S. *Ibid.* **1973**, *95*, 4755. Snyder, J. P.; Carlsen, L. *Ibid.* **1977**, *99*, 2931.
41. Bastian, E. J. Jr., Martin, R. B. *J. Phys. Chem.* **1973**, *77*, 1129. Sugeta, H.; Go, A.; Miyazawa, T. *Bull. Chem. Soc. Jpn.* **1973**, *46*, 3407. Van Wart, H. E.; Lewis, A.; Scheraga, H. A.; Saeva, F. D. *Proc. Nat. Acad. Sci. USA*, **1973**, *70*, 2619. Van Wart H. E.; Shipman, L. L.; Scheraga, H. A. *J. Phys.* Chem. **1974**, *78*, 1848. Van Wart, H. E.; Scheraga, H. A. *J. Phys. Chem.* **1976**, *80*, 1812, 1832.
42. Steudel, R. *Top. Curr. Res.* **1982**, *102*, 149.
43. Steudel, R. *Stud. Inorg. Chem.* **1984**, *5*, 3.
44. Steudel, R. *Phosphorus Sulfur* **1985**, *23*, 33.
45. Steudel, R.; Laitinen, R. *Top. Curr. Res.* **1982**, *102*, 177.
46. Schweig, A.; Thon, N. *Chem. Phys. Lett.* **1976**, *38*, 482. Müller, C.; Schäfer, W.; Schweig, A.; Thon, N.; Vermeer, H. *J. Am. Chem. Soc.* **1976**, *98*, 5440. Friege, H.; Klessinger, M. *J. Chem. Res. S* **1977**, 208. Schweig, A.; Thon, N.; Vermeer, H. *J. Am. Chem. Soc.* **1979**, *101*, 80. Honegger, E.; Heilbronner, E. *Chem. Phys. Lett.* **1981**, *81*, 615.
47. Klessinger, M.; Rademacher, P. *Angew. Chem. Int. Ed. Engl.* **1979**, *18*, 826.
48. Brown, R. S.; Jørgensen, F. S. In "Electron Spectroscopy: Theory, Techniques and Applications," Vol. 5, Brundle, C. R.; Baker, A. D. Eds., Academic Press: London, 1984, pp 1–122.
49. Martin, H.-D.; Mayer, B. *Angew. Chem. Int. Ed. Engl.* **1983**, *22*, 283.
50. Gleiter, R.; Spanget-Larsen, J. *Top. Curr. Chem.* **1979**, *86*, 140.
51. Rao, C. N. R.; Basu, P. K. In "Physical Methods in Heterocyclic Chemistry," Gupta, R. R., Ed. Wiley: New York, 1984, pp 231–279.
52. For a general introduction to photoelectron spectroscopy see Rabalais, J. W. "Principles of Ultraviolet Electron Spectroscopy." Wiley: New York, 1977.
53. Riga, J.; Verbist, J. J. *J. Chem. Soc., Perkin Trans. II*, 1983, 1545.
54. Koopmans, T. *Physica* **1934**, *1*, 104.
55. Hoffmann, R. *Acc. Chem. Res.* **1971**, *4*, 1.
56. Boer, F. P.; Lipscomb, W. N. *J. Chem. Phys.* **1969**, *50*, 989. Sakai, H.; Yamabe, T.; Kato, H.; Nagata, S.; Fukui, K. *Bull. Chem. Soc. Jpn.* **1975**, *48*, 33. Guest., M. F.; Rodwell, W. R. *Mol. Phys.* **1976**, *32*, 1075.; Mollere, P. D.; Houk, K. N. *J. Am. Chem. Soc.* **1977**, *99*, 3226.
57. Sweigart, D. A.; Turner, D. W. *J. Am. Chem. Soc.* **1972**, *94*, 5599.
58. Bock, H.; Wagner, G. *Angew. Chem. Int. Ed. Engl.* **1972**, *11*, 150.
59. Wagner, G.; Bock, H. *Chem. Ber.* **1974**, *107*, 68.
60. Guimon, M.-F.; Guimon, C.; Pfister-Guillouzo, G. *Tetrahedron Lett.* **1975**, 441.
61. Guimon, M.-F.; Guimon, C.; Metras, F.; Pfister-Guillouzo, G. *J. Am. Chem. Soc.* **1976**, *98*, 2078.
62. Setzer, W. N.; Coleman, B. R.; Wilson, G. S.; Glass, R. S. *Tetrahedron* **1981**, *16*, 2743.
63. Coleman, B. R.; Glass, R. S.; Setzer, W. N.; Prabhu, U. D. G.; Wilson, G. S. *Adv. Chem. Ser.* **1982**, *201*, 417.
64. Jørgensen, F. S.; Snyder, J. P. *J. Org. Chem.* **1980**, *45*, 1015. Rindorf, G.; Jørgensen, F. S.; Snyder, J. P. *Ibid.* **1980**, *45*, 5343.
65. Bocian, D. F.; Strauss, H. L. *J. Am. Chem. Soc.* **1977**, *99*, 2866.
66. Bixon, M.; Lifson, S. *Tetrahedron* **1967**, *23*, 769.
67. Hendrickson, J. B. *J. Am. Chem. Soc.* **1967**, *89*, 7036.
68. Dale, J. *Acta Chem. Scand.* **1973**, *27*, 1115.

69. Lightner, D. A.; Docks, E. L. *Tetrahedron* **1979**, *35*, 713.
70. Kao, J.; Allinger, N. L. *Inorg. Chem.* **1977**, *16*, 35.
71. Kawada, I.; Hellner, E. *Angew. Chem. Int. Ed. Engl.* **1970**, *9*, 379.
72. Steudel, R.; Reinhardt, R.; Schuster, F. *Angew. Chem. Int. Ed. Engl.* **1977**, *16*, 715.
73. Steudel, R.; Schuster, F. *J. Mol. Struct.* **1978**, *44*, 143.
74. Setzer, W. N.; Wilson, G. S.; Glass, R. S. *Tetrahedron* **1981**, *37*, 2735.
75. Schmidt, M.; Bender, R.; Burschka, Ch. *Z. Anorg. Allg. Chem.* **1979**, *454*, 160.
76. Cook, M. J.; Ghaem-Maghami, G.; Kaberia, F.; Bergensen, K. *Org. Magn. Resn.* **1983**, *21*, 339.
77. Rys. B.; Duddeck, H. *Tetrahedron* **1985**, *41*, 889.
78. Lindner, H. J.; Kitschke, B. *Chem. Ber.* **1978**, *111*, 2047.
79. Schmidt, H.-J.; Ziegler, M. L. *Z. Naturforsch.* **1973**, *28B*, 508.
80. Moriarty, R. M.; Ishibe, N.; Kayser, M.; Ramey, K. C.; Gisler, H. J., Jr. *Tetrahedron Lett.* **1969**, 4883.
81. Nishikawa, M.; Kamiya, K.; Kobayashi, S.; Morita, K.; Tomie, T. *Chem. Pharm. Bull.* **1967**, *15*, 756.
82. Kabuss, S.; Lüttringhaus, A.; Friebolin, H.; Mecke, R. *Z. Naturforsch.* **1966**, *21B*, 320.
83. Friebolin, H.; Mecke, R.; Kabuss, S.; Lüttringhaus, A. *Tetrahedron Lett.* **1964**, 1929.
84. Kabuss, S.; Lüttringhaus, A.; Friebolin, H.; Schmid, H. G.; Mecke, R. *Tetrahedron Lett.* **1966**, 719.
85. Kojić-Prodić, B.; Ruzić-Toros, Z.; Sunjć, V.; Decorte, E.; Moimas, F. *Helv. Chim. Acta* **1984**, *67*, 916.
86. Bandoli, G.; Nicolini, M.; Tollenaere, J. P. *J. Crystallogr. Spectrosc. Res.* **1984**, *14*, 401.
87. Bandoli, G.; Nicolini, M. *J. Crystallogr. Spectrosc. Res.* **1982**, *12*, 425.
88. Koch, M. H. J.; Evrard, G. *Acta Crystallogr.* **1974**, *B30*, 2925.
89. Aschwanden, W.; Kyburz, E.; Schönholzer, F. *Helv. Chim. Acta* **1976**, *59*, 1245.
90. Jaunin, A.; Petcher, T. J.; Weber, H. P. *J. Chem. Soc., Perkin Trans. II* **1977**, 186.
91. Bandoli, G.; Nicolini, M. *J. Crystallogr. Spectrosc. Res.* **1983**, *13*, 293.
92. Guttenberger, H. G.; Bestmann, H. J.; Dickert, F. L.; Jørgensen, F. S.; Snyder, J. P. *J. Am. Chem. Soc.* **1981**, *103*, 159.
93. Rahman, R.; Safe, S.; Taylor, A. *J. Chem. Soc., Q. Rev.* **1970**, *24*, 108.
94. Morris, J. L.; Rees, C. W. *Chem. Soc. Rev.* **1986**, *15*, 1.
95. Daley, S. T. A. K.; Rees, C. W.; Williams, D. J. *J. Chem. Soc. Chem. Commun.* **1984**, 55.
96. Jones, R.; Morris, J. L.; Potts, A. W.; Rees, C. W.; Rigg, D. J.; Rzepa, H. S.; Williams, D. J. *J. Chem. Soc., Chem. Commun.* **1985**, 398.
97. Srinavasan, R.; Srikrishnan, T. *Tetrahedron* **1971**, *27*, 1009.
98. Anet, F. A. L.; Anet, R. In "Dynamic Nuclear Magnetic Resonance Spectroscopy," Jackman, L. M.; Cotton, F. A., Eds. Academic Press: New York, 1975, Chap. 14.
99. Anet, F. A. L.; Basus, V. J. *J. Am. Chem. Soc.* **1973**, *95*, 4424.
100. Meiboom, S.; Hewitt, R. C.; Luz, Z. *J. Chem. Phys.* **1977**, *66*, 4041.
101. Allinger, N. L.; Hirsch, J. A.; Miller, M. A.; Tyminski, J. J.; Van Catledge, F. A. *J. Am. Chem. Soc.* **1968**, *90*, 1199.
102. Anet, F. A. L.; Krane, J. *Tetrahedron Lett.* **1973**, 4029.
103. Dunitz, J. D. In "Perspectives in Structural Chemistry," Vol. II, Dunitz, J. D.; Ibers, J. A., Eds. Wiley: New York, 1968, pp 1–70.
104. Groth, P. *Acta Chem. Scand.* **1965**, *19*, 1497.
105. Srikrishnan, T.; Srinavasan, R.; Zand, R. *J. Cryst. Mol. Struct.* **1971**, *1*, 199.
106. Egmond, J. V.; Romers, C. *Tetrahedron* **1969**, *25*, 2693.
107. Miller, R. W.; McPhail, A. T. *J. Chem. Soc., Perkin Trans. II* **1979**, 1527.
108. Anet, F. A. L.; Degen, P. J. *J. Am. Chem. Soc.* **1972**, *94*, 1390.
109. Dale, J.; Ekeland, T.; Krane, J. *J. Am. Chem. Soc.* **1972**, *94*, 1389.
110. Dale, J.; Ekeland, T. *Acta Chem. Scand.* **1973**, *27*, 1519.
111. Warren, B. E.; Burwell, J. T. *J. Chem. Phys.* **1935**, *3*, 6.
112. Abrahams, S. C. *Acta Crystallogr.* **1955**, *8*, 661.
113. Sands, D. E. *J. Am. Chem. Soc.* **1965**, *87*, 1395.
114. Watanabe, Y. *Acta Crystallogr., Sect. B* **1974**, *B30*, 1396.
115. Coppens, P.; Yang, Y. W.; Blessing, R. H.; Cooper, W. F.; Lawsen, F. K. *J. Am. Chem. Soc.* **1977**, *99*, 760.
116. Boschi, R.; Schmidt, W. *Inorg. Nucl. Chem. Lett.* **1973**, *9*, 643.
117. Johnson, S. M.; Paul, I .C. *Tetrahedron Lett.* **1969**, 177.
118. Johnson, S. M.; Maier, C. A.; Paul, I. C. *J. Chem. Soc., Sect. B* **1970**, 1603.
119. Olmstead, M. M.; Williams, K. A.; Musker, W. K. *J. Am. Chem. Soc.* **1982**, *104*, 5567.

120. Doi, J. T.; Kessler, R. M.; de Leeuw, D. L.; Olmstead, M. M.; Musker, W. K. *J. Org. Chem.* **1983**, *48*, 3707.
121. Hill, N. L.; Hope, H. *Inorg. Chem.* **1974**, *13*, 2079.
122. Musker, W. K. University of California at Davis, private communication.
123. Hope, H.; Nichols, B. G. *Acta Crystallogr., Sect. B* **1981**, *B37*, 158.
124. Musker, W. K.; Olmstead, M. M.; Goodrow, M. H. *Acta Crystallogr., Sect. C* **1983**, C39, 887.
125. Musker, W. K.; Olmstead, M. M.; Kessler, R. M. *Inorg. Chem.* **1984**, *23*, 1764.
126. Musker, W. K.; Olmstead, M. M.; Kessler, R. M. *Inorg. Chem.* **1984**, *23*, 3266.
127. Olmstead, M. M.; Musker, W. K.; Kessler, R. M. *Acta Crystallogr., Sect. C* **1984**, *C40*, 1172.
128. Olmstead, M. M.; Musker, W. K. *Acta Crystallogr., Sect. B* **1981**, *B37*, 261.
129. Anet, F. A. L.; Ghiaci, M. *J. Org. Chem.* **1980**, *45*, 1224.
130. Frank, G. W.; Degen, P. J.; Anet, F. A. L. *J. Am. Chem. Soc.* **1972**, *94*, 4792.
131. Frank, G. W; Degen, P. J. *Acta Crystallogr., Sect. B* **1973**, *B29*, 1815.
132. Abel, E. W.; King, G. D.; Orrell, K. G.; Sik, V.; Cameron, T. S.; Jochem, K. *J. Chem. Soc., Dalton Trans.* **1984**, 2047.
133. Abel, E. W.; King, G. D.; Orrell, K. G.; Sik, V. *Polyhedron* **1983**, *2*, 1363.
134. Bhatt, M. V.; Atmaram, S.; Baliah, V. *J. Chem. Soc., Perkin Trans. II* **1976**, 1228.
135. Grandjean, D.; Leclaire, A. *Compt. Rend. Acad. Sci. Paris, Ser. C* **1967**, *265*, 795.
136. Dräger, M. *Z. Anorg. Allg. Chem.* **1985**, *527*, 169.
137. Dräger, M.; Guttmann, H.-J. *J. Organomet. Chem.* **1981**, *212*, 171.
138. Dräger, M.; Schmidt, B. M. *J. Organomet. Chem.* **1985**, *290*, 133.
139. Martin, J.; Robert, J.-B. *Acta Crystallogr., Sect. B* **1979**, *B35*, 1623.
140. Ceré, V.; Pollicino, S.; Sandri, E.; Fava, A. *J. Am. Chem. Soc.* **1978**, *100*, 1516.
141. Calderoni, C.; Ceré, V.; Pollicino, S.; Sandri, E.; Fava, A.; Guerra, M. *J. Org. Chem.* **1980**, *45*, 2641.
142. Ollis, W. D.; Stoddart, J. F.; Sutherland, J. O. *Tetrahedron* **1974**, *30*, 1903.
143. Crossley, R.; Downing, A. P.; Nógrádi, M.; Bruga de Oliveira, A.; Ollis, W. D.; Sutherland, I. O. *J. Chem. Soc., Perkin Trans. I* **1973**, 205.
144. Medard, J.-M.; Rodier, N.; Reynauld, P.; Brion, J. D.; Xuong, N. T. *Acta Crystallogr.* **1983**, *C39*, 1136.
145. Guise, G. B.; Ollis, W. D.; Peacock, J. A.; Stephanatou, J. S.; Stoddart, J. F. *J. Chem. Soc., Perkin Trans. I* **1982**, 1637.
146. Gafner, G.; Admiral, L. J. *Acta Crystallogr., Sect. B* **1969**, *B25*, 2114.
147. Goodrow, M. H.; Olmstead, M. M.; Musker, W. K. *Tetrahedron Lett.* **1982**, *23*, 3231.
148. Kopf, J.; von Deuten, K.; Nakhdjavan, B.; Klar, G. *Z. Naturforsch.* **1979**, *34B*, 48.
149. Capasso, S.; Mattia, C.; Mazzarella, L.; Puliti, R. *Acta Crystallogr., Sect. B* **1977**, *B33*, 2080.
150. Capasso, S.; Mazzarella, L.; Tancredi, T. *Biopolymers* **1979**, *18*, 1555.
151. Chandrasekharan, R. *Proc., Indian Acad. Sci., Sect. A* **1968**, *68*, 13.
152. Chandrasekharan, R.; Balasubramanian, R. *Biochim. Biophys. Acta* **1969**, *188*, 1.
153. Ernest, I.; Holick, W.; Rihs, G.; Schomburg, D.; Shoham, G.; Wenkert, D.; Woodward, R. B. *J. Am. Chem. Soc.* **1981**, *103*, 1540.
154. Wahl, G. H., Jr.; Bordner, J.; Harpp, D. N.; Gleason, J. G. *J. Chem. Soc., Chem. Commun.* **1972**, 985; *Acta Crystallogr., Sect. B* **1973**, *B29*, 2272.
155. Anet, F. A. L.; Wagner, J. J. *J. Am. Chem. Soc.* **1971**, *93*, 5266.
156. Anet, F. A. L.; Krane, J. *Isr. J. Chem.* **1980**, *20*, 72.
157. Rustad, S.; Seip, H. M. *Acta Chem. Scand.* **1975**, *A29*, 378.
158. For a definition of this nomenclature, see Ref. 68.
159. Samuel, G.; Weiss, R. *Tetrahedron Lett.* **1969**, 3529.
160. Groth, P. *Acta Chem. Scand.* **1967**, *23*, 1311.
161. Borgen, G.; Dale, J. *J. Chem. Soc., Chem. Commun.*, **1970**, 1105.
162. Borgen, G.; Dale, J.; Anet, F. A. L.; Krane, J. *J. Chem. Soc., Chem. Commun.*, **1974**, 243.
163. Bryan, R. F.; Dunitz, J. D. *Helv. Chim. Acta* **1960**, *43*, 3.
164. Dahl, S. G.; Groth, P. *Acta Chem. Scand.* **1971**, *25*, 1114.
165. Glass, R. S.; Wilson, G. S.; Setzer, W. N. *J. Am. Chem. Soc.* **1980**, *102*, 5068.
166. Weighardt, K.; Küppers, H.-J.; Weiss, J. *Inorg. Chem.* **1985**, *24*, 3067.
167. Setzer, W. N.; Ogle, C. A.; Wilson, G. S.; Glass, R. S. *Inorg. Chem.* **1983**, *22*, 266.
168. Küppers, H.-J.; Neves, A.; Pomp, C.; Ventur, D.; Wieghardt, K.; Nuber, B.; Weiss, J. *Inorg. Chem.* **1986**, *25*, 2400.
169. Ashby, M. T.; Lichtenberger, D. L. *Inorg. Chem.* **1985**, *24*, 636.
170. Wieghardt, K.; Küppers, H.-J.; Raabe, E.; Krüger, C. *Angew. Chem. Int. Ed. Engl.* **1986**, *25*, 1101.

171. Blake, A. J.; Gould, R. O.; Holder, A. J.; Hyde, T. I.; Lavery, A. J.; Odulate, M. O.; Schröder, M. *J. Chem. Soc. Chem. Commun.* **1987**, 118.
172. Hart, S. M.; Boeyens, J. C. A.; Michael, J. P.; Hancock, R. D. *J. Chem. Soc. Dalton Trans.* **1983**, 1601.
173. Boeyens, J. C. A.; Dobson, S. M.; Hancock, R. D. *Inorg. Chem.* **1985**, *24*, 3073.
174. Sato, T.; Uno, K. *J. Chem. Soc., Perkin Trans. I* **1973**, 895.
175. Harris, M. M.; Patel, P. K. *J. Chem. Soc., Perkin Trans. II* **1978**, 304.
176. Dunitz, J. D.; Venkatesan, K. *Helv. Chim. Acta* **1961**, *44*, 2033.
177. Huber-Buser, E.; Dunitz, J. D. *Helv. Chim. Acta* **1960**, *43*, 760; **1961**, *44*, 2027.
178. Dunitz, J. D.; Weber, H. P. *Helv. Chim. Acta* **1964**, *47*, 951.
179. Mladeck, M. H.; Nowacki, W. *Helv. Chim. Acta* **1964**, *47*, 1280.
180. Reinhardt, R.; Steudel, R.; Schuster, F. *Angew, Chem. Int. Ed. Engl.* **1978**, *17*, 57.
181. Setzer, W. N.; Glass, R. S. Unpublished results.
182. Goodrow, M. H.; Olmstead, M. M.; Musker, W. K. *Phosphorus Sulfur* **1983**, *16*, 299.
183. Valle, G.; Piazzesi, A.; Del Pra, A. *Cryst. Struct. Commun.* **1972**, *1*, 289.
184. Valle, G.; Piazzesi, A.; Busetti, V. *Cryst. Struct. Commun.* **1972**, *1*, 293.
185. Valle, G.; Zanotti, G. *Cryst. Struct. Commun.* **1977**, *6*, 651.
186. Feigenbaum, A.; Lehn, J.-M. *Bull. Soc. Chim. Fr.* **1969**, 3724; **1973**, 198.
187. Ceré, V.; Dalcanale, E.; Paolucci, C.; Pollicino, S.; Sandri, E.; Lunazzi, L.; Fava, A. *J. Org. Chem.* **1982**, *47*, 3540.
188. Keehn, P. M. In "Cyclophanes," Keehn, P. M.; Rosenfeld, S. M., Eds. Academic Press: New York, 1983, Chap. 3. Mitchell, R. H. *Ibid.*, Chap. 4.
189. Mitchell, R. H.; Boeckelheide, V. *J. Heterocycl. Chem.* **1969**, *6*, 981.
190. For a review of related studies that illustrate the smaller steric size of a lone pair of electrons on a pyridine nitrogen as compared with the benzene C—H moiety see Förster, H.; Vögtle, F. *Angew. Chem. Int. Ed. Engl.* **1977**, *16*, 429.
191. Kiryu, S.; Nowacki, W. *Z. Kristallogr.* **1975**, *142*, 99.
192. Kiryu, S.; Nowacki, W. *Z. Kristallogr.* **1975**, *142*, 108.
193. Vögtle, F.; Schafer, R.; Schunder, L.; Neumann, P. *Justus Liebig's Ann. Chem.* **1970**, *734*, 102.
194. Bottino, F.; Foti, S.; Pappalardo, S.; Finocchiaro, P.; Ferrugia, M. *J. Chem. Soc. Perkin Trans. I* **1979**, *198*.
195. Anker, W.; Bushnell, G. W.; Mitchell, R. H. *Can. J. Chem.* **1979**, *57*, 3080.
196. Pahor, N. B.; Calligaris, M.; Randaccio, L. *J. Chem. Soc., Perkin Trans. II* **1978**, 42.
197. Kamenar, B.; Prout, C. K. *J. Chem. Soc.* **1965**, 4838.
198. Winberg, H. E.; Fawcett, F. S.; Mochel, W. E.; Theobald, C. W. *J. Am. Chem. Soc.* **1960**, *82*, 1428. Fletcher, J. R.; Sutherland, I. O. *J. Chem. Soc. Chem. Commun.* **1969**, 1504.

Phosphorus Heterocycles

Louis D. Quin
UNIVERSITY OF MASSACHUSETTS,
AMHERST, MA

5.1. SOME GENERAL CONSIDERATIONS

In 1977, a valuable review of the stereochemistry of heterocyclic phosphorus compounds was published by Gallagher.[1] Of the 443 references cited in this review, only five pertained to ring systems with seven or more members, and in these references only six different ring systems were described. During the following decade, enormous progress was made throughout the field of heterocyclic phosphorus chemistry, and in the literature survey conducted for the present review, some 60 references were found for conformational studies on the medium heterocycles alone, in which 39 different ring systems are described. Nevertheless, even with this heightened activity, many fundamental ring systems have been completely ignored, and severe gaps remain in our understanding of conformations of the medium rings. Many opportunities for research can be found in this area; there is, for example, not a single reference to conformational preferences expressed by either the seven- or eight-membered rings containing only carbon atoms with a single phosphorus atom. Much of the conformational work has been conducted with rings containing the O—P—O unit, frequently with fused benzo groups, since these are generally easier to synthesize.

A summary of the rings that have been studied from the conformational standpoint is given in Table 5-1. In the sections that follow, the available information on these rings will be presented. As with other heterocyclic families, it is convenient to refer to the carbocyclic rings for the basic shapes that can be expected for these systems.[2] In many cases, it will be seen that the resemblance is strong, even though the bonds associated with phosphorus are considerably

TABLE 5-1. Medium-Ring Phosphorus Heterocycles Subjected to Conformational Analysis

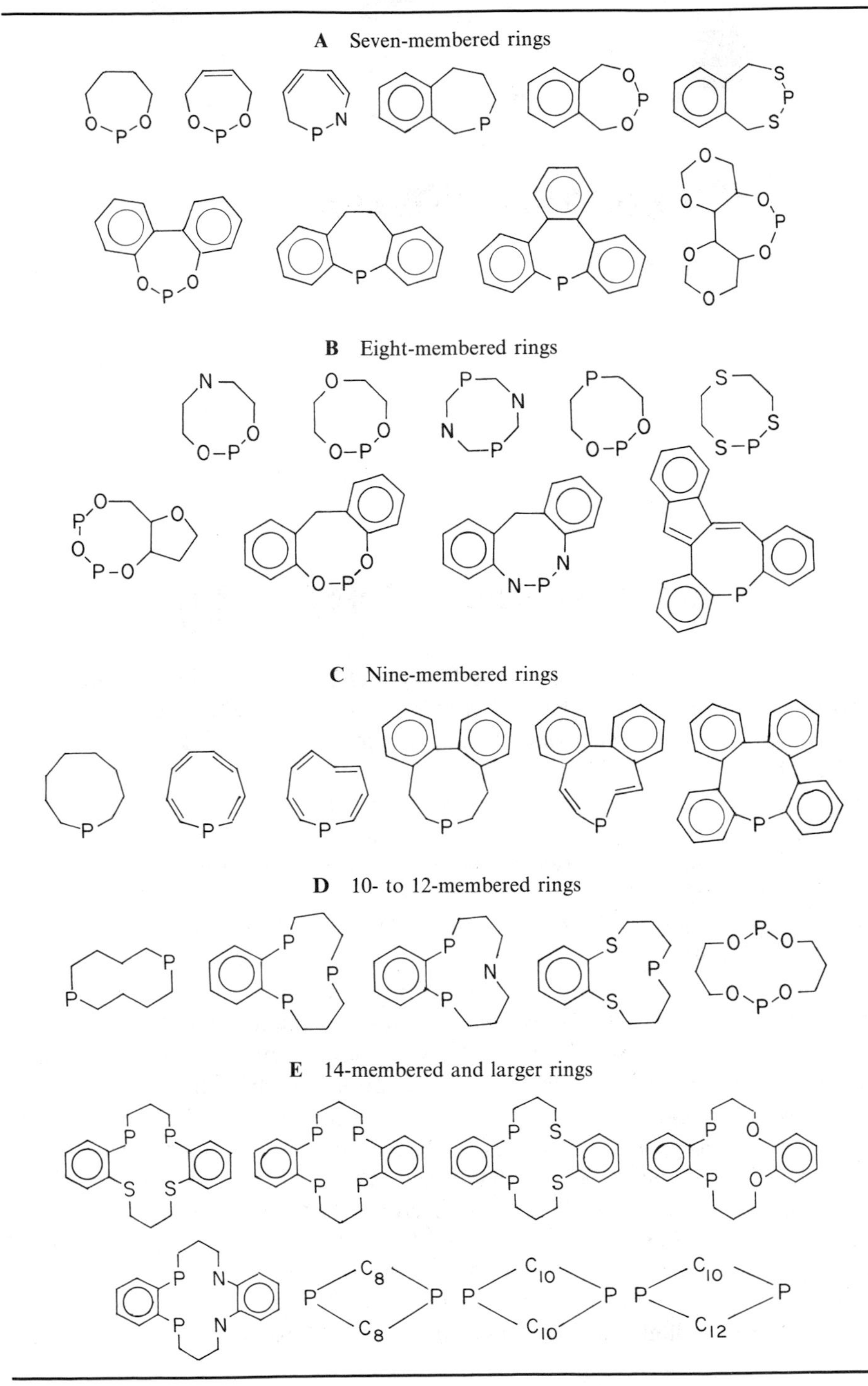

longer[3] (e.g., C—P, 1.8 Å; compared with C—C, 1.54 Å) and the angles at phosphorus can vary from 100° to 110°, depending on the coordination state [with P(III) compounds at the lower end and phosphoryl derivatives at the higher]. These differences necessarily introduce some distortions in the ring, but the basic shape remains discernible. Thus, it will be seen that in the solid state the familiar chair (twisted) characterizes the seven-membered ring, the crown the eight-membered ring, and the twist–chair–chair the nine-membered ring.

When the common ring shapes are adopted, several invertomeric (from ring pseudorotation) forms are possible, as a result of the preference of the phosphorus atom, or other heteroatoms that may be present, for particular locations. An excellent example is provided by the 1,3,2-dioxaphosphepane system; two different chair conformations are shown here, and evidence will be presented in section 5.2 that both forms (twisted) may be found among various substituted compounds.

Phosphorus chemistry is made very rich by the large number of functional groups that can be developed around this atom. This fact complicates the conformational picture, however, as quite different preferences for substituents on the ring can be found for a given ring system where variation in the coordination state can be effected. This is an especially well-known effect among the six-membered phosphorinanes; for example, in phosphines, the P-substituent can be largely axial on a chair conformation, but in the corresponding oxide, the same substituent is largely equatorial. This same possibility can exist in the larger rings, and examples will be discussed. In some cases, the substituents can dictate the shape adopted by the ring; thus, *cis,trans* isomers can have different conformations to alleviate the nonbonded interactions.

Many phosphorus compounds have been studied by X-ray diffraction of single crystals, and no special problems are encountered on using this most powerful structure-determination technique. After the establishment of the exact conformation adopted in the solid state, the usual problem remains of determining whether this same conformation fully characterizes the substance when in solution, or if it is the major contributor to a conformational equilibrium. It is generally the case that the solid-state conformation does indeed prevail in solution,[4] but by no means can this be assumed about a compound without an appropriate experimental verification. Indeed, instances will be mentioned where quite different conformations were adopted in the solid versus the solution. Several techniques are available for the study of solution conformations, some that are common to any heterocyclic family [e.g., the dihedral angle control of vicinal (three-bond) H—H coupling], but there are some that are very specially associated with the magnetic characteristics of the ^{31}P nucleus. This nucleus has a spin quantum number of $\frac{1}{2}$, and because it is the only naturally occurring isotope, it splits the signals of common nuclei like ^{1}H and ^{13}C to doublets. The magnitude of the coupling constants can be controlled by steric relations of the coupled nuclei in very special and extremely useful ways. Thus, geminal (two-bond) coupling, while of limited value in proton–proton coupling, can be extremely informative in both the ^{1}H and ^{13}C spectra of phosphorus compounds based on three-coordinate phosphorus.

The empirical relation is that two-bond coupling to both ^{1}H and ^{13}C is large when the lone pair on ^{31}P is oriented close to these nuclei, and can be small, even zero, when remote. These effects are found in many structural types, and are adequately discussed elsewhere.[5,6] The effect has not been employed widely in the case of medium-ring systems, but the potential for its use is there. For P(IV) compounds, a valuable relation like that of Karplus for three-bond H—H coupling exists. For coupling of ^{31}P to both ^{1}H[5] and ^{13}C[6], maximal values are achieved in structures where the dihedral angles relating the coupled nuclei are 0° or 180°, with small or negligible coupling at 90°. These relations must be used with caution when applied to the medium rings, however, since plots developed mostly from smaller rings do not necessarily hold for these rings, as will be discussed in later sections. Three-bond coupling to ^{13}C in P(III) compounds also is under a dihedral-angle control,[7] but there can be considerable distortion of the plot from the fairly regular plot so familiar from H—H coupling. Furthermore, lone-pair orientation effects can be present here also, and if the lone pair is remote from the coupled H or P, there may be negligible coupling regardless of the dihedral-angle relation of the coupled nuclei.

In a review that is devoted to conformational matters, it is not possible to provide details on the synthetic methods used to prepare the compounds being studied. These methods are usually provided in the original literature cited. Some review literature is available that is helpful in introducing and organizing the information.[8–10] Reviews are also useful in providing background on conformational properties of the smaller ring systems,[1,9–11] which receive considerable comment in the present review.

5.2. Rings with Seven Members (Phosphepanes)

Only one seven-membered ring system, 1,3,2-dioxaphosphepane, has received broad conformational consideration, and this no doubt because of relatively simple access from the reaction of phosphorus halides and dihydroxy compounds. An important contribution was made early in this work by Coulter,[12] who performed an X-ray diffraction analysis of the four-coordinate acid **1**. The ring was basically in the chair shape, with some twisting from the ideal. The phosphorus and one oxygen were located below the plane established by O-3, C-4, C-6, and C-7, with C-5 above the plane (as in **1a**). The twisting occurs between the O—C bond and the C—C bond of the plane. The twisting may be assumed to arise from relief of the eclipsing interaction between the protons on C-6 and C-7 of the plane.

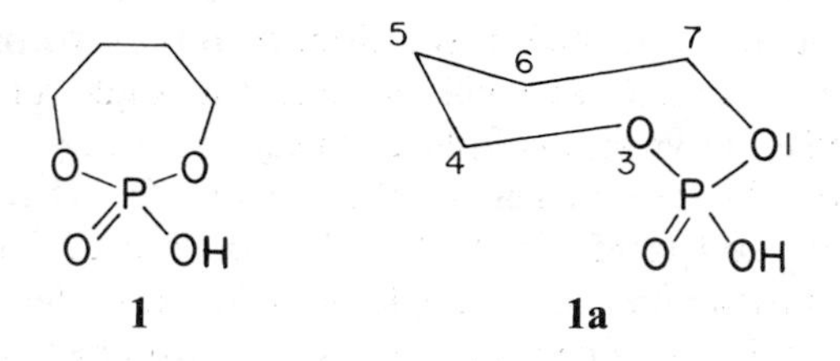

Some important structural data from the analysis are summarized in Table 5-2. By adopting this ring shape, the molecule easily accommodates the preferred tetrahedral angle of phosphate phosphorus. The twist chair is also known as a low-energy conformation of cycloheptane, and in this sense the result for **1** is not surprising. However, there are several invertomeric forms possible that would arise from placement of the heteroatoms at different orientations of the twist chair, and the unsymmetrical invertomeric form actually adopted might not have been easily predicted. No data are available on the fate of this conformation when **1** is placed in solution. However, solution studies have been performed on several P-substituted derivatives of **1**. As pointed out by Guimaraes

TABLE 5-2. SOME STRUCTURAL DATA FOR TETRAMETHYLENE PHOSPHORIC ACID[a]

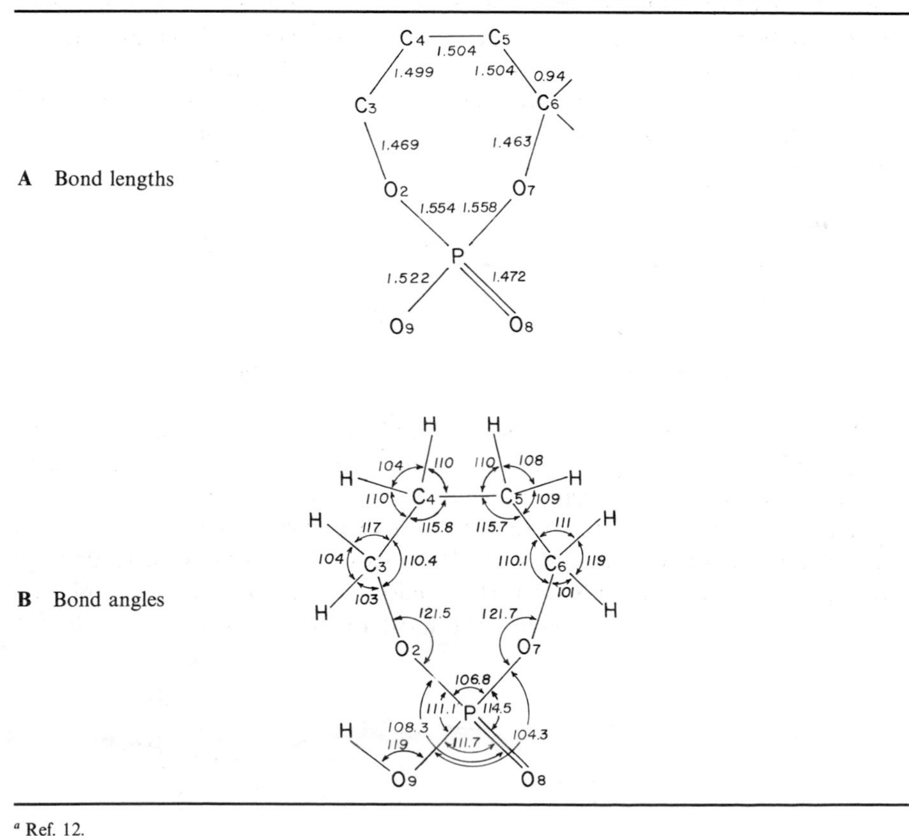

[a] Ref. 12.

et al.,[13] if this structure persisted in solution, then the two protons on C-4, as well as those on C-7, would have different dihedral-angle relations to ^{31}P and, therefore, have different vicinal coupling constants. Using the data of Coulter, they calculated the dihedral angles for the P—O—C—H units in twist chair **1a** to be about 31° and 150.5° for C-4, and 31.6° and 151° for C-7. Were the ring frozen in this twist chair, these angles should cause the $^{3}J_{POCH}$ values at a CH_2 to be quite different. From information on another ring system that is strongly biased, to be discussed, such angles should result in values very roughly around 10 and 25 Hz respectively. In fact, several derivatives having structure **2** (R=Cl, OC_6H_5, C_6H_5, CH_3, NMe_2) have very similar values for the two couplings.

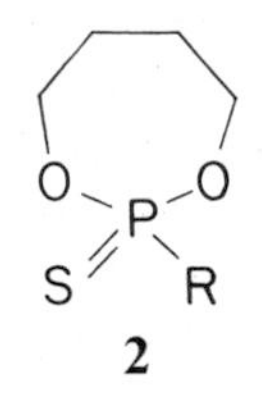

2

A reasonable explanation is that there is a rapid equilibration between mirror-image twist chairs with the same invertomer structure that results in an exchange of the two protons on C-4 and C-7. Indeed, the experimental value can be related quite precisely to the averaged value for the two $^{3}J_{POCH}$ when the exchange occurs between dihedral angles of 30° and 150°. For example, for **2**, R=Cl, the experimental value for $^{3}J_{POCH}$ as well as the calculated value is 21.5 Hz; the equilibrium between **2a** and **2b** seems indicated.

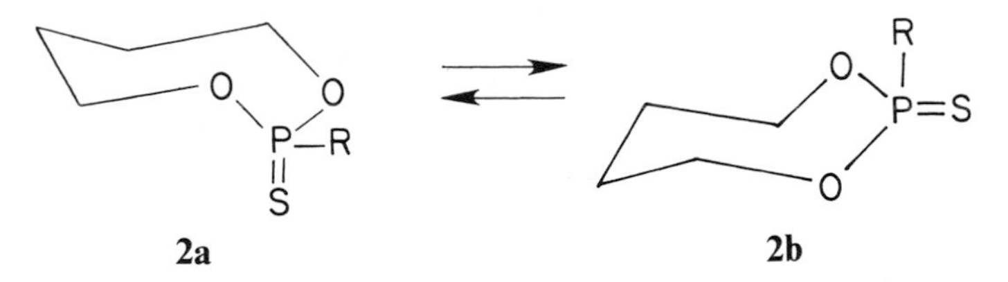

2a **2b**

However, other explanations can be found for the observed identity of the $^{3}J_{POCH}$ values. If the ring were in the symmetrical chair conformation, then a rapid equilibrium of two equally populated conformations would give the same coupling constant for a CH_2 group. But this explanation ignores the fact that nonbonded interactions involving the P-substituent (R) should disfavor a 1:1 equilibrium mixture (and also that this conformation would have eclipsed hydrogens on C-5 and C-6). Support for the twist-chair of **2** comes from the ^{13}C

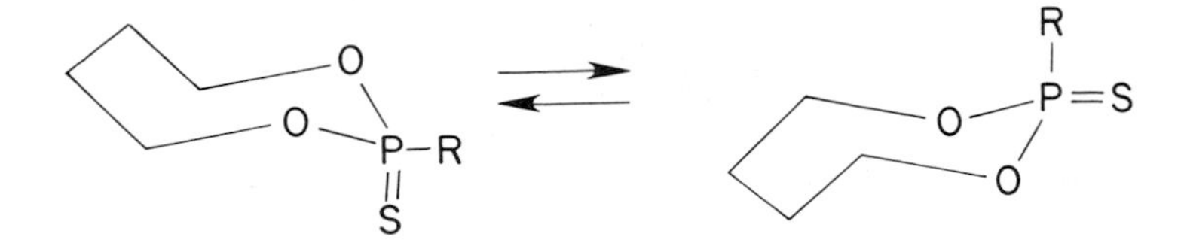

NMR data. In the structure of Coulter (**1a**), the dihedral angle relating ^{31}P to C-5 (and through the equilibrium to C-6) is about 90°; this leads to an expected value for $^3J_{POCC}$ near zero, and indeed no splitting of the C-5 or C-6 signal is observed. This does not necessarily eliminate other forms from consideration, however, since this dihedral angle may not be unique to the twist–chair. Another technique was also employed in this study to support the twist–chair structure; proton relaxation times (T_1) were in good agreement with those calculated for the twist–chair of Coulter, and not those calculated for the symmetrical chair. To complicate the picture, a solvent influence on the $^3J_{POCH}$ values was noted, possibly suggesting that other minor conformations were present in the solution.

A similar situation probably holds for compounds of structure **3**, where oxygen replaces sulfur at phosphorus in the same ring system.[14] Thus, for **3a**, nearly identical ^3J values (16.8, 16.2 Hz) were observed, as was also true of **3b** (16.5, 16.8 Hz).

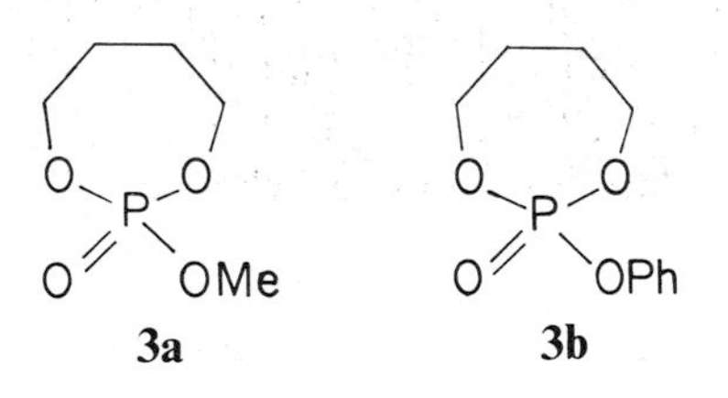

3a **3b**

A very different experimental approach was used in the study of phosphoryl derivative **4**.[15] The IR and Raman spectra of **4** were obtained for the crystal, the

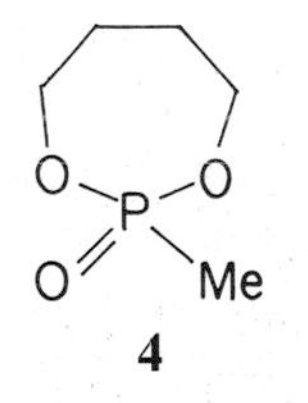

4

melt, and solutions (CS_2 or CH_3CN); the important observations were made that, relative to the crystal, no new bands appeared or intensity ratio changes occurred, in the melt or in the solutions. This is strong evidence that the conformation of the ring remains the same in all media, either as a single invertomer or as an average from equilibration between mirror antipodes. Other features of the spectra pointed to an unsymmetrical invertomeric form of the twist chair, which is, of course, the same proposal made to interpret the proton NMR spectra, and was further supported by the observation of a stretching frequency for P=O that is between the values expected for a purely axial or purely equatorial disposition. It is informative to make a comparison here with the six-membered (1,3,2-dioxaphosphorinane) analogue; this system exhibits strikingly different behavior, in that the vibration spectra showed marked differences as the media were changed. This implies an equilibrium to be present that involves chair conformations containing axial and equatorial P=O in a ratio controlled by the medium.

Removal of the phosphoryl oxygen from **3b** (forming phosphite **5**) causes a change in the conformational characteristics of the seven-membered ring.[16] The protons on C-4 (or C-7) of **5** no longer have identical $^3J_{PH}$ values; instead, values of 13.5 and 9.5 Hz were observed at 20°. Furthermore, the coupling proved to be temperature sensitive, as seen in the following table.

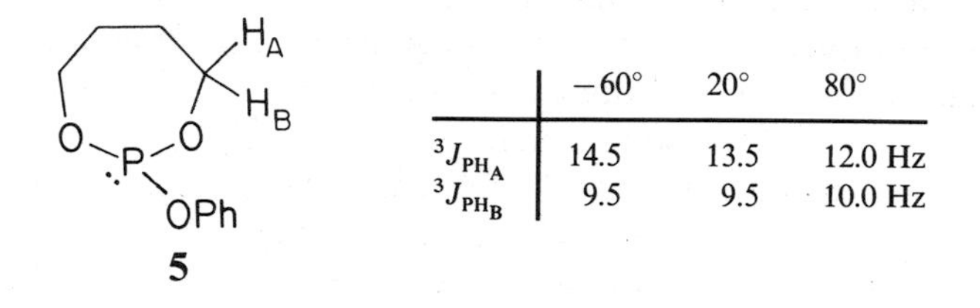

	−60°	20°	80°
$^3J_{PH_A}$	14.5	13.5	12.0 Hz
$^3J_{PH_B}$	9.5	9.5	10.0 Hz

These data indicate that the system no longer can be described by an equilibrium between two forms of equal energy. That the difference in $^3J_{PH}$ increases as the temperature is lowered indicates that the equilibrium shifts to favor a form with quite different dihedral angles for the CH_2 group, as would be found in a symmetrical conformation such as **5a** (where eclipsing of protons on C-6,7 is introduced).

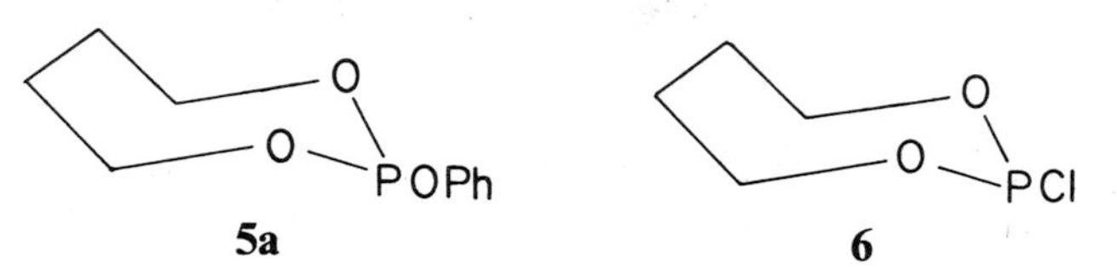

5a **6**

The P(III) molecule **6** was studied by IR and Raman spectroscopy,[17,18] which gave spectra having more frequencies than would be expected for a single conformation. Furthermore, these spectra were temperature dependent, indicating the presence of a mixture of different conformers. At 25°, two conformations seemed to be nearly equally populated. One of these is suggested to have a twisted ring as in **6a**, which by calculation is found to be the more favorable. As the temperature is lowered, the intensity of the other form diminishes. This less stable form is the more symmetrical one, as indicated in the Raman spectrum by the presence of several depolarized vibrations, and is suggested to be closer to the symmetrical chair structure with axial chlorine (**6b**). Other conformations related to this may also be present.

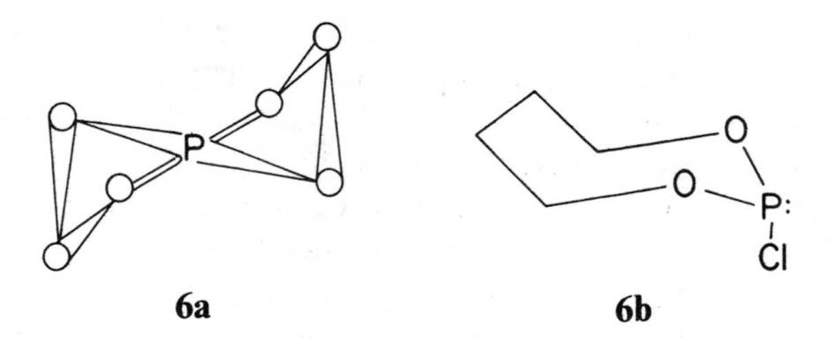

6a **6b**

The introduction of methyl substituents on this ring at the 4,7-positions creates isomers,[17] and if both methyls are to be placed in the less crowded equatorial positions, then conformational adjustments must ensue. Structure **7** is that for the *trans* isomer, and the Raman spectrum can be interpreted on this basis. The spectrum of the *cis* isomer is quite different, and suggests less conformational homogeneity. The structure may be more like the symmetrical chair.

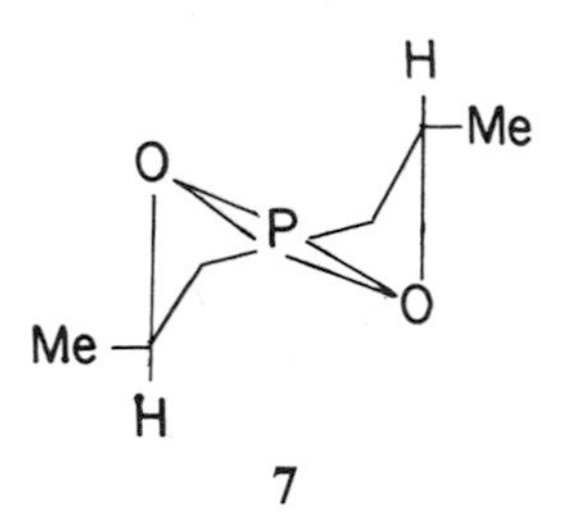

Installation of a double bond at C-5,6 imparts planarity to that region of the molecule, and simplifies the conformational problem. As for cycloheptene, the following three conformations are suggested to be the most energetically favorable.

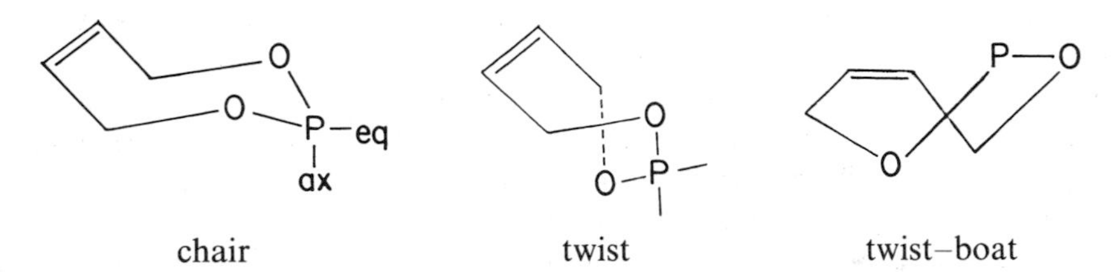

There seem to be no X-ray studies on crystals of such substances, but proton NMR studies on **8a–e** have proved helpful in establishing the conformational tendencies in solution.[13] The methylene protons have quite different chemical shifts and $^3J_{PH}$ values (e.g., **8b** in CCl_4, δ 4.86, $J = 10.5$ Hz; δ 4.43, $J = 27.0$). This has been interpreted on the basis of an equilibrium highly biased toward a chair conformation (**8f**). There is some dependence of these parameters on solvent polarity, perhaps due to a stabilization of one of the participants in the equilibrium.

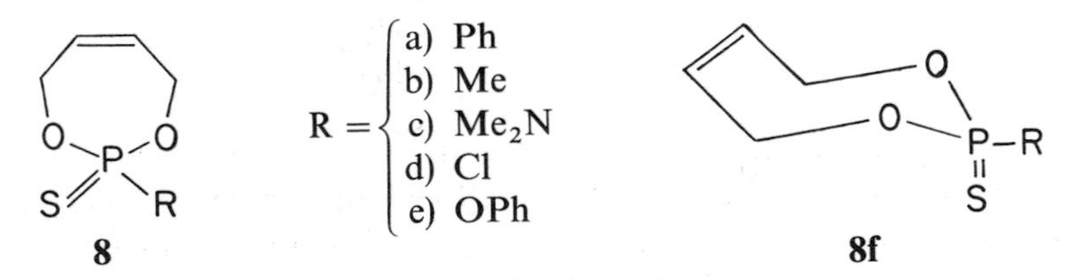

For **8d,e**, however, only one CH_2 signal is observed (e.g., **8a**, δ 4.30, $J =$ 21.3 Hz), which could result from the preference for a twist conformation where the protons are equivalent, or a 1:1 chair–chair equilibrium, or even accidental equality of chemical shifts that would give an observed J value that is the mean of the two couplings. Based on the definite preferences expressed in the phosphorinanes, it is likely that CH_3, C_6H_5, and $N(CH_3)_2$ will prefer the equatorial position, and Cl and OC_6H_5 will prefer the axial. With compound **9**, a somewhat different result has been obtained.[19] A conformational equilibrium preference seems to be expressed from the fact that the protons become more similar in shift and coupling (δ 5.12, $J = 15.2$ Hz; δ 4.99, $J = 18.7$ Hz).

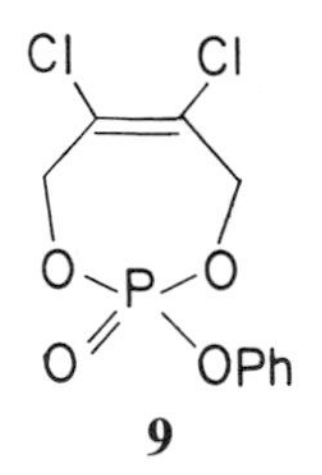

The data were interpreted to imply that the equilibrium existed between twist–boats that differed in axial or equatorial disposition of the P-substituent. In the twist–boat, the dihedral angles are equivalent for the two methylene protons, and ring inversion causes equivalence of the two methylenes, so that no large difference in $^3J_{PH}$ is expected. This explanation was supported by dipole moment and Kerr effect studies, and is preferred over an explanation based on a chair–chair (1:1) equilibrium. However, the problem of the axial–equatorial preference of the P-substituent is not addressed. No information bears on the origin of the conformational difference from **8a–e**, and it is not known whether the chlorine atoms or the oxygen on phosphorus are responsible.

In the absence of the fourth atom on phosphorus, as in phosphite **10**, a large difference in the methylene protons is observed (e.g., δ 4.66, $J = 13.0$ Hz; δ 4.18, $J = 7.0$), and these values were found to remain constant over a wide temperature range (-60 to $+80°$).[16]

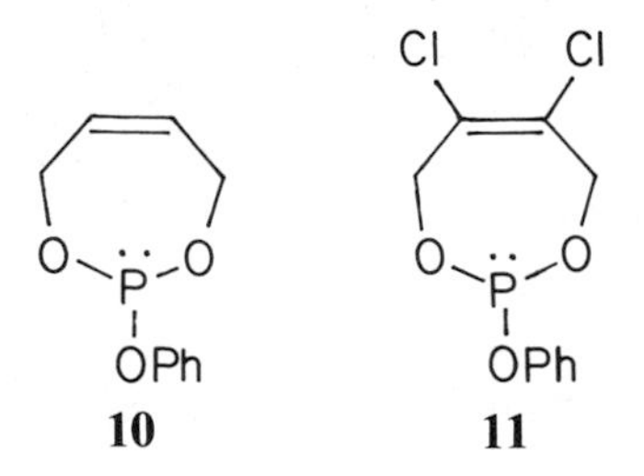

A similar observation was made[19] for the 5,6-dichloro derivative **11**; on varying the temperature, no change in the IR or Raman spectra or in the proton NMR characteristics of this compound occurred. A rigid chair (**10a**) or boat seems indicated, since, as noted, a twist form would give $^3J_{PH}$ values that were more alike. The larger J can be assigned to the equatorial H, and the smaller to the axial. Following the pattern set in six-membered rings, it was suggested[16] that the oxygen atom is axial. The ^{13}C NMR data reported in this paper[16] can be used to support this assignment; the two-bond coupling of ^{31}P to CH_2 is

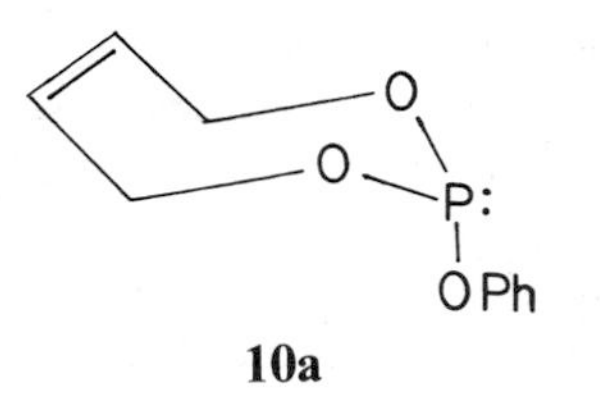

zero, which implies a remote (equatorial) orientation of the lone pair on phosphorus to this carbon. In phosphorinanes, derivatives with axial substituents have small ^2J values, while equatorial substitution leads to quite sizable values.[6]

A similar assignment,[20] again based on the stereospecificity of the three-bond coupling to protons, was reached for **12a** and **12b**. Here an equilibrium is present where the chair form (with equatorial NR_2) dominates over a twist by about 3:1.

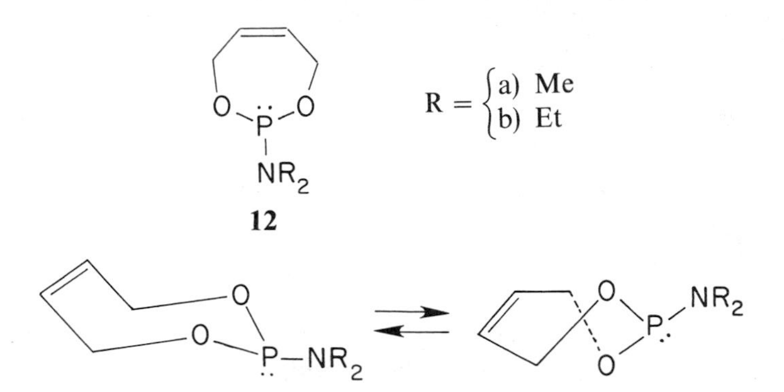

Dipole moment data[21] are in support of this proposal, since the experimental value (e.g., 1.87*D* for **12a**) falls in between the calculated values for the chair (2.14*D* if equatorial, 2.42 if axial) and the twist conformation (1.05). On the other hand, IR–Raman studies of **12a,b** failed to give any indication of a second conformation; the recorded number of strong lines and bands did not exceed that theoretically expected for one form. Also, there was no appreciable change in these spectra when the temperature was changed.

In the benzo series, X-ray analysis provides a starting point for conformational considerations. For compounds **13a,b**, a chair conformation (like **10a**) is present with axial P=S and equatorial P—R.[22]

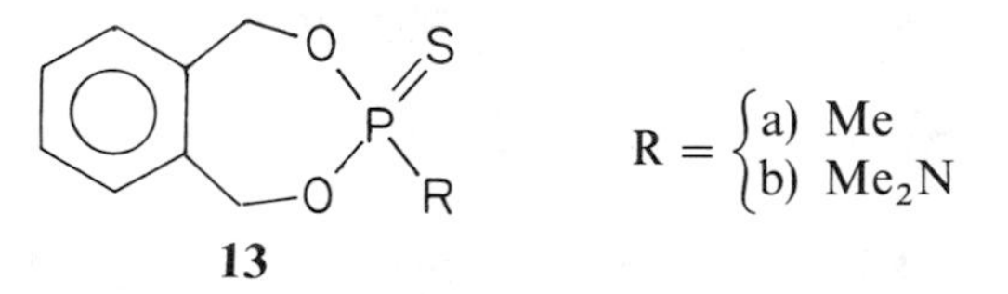

Were this conformation to prevail in solution, one would again expect large differences in proton NMR shifts, as well as $^3J_{PH}$, and indeed these effects are observed[13] (e.g., **13a**, δ 5.92, $J = 12.3$, and δ 4.58, $J = 25.9$ Hz). Small changes in these values with solvent suggest that an equilibrium may be present, and this has also been proposed to explain the observation[14] that $Eu(fod)_3$ can induce a considerable change in the ^3J values of some compounds. This is attributed to a shift in the equilibrium to favor the conformation with axial P=O (expressed by the authors as resulting from the usual equatorial preferences of P=O being diminished by complexation to Eu, through modification of the nature of P=O bonding). Compounds **14a,b** were employed in this study.

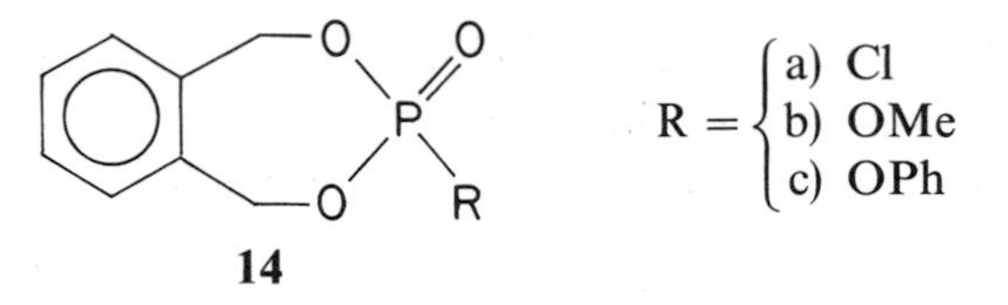

Compound **14c** showed no temperature dependence of $^3J_{PH}$ and exists in the chair form with P=O thought to be equatorial from $\nu_{P=O} = 1320\ \text{cm}^{-1}$, a value typical for this structural feature. The Eu complexation then shifts the equilibrium to the axial P=O side.

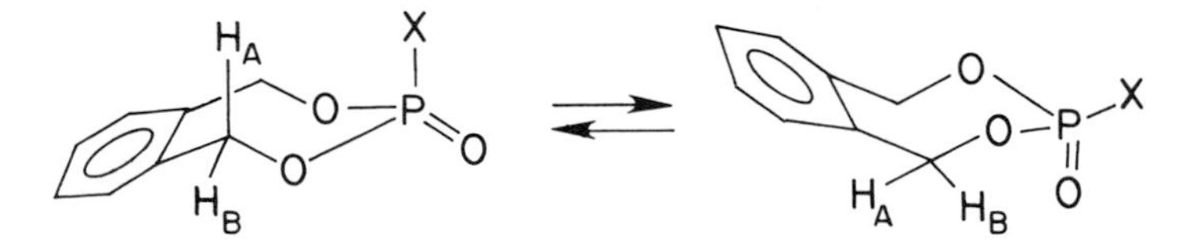

This is indicated, for example, by a greater downfield shift for the axial proton over the equatorial, which would result from proximity of Eu to the axial H by complexation with axial P=O. For **14b**, a mixture of chair forms with axial and equatorial OMe appears to be present (from the similar $^3J_{PH}$ values), and this equilibrium is shifted by Eu to favor the axial P=O conformer.

IR–Raman studies of **13b** [and other P(IV) compounds] also show that the spectra can be interpreted by the presence of a conformational equilibrium strongly dominated by one form, presumed to be the chair with equatorial NR_2.[21] A small amount (up to 10%) of twist form may also be present in solution.

For the P(III) system,[16] proton NMR data are available to indicate that the CH_2 protons have very similar environments. Thus, for **15**, the $^3J_{PH}$ values differ by only about 1 Hz, and this excludes a rigid chair or boat.

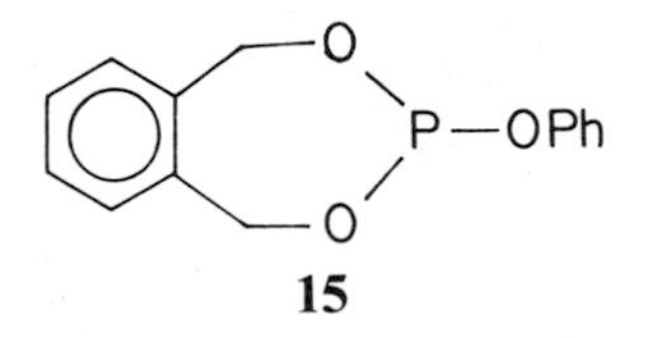

A 1:1 chair–chair equilibrium ignores the axial preference of OC_6H_5, which would modify this ratio. An increase in the coupling difference occurs on lowering the temperature, indicating that a conformational equilibrium is present. At the temperature of $-60°$, the values are 11.5 and 8.5 Hz, while at 20°, they are 10.5 and 9.5 Hz. The authors propose an equilibrium between the chair (the more stable) and the twist, probably with axial OC_6H_5. Support for the axial dominance comes from the stereospecificity of $^2J_{PC}$; the remote lone pair is associated with the small value of about 2 Hz.

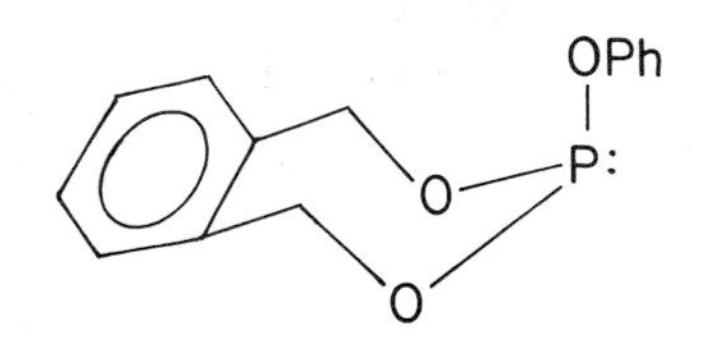

For compound **16**, R=Me, it was convincingly demonstrated[23] by freezing out the two forms that an equilibrium between two conformers was in effect.

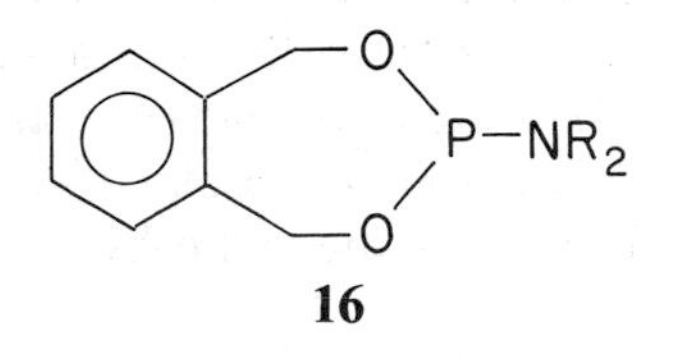

16

At 149°K, two sets of signals for both the NMe_2 and CH_2 groups were present, but these vanished as the temperature was raised.

	δMe	δH_A	δH_B	$^3J_{PH_A}$	$^3J_{PH_B}$
288°K	2.60	4.98	4.81	7.3	18.3
149°K	2.69	5.12	4.49	1.5 ± 1.5	24.1
	2.35	4.98	4.72	10.3	15.5 ± 1.5

That conformer giving the most downfield shift (δ 5.12) for a CH_2 proton has greatly different $^3J_{PH}$ values, consistent with a chair conformation. As for phosphorinanes, the P–NMe_2 group is assumed to be equatorial. The other frozen-out conformer has quite similar coupling constants, which allows assignment of the twist structure. The equilibrium constant for chair ⇄ twist is 1.8 ± 0.2 (at 149°K). The IR–Raman spectra of this compound, as well as for the NEt_2 derivative, show the extra lines expected for an equilibrating system, and they also display a temperature dependence.[21] In good agreement with the NMR result, peak intensity measurements gave a conformer ratio of about 1:1 for the same chair and twist forms.

Dibenzo-dioxaphosphepanes have also been studied, but here there are no convenient proton signals to employ in the conformational analysis. Using dipole moments, Arbuzov et al.[24] were led to suggest that the molecule **17** adopted a twisted conformation, with the angle of twist between the rings of the diphenyl moiety being 41–48°. The twisting probably results from the repulsions between ortho hydrogens, as in biphenyl itself, which has an angle of 41.6°. With this twisted conformation (**17a**), a good match of experimental and calculated values was obtained.

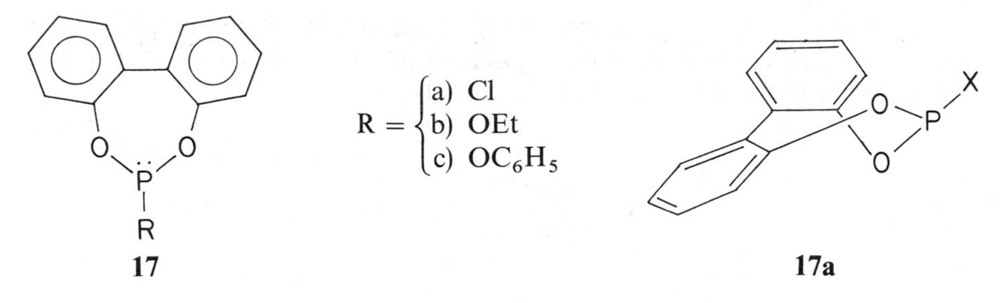

A substituted derivative of this ring system (**18**) was examined by both proton and carbon NMR,[25] but no information about the conformation of the ring was obtained. The ortho-t-butyl groups gave identical signals, as did those ring carbons bearing H. This is consistent with a rapid flipping of the ring between two equivalent twisted conformations, perhaps like structure **17a**. A symmetrical conformation with both benzo groups in a plane would give the same spectral data, but seems unlikely because of the ortho-hydrogen repulsions.

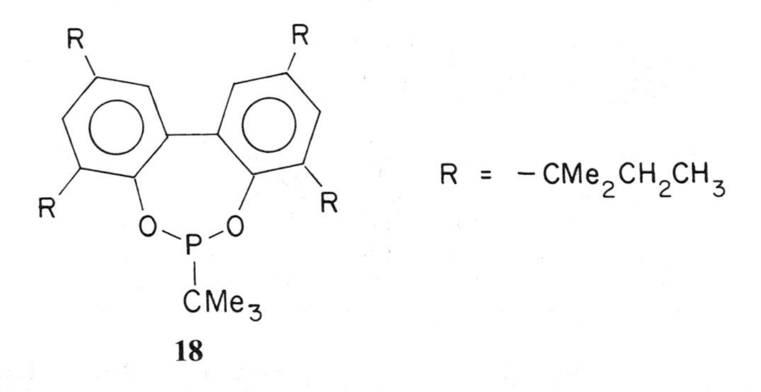

The effect of replacing both oxygens of the 4,5-benzo ring system by sulfur, so as to give **19**, has also been studied.[26] The X-ray analysis of one such derivative (R = *p*-chlorophenyl) revealed that the ring is in the shape of a symmetrical, slightly flattened chair, with OR axial and P=O equatorial. The NMR signals for the CH_2 groups were reported only as multiplets, and thus provided no information on the solution conformation.

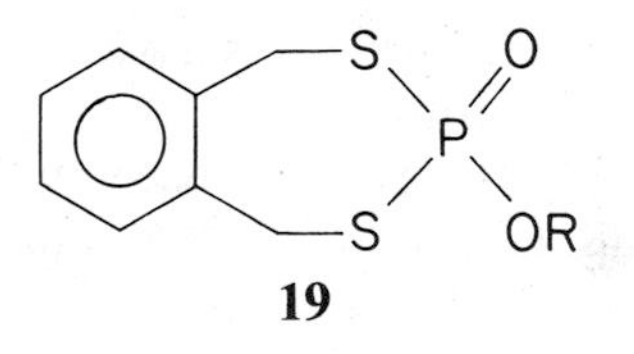

Aza derivatives of the phosphepane ring system are known,[27] but it appears that the only conformational consideration accorded such compounds has been the determination by X-ray analysis of the structure of the unsaturated derivative **20**. The ring adopts a nonplanar, unsymmetrical shape to accommodate the planarity at the four sp^2 carbons and at nitrogen.

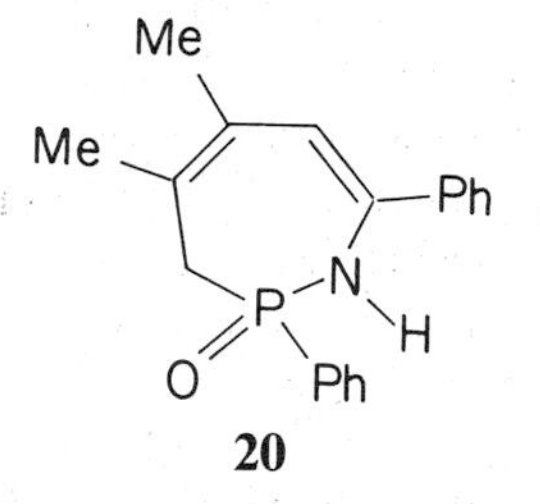

20

As mentioned in section 4.1, no conformational studies have been performed on the parent ring containing only carbon and a single phosphorus atom. However, the shapes of mono-, di-, and tribenzo derivatives have been determined by X-ray analysis.

The monobenzo derivative **21** was found[28] to have a twist–chair conformation.

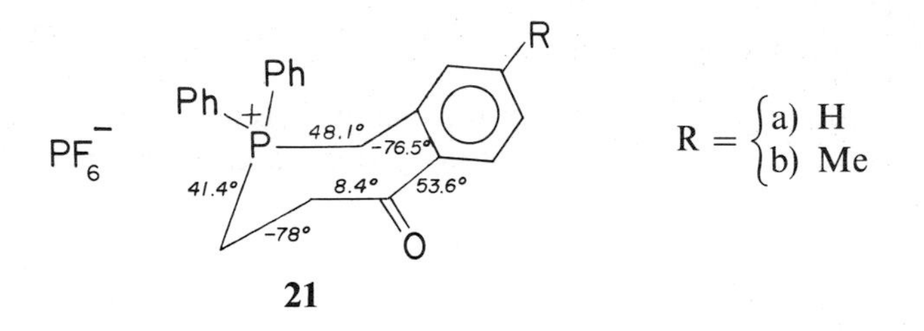

21

The internal torsion angles in the ring help define the shape; these are noted on the structure at the appropriate bond. The proton NMR spectrum of this molecule shows only slight line broadening at the lower end of the wide range +150° to −110°, indicating that the ring has little flexibility and is not involved in conformational equilibration. The proton NMR spectrum was not adequately resolved to allow a full study of the usually valuable P—H coupling effects. The X-ray analysis gave a C-3, C-4 torsion angle of −78°; were it retained in solution, such an angle would allow very little coupling of P to ring carbon 5, and indeed this signal was reported to have δ 200.04 with unresolved coupling. The ring shape allows P to have nearly tetrahedral bond angles.

The dibenzo derivative **22** adopts a "butterfly" shape (**22a**), with benzo groups inclined at 56.8° to each other.[29] Carbons 4,5 are almost coplanar with the ring to which they are attached, but are staggered and have the attached carbons (3 and 6) at a torsion angle of 67.4°. The P nucleus is tetrahedral (107.2°); the

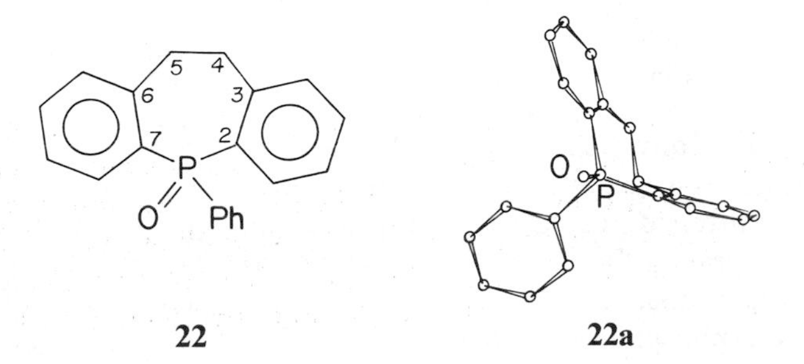

22 **22a**

oxygen substituent is nearly coplanar with C-2 and C-7, and the phenyl group is oriented in a pseudoaxial position.

A similar conclusion about the conformation of **22** was developed independently from its proton NMR spectrum.[30] The single observation that the ortho-hydrogens were strongly deshielded (*m*, δ 8.18–8.52, $^3J_{PH} = 12$ Hz) was attributed to their being in close proximity to P=O, and this was confirmed by the strong effect of Eu shift reagent at this position. To achieve this H—O proximity, the same ring shape that was defined by the X-ray analysis was proposed. In some derivatives (R = Cl, OMe), the two methylene protons were found not to be interconverting by rapid inversion of the ring, and this resulted in an AA′BB′ spectrum. In other derivatives (R = C_6H_5), a singlet for the four protons was obtained, meaning that rapid equilibration was occurring. From the interpretation of the AA′BB′ spectrum, the values $^3J_{1,2} = 6.0$, $^3J_{1,4} = 7.9$, $^3J_{1,3} = -15.8$ Hz were found. These data led to a calculated dihedral angle in the CH_2CH_2 unit, and thence to an angle for the carbons of the attached rings of 47°, in reasonable agreement with the X-ray result of 56.8°. When a double bond is placed in the bridge, the butterfly shape is retained but the central ring adopts a pseudoboat shape that inverts by boat–boat interconversion.

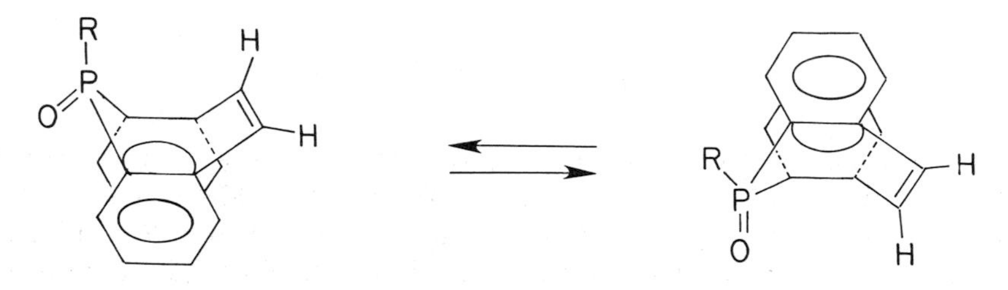

If the double bond is part of a benzo group, as in **23**, the barrier to this inversion is so large that the individual conformers are stable at room temperature and can be isolated as crystalline solids.[31]

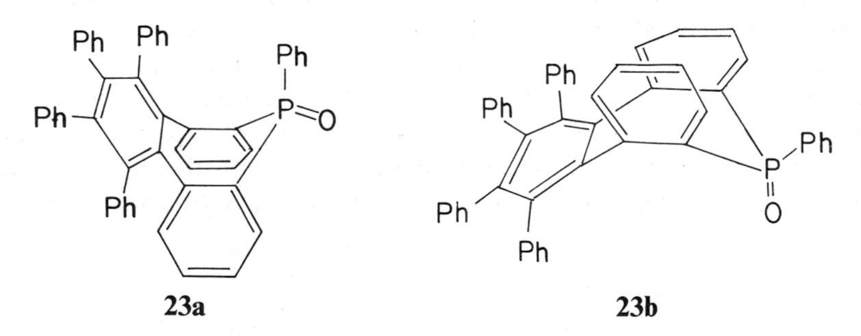

Isomer **23a** is converted to **23b** at 140°, with $\Delta G^{\ddagger}$ of 31 kcal/mol. The phosphines formed from deoxygenation of the oxides had a similar energy barrier, but here the inversion could be accomplished merely by isomerization of the phosphorus pyramid. In any event, the large barrier allowed the determination of the physical properties of the individual isomers in solution. In the ^{31}P NMR spectrum, the phosphines differed in their shifts by 10 ppm while the oxides differed by 1 ppm. The interconversion of the phosphines could be conveniently followed by measuring the change in the ^{31}P spectrum with temperature. The thermal equilibration of the oxides was studied with the aid of a chromatographic technique. The more stable oxide conformer appeared to be that with the phenyl group in the axial position, based on comparisons of the NMR effects with monocyclic compounds, but in the phosphines phenyl was pseudo-equatorial. In a subsequent study,[32] the same authors determined the structure of the phosphines by X-ray analysis, and confirmed the boat shape with equatorial phenyl.

Sugar derivatives where a phosphate group is attached to the 2,5-carbons can formally be considered as fused 1,3,2-dioxaphosphepanes. The structures of two such compounds have been determined by X-ray analysis, one an arabinosylcytidine phosphate (**24**)[33] and the other a dimethylene mannitol phosphoramidothionate (**25**).[34]

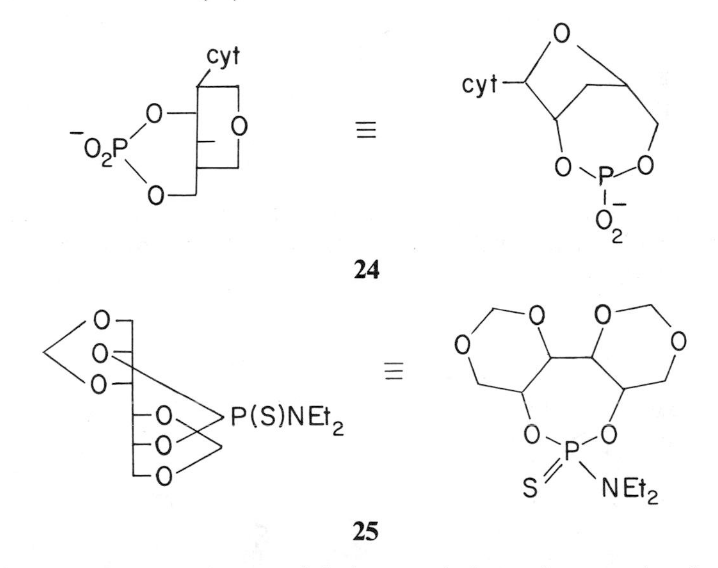

The former, of course, has special characteristics and constraints because its ring has a bridging group. The latter is more akin to the substances under review. Here the seven-membered ring is present as a distorted chair, represented by the authors with structure **25a**. Sulfur is pseudoequatorial, and the nitrogen function is pseudoaxial.

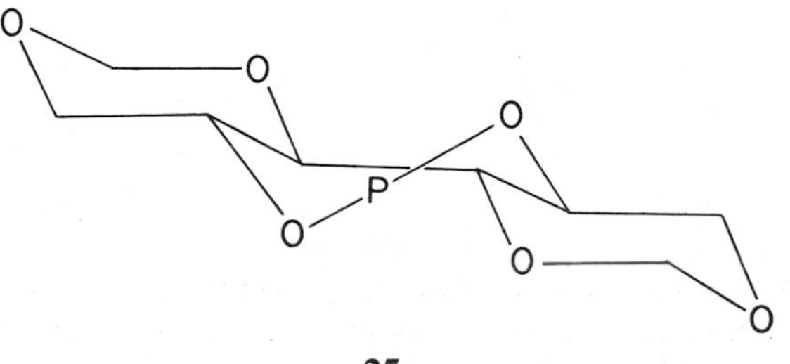

25a

5.3. Rings with Eight Members (Phosphocanes)

Synthetic limitations have dictated the direction of research on eight-membered rings into a somewhat different path than that taken with seven-membered rings. As for the latter, there is no published information on the conformational preferences of the system with one phosphorus and seven carbon atoms. The most popular system for study is that where 1,3,6-trihetero-2-phospha substitution is present, a system that attracts special attention because of the possibility that transannular interaction between the 6-heteroatom and phosphorus might be present (as in **26**) and lead to conformational biasing. The only other monocyclic system that has received consideration is that with 1,5-diaza-3,7-diphospha substitution.

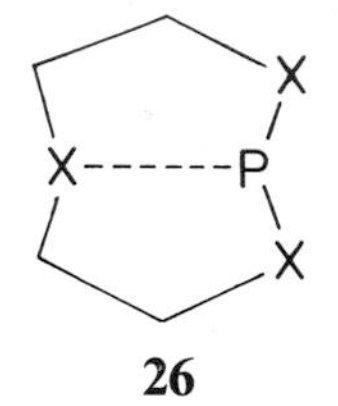

26

The crown conformation, a low-energy form important in cyclooctanes, has been established by X-ray analysis for compounds **27**–**31** shown in the proper conformational perspective.

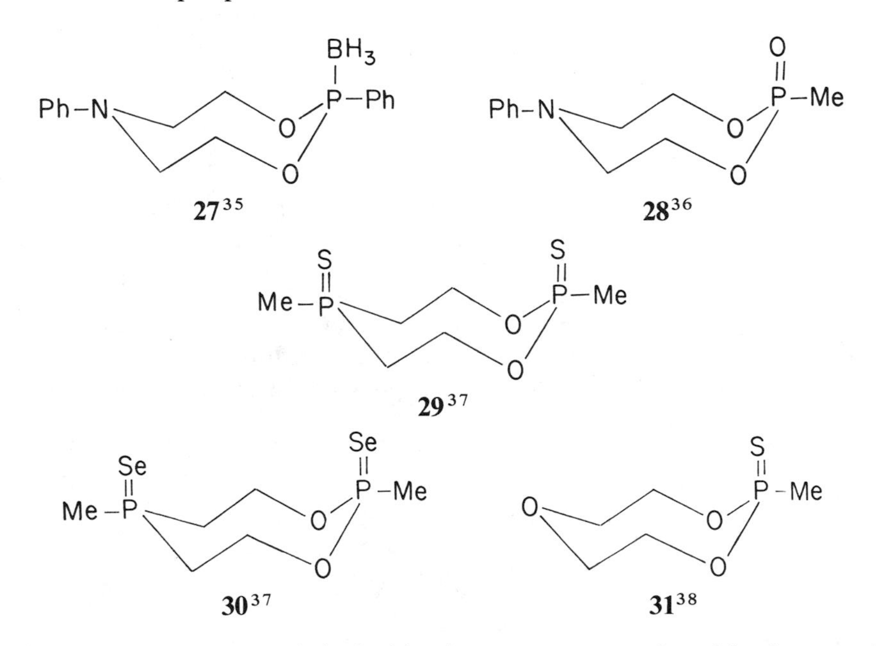

An approximate boat–chair, lacking in symmetry, was found in the crystals of **32**–**34**. An asymmetric structure, resembling a twist–boat–chair with axial P=S, was established[41] for **35**.

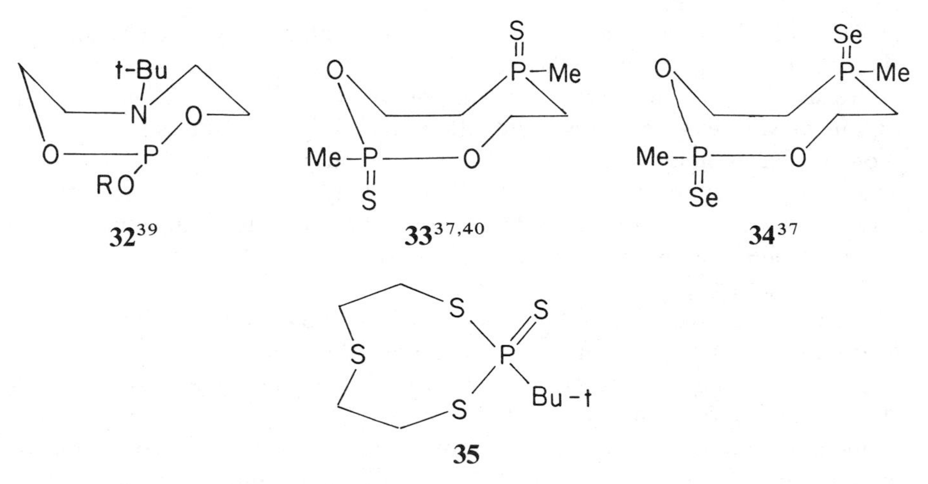

The interatomic distance for P and atom-6 in the ring is small and indicative of transannular electronic interaction for **31** and **32**. The distance is too great in **29**, **30**, **33**, **34**, and **35**.

In all of the P(IV) structures, the P-substituent is found as usual in the least crowded pseudoequatorial position, leaving the smaller O, S, or Se to occupy the pseudoaxial. The P(III) compound **32** has its exocyclic substituent in the pseudo-equatorial position. The preference of methyl for the equatorial position may be responsible for imparting different shapes to the *cis,trans* pairs **29**–**33** and **30**–**34**. Were the same ring shape present in a *cis,trans* pair, one methyl would have to be in a more crowded pseudoaxial position.

The crown conformation was also established by X-ray analysis for some derivatives of the 1,5-diaza-3,7-diphospha system (**36**, **37**). A distorted boat was found[44] for **38**.

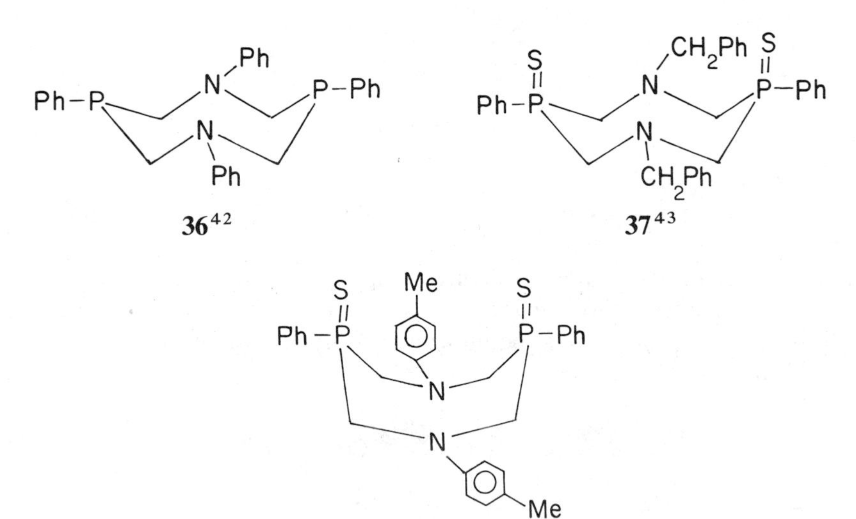

The question of the conformation adopted in solution by heterocycles with the substitution patterns typified by **27–31** and **36–37** has been specifically addressed for some representative compounds. In such studies, the determination of various stereo-dependent NMR coupling constants and of dipole moments has been found valuable. Using the NMR parameters of $^3J_{HH}$, $^3J_{POCH}$, and $^3J_{PCCH}$, Piccinni-Leopardi et al.[37] were able to show that the symmetrical crown shape for the *cis* isomers **29** and **30** is largely retained in solution. The torsion angle established in the crystal allowed prediction of the three-bond couplings, and a good match with experimental values was obtained. Some distortions occurred with the *trans* isomers **33** and **34**. Two enantiomeric conformations are present in the solid, and the NMR data were interpreted as implying a rapid equilibration of the two forms. The data suggest that the solid-state conformation is largely retained in solution, although for both *cis* and *trans* isomers some other minor conformations could also be present. This is indeed a significant discovery, as the number of shapes and rotameric forms possible for such highly substituted ring systems is large and could lead to solution behavior of great complexity.

Compound **31** was examined by NMR long before the crystal structure was reported.[45] It was deduced from the $^3J_{POCH}$ values and the torsion angles indicated by them that only the fixed crown or fixed boat–boat (**39**) could be present in solution. Thus, ΔJ was 17 Hz, compatible with angles of 180° and 60°. However, the boat–boat would have severe nonbonded interactions, and a preference was therefore expressed for the crown (**31**). This received support from the $^3J_{POCC}$ value, which was only 1.3 Hz and thus indicative of a dihedral angle close to 90°. The boat–boat would have an angle of 60°. The structure deduced is, of course, the same as conclusively found for the solid.

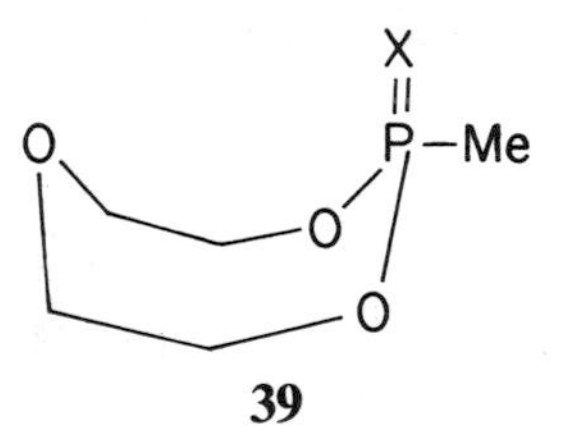

39

The solution conformation of **28** was also deduced to be solely that of the crown, based on dipole moment and spectral (NMR, IR) measurements.[46] The P-methyl and P-butyl analogues of **28**, however, gave indications of the presence of a conformational equilibrium, involving the crown and chair–boat conformations. Proton NMR evidence, using the usual coupling effects, also indicated[47] a conformational equilibrium to prevail for several derivatives of general formula **40**.

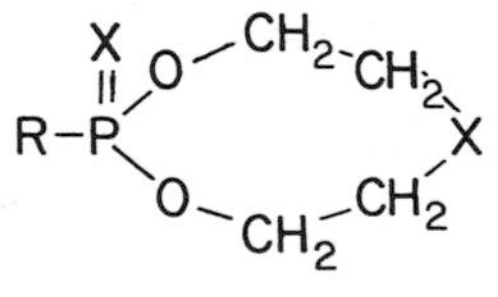

40

Derivatives of the 1,3,6,2-trioxaphosphocane ring system with phosphorus in the tricoordinate condition have also been examined.[45] The same crown shape seems to be present, as suggested from the large difference in $^3J_{POCH}$ values (23.8 and 5.6 Hz). These values are also indicative of the orientation of the lone pair; the resemblance of the former value to that found in a P-methyl 1,3,2-dioxaphosphorinane with an equatorial substituent (25 Hz) suggested that the methyl might have the same orientation in the eight-membered ring. Support for this suggestion comes from the $^2J_{PC}$ value of 12.0 Hz; a value of this magnitude implies that the lone pair is close to (hence axial), not remote from, the ring carbon.[6]

The 1,5,3,7-diazadiphosphocane ring system (in **36**) was examined by NMR techniques,[48] and the crown conformation was indicated by the appearance of the CH_2 groups in the proton spectrum. There was only one signal, which had two different two-bond couplings to ^{31}P. This is reasonably explained by the P-substituent being equatorial on the symmetrical crown conformation; this places the lone pair oriented axially, and hence *gauche*, to one proton (equatorial, $^2J = 12.5$ Hz) and anti to the other (axial H, $^2J = 4.5$ Hz). This is the same conformational assignment as later established by X-ray analysis.[49] Dipole moment measurements on benzene solutions of **36** also indicated retention of the solid-state conformation in the solution; good agreement existed between experimental data and values calculated from this conformation. On the other hand, the dipole moment measurement[49] on the disulfide **37** gave a value that differed significantly from that calculated for the boat–boat structure established by X-ray analysis. Better agreement was obtained for a crown conformation, implying a change in structure when the solid was placed in solution.

The 1,3,6,2-trioxaphosphocane with tricoordinate phosphorus (as in **41**) gave a temperature-dependent NMR spectrum, whose analysis indicated a chair–chair conformation to be in predominance at room temperature.[50,51] In this study, it was observed that the substituents Me, Ph, and OMe exerted the same equatorial preference as found for Cl and NMe_2; the former substituents normally take the axial position in the dioxaphosphorinanes, an observation that calls for great caution in extending ideas developed for one ring system to a larger one.

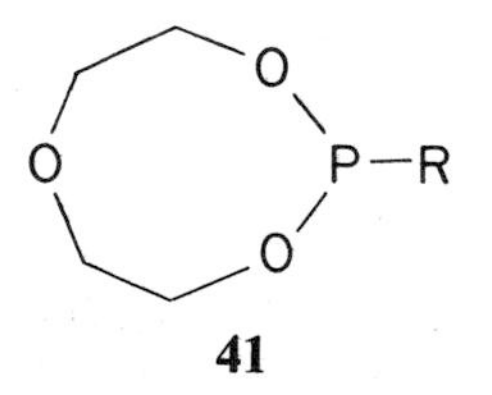

41

An application of ^{31}P NMR spectroscopy aided the conformational analysis of the 1,3-dioxa-6-thia-2-phosphocane **42**; the single peak observed at room temperature (+183.1) was split into two signals of unequal intensity at $-140°$ (δ +188.5 and +172.0), implying that two different conformations have been frozen out.[51] The major conformation at room temperature was proposed to be the crown.

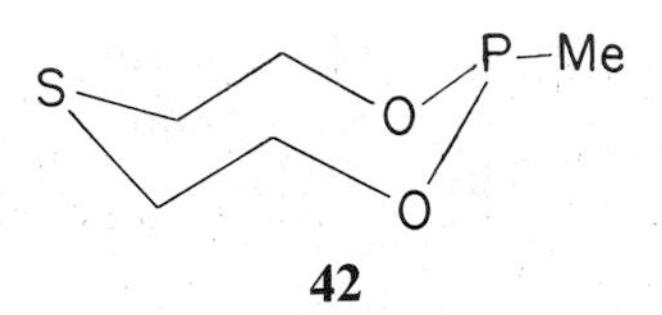

42

The only other type of heterocyclic system with eight ring members to receive significant attention is the dibenzo derivative of the 1,3,2-dioxaphosphocane ring, again because it is an easily synthesized system. Here the dibenzo rings drastically restrict the rotational possibilities for the central ring, and X-ray analysis of **43** proved that the boat–chair conformation (**43a**) with C_s symmetry was adopted.[52]

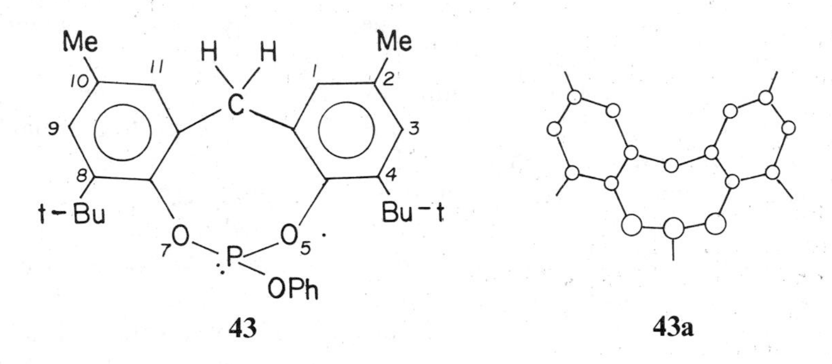

43 **43a**

The phenoxy group on phosphorus was in the equatorial position, again emphasizing that the group preference rule found for six-membered analogues does not apply in the eight-membered rings.

Some P—C coupling data were reported[53] for compound **43** that bear on its solution conformation. The two-bond coupling to C-4a,7a was 7.1 Hz, which is a value more indicative of an axially oriented lone pair, as consistent with the structural parameters found in the X-ray analysis, than an equatorial lone pair (remote from the coupled carbon). Of greater significance is the nearly identical (3.7, 3.9 Hz) three-bond coupling to the benzo ring carbons (C-11a,12a and C-4,C-8) ortho to the oxygen atoms. This near-identity implies that the dihedral angle and the lone-pair orientation on P are probably similar for these carbons, and this is the relation that is present in the solid-state boat–chair conformation. A similar conclusion was reached[54] for the P-chloro analogue **44**.

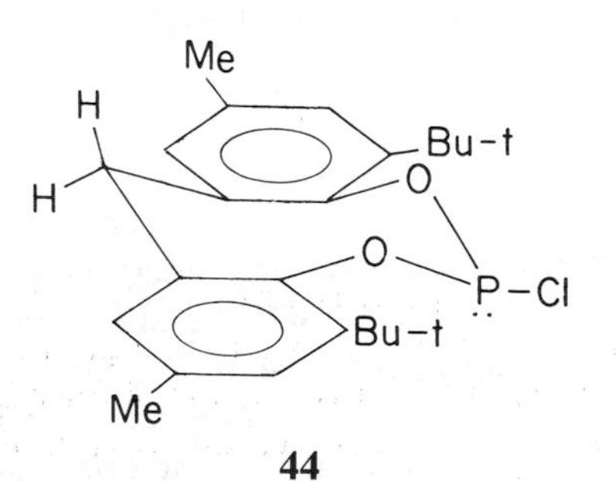

44

Quantum mechanical calculations indicated the boat–chair with equatorial Cl to be the most stable, and then, in order, the boat with equatorial Cl, the twist–boat, and the boat–chair with axial Cl. The UV spectrum supported the conformation **44**; there was a lack of conjugation between the oxygens and the benzene rings that implies an orthogonal relation of the *p*-orbitals on oxygen and the π-orbitals of the ring, in accord with this conformation. However, dipole moment data were in accord with a boat–chair with equatorial Cl in equilibrium (1:1) with another conformation, possibly the boat. Equilibria with other forms could also be present.

In the first study[55] of the proton NMR spectra of this family of compounds, it was observed that the two hydrogens of the methylene group were in distinctly different environments; they gave well-separated signals (e.g., for **43**, δ 3.42 and 4.16, $^2J_{HH} = 12.7$ Hz). Furthermore, the important observation was made that the down-field signal was coupled to ^{31}P ($J = 2.9$ Hz) but the up-field signal was not coupled. When an alkyl group replaces one H of the methylene group, *cis,trans* isomerism is present,[56] and spectral evidence for the two isomers was obtained in the case of, for example, **45** by the presence of proton signals for two different methyl doublets (δ 1.62 and 1.84).

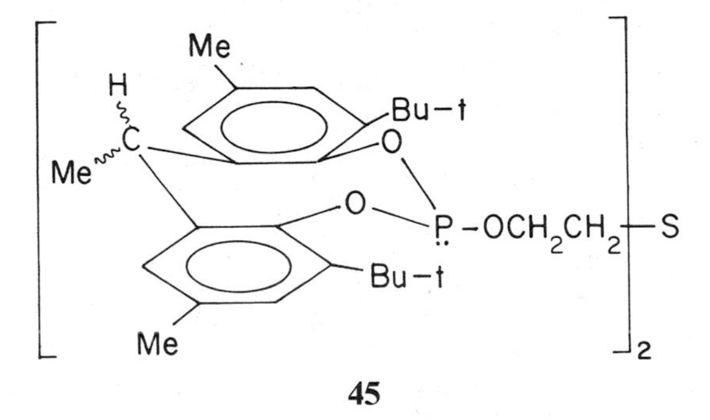

45

The nonequivalence of the methylene protons is apparent in the spectra of derivative **46** with alkyl groups directly attached to phosphorus,[57] and also in derivatives where phosphorus is pentacoordinate, as in **47**.[58] Here the internal ring is probably attached to phosphorus in equatorial positions of the trigonal

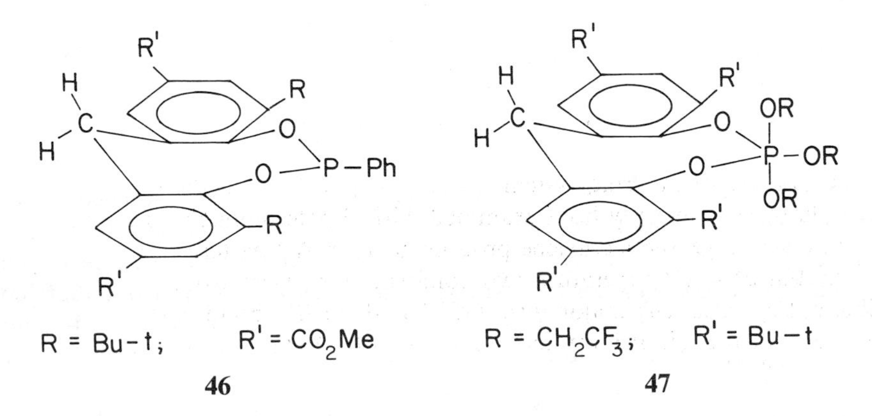

R = Bu−t; R' = CO_2Me

46

R = CH_2CF_3; R' = Bu−t

47

bipyramid. The long-range coupling is detectably different for the two protons at $-27°$ (1.9 and 2.8 Hz) but is the same at 100° (2.7 Hz). Conformational mobility of the central ring seems to be implied by this observation.

A low-temperature proton NMR study[54] of the P-chloro compound **44** revealed that two forms contributing to a conformational equilibrium could be frozen out and independently observed. The data are summarized in the following table, where it will be seen that a major difference was the presence of the five-bond coupling in one form but not in the other. At room temperature, this value appeared to be averaged from these contributors (2.4 Hz).

	δ_A	δ_B	$\Delta\delta$	$^2J_{HH}$	$^5J_{PH}$	%	$\Delta G^{\ddagger}$
Form I	3.38	4.26	0.88	12.4	3.9	56	45.5 ± 3 kJ/mol
Form II	4.16	3.31	-0.85	12.6	0	44	

The presence of the five-bond coupling first led[56,57] to the proposal that the room-temperature spectrum could be explained by conformation **48**, where P and one H are in close proximity. However, this is the boat–boat conformation, which has been calculated by Arshinova et al.[54] to be of relatively high energy and not in accord with the carbon NMR spectrum. In fact, it may not be necessary to have spatial proximity of the coupled nuclei, as the effect may be transmitted through the bonds, not space. This would imply that there is stereospecificity in the five-bond coupling, just as is so well established for $^{2-4}J_{PH}$.[5] Nothing seems to be known about this, and it remains speculative until studied with model compounds having rigid structures. In any case, the $^5J_{PH}$ effect seen here clearly indicates a considerable difference in the conformations of the forms that contribute to the equilibrium observed for the P-chloro compound **44**. Most probably, the form showing the five-bond coupling is the boat–chair with equatorial Cl; the other remains uncertain.

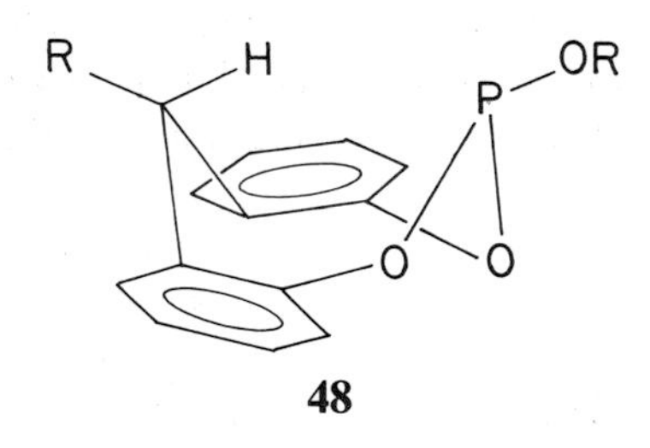

48

Derivatives of this ring system where phosphorus is in the four-coordinate state have very recently been examined with dipole moment and NMR techniques.[59] For **49**, the methylene protons gave an AB pattern at room temperature, but at low temperature two conformations were frozen out that have different spectra. The major form (86%) had no five-bond coupling to phosphorus, but the minor form had a coupling of 3.4 Hz. For another compound

(**50**), the room-temperature spectrum displayed the five-bond coupling (3.4 Hz), implying that its conformation resembled that of the minor conformer of **49**.

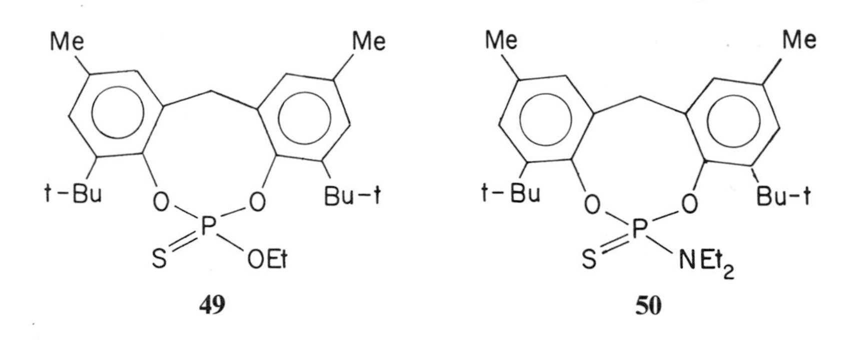

49 **50**

This is a significant observation, as it shows that in eight-membered rings the nature of the substituent on phosphorus can control the conformation of the ring. For **50**, the dipole moment measurements suggested the boat–chair with equatorial substituent to be present. For **49**, the major contributor may be the boat with equatorial substituent, and the minor may be the boat–chair with equatorial substituent. These authors also pointed out that the $^2J_{HH}$ values may reveal conformational tendencies. Thus, the boat–chair form may be associated with values around -12.5 Hz, while the boat may have -15.5 to -16.5. The former value is the only one observed for the two conformers with three-coordinate phosphorus (**43**, **44**), and if the relation holds here, then it is implied that the boat form makes no notable contribution to the equilibria for such compounds. Continued study of this ring system should provide clarification of the conformational picture.

Two other eight-membered ring systems have been subjected to X-ray analysis, but no information is available on solution conformations. In compounds **51**[60] and **52**,[61] the nitrogen atoms are planar. The shape for the latter is shown in **52a**. The unusual compound **53** is of interest as the only eight-membered ring derivative with phosphorus and carbon composing the ring system. A boat shape (**53a**) with quasi-axial phenyl was deduced from the X-ray data.[62]

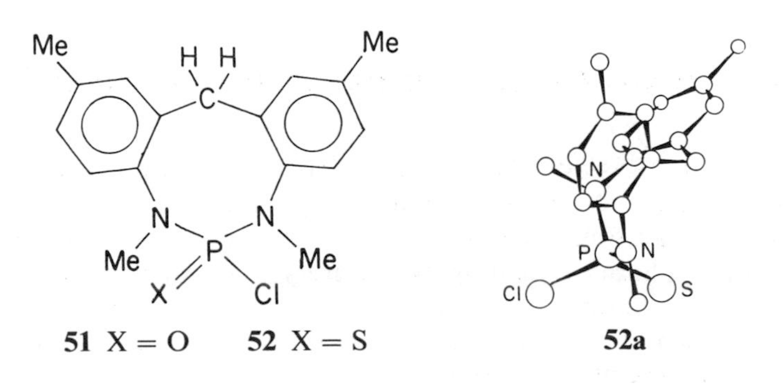

51 X = O **52** X = S **52a**

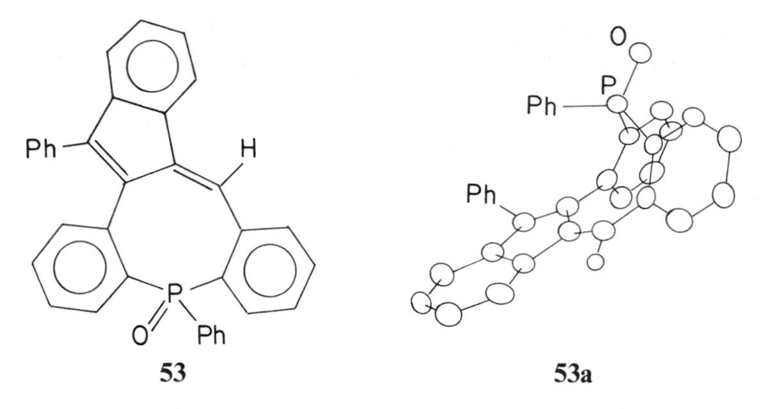

53 **53a**

In a 3′,5′-cyclopyrophosphate of a nucleoside, the 1,3,5,2,4-trioxadiphosphocane ring system is present, fused to a tetrahydrofuran ring. Such compounds, while of great biochemical interest, are of a highly specialized nature and will be mentioned only briefly here. An example of this structure (**54**) has been examined with proton and phosphorus NMR, and also subjected to conformational analysis by the method of atom–atom potentials.[63] The compound is described as undergoing equilibration between two nearly equally populated conformations that were calculated to be of low energy. Both possessed a crown shape for the phosphocane ring, and different shapes for the tetrahydrofuran ring. The stereospecificity of $^{3}J_{HH}$ and $^{3}J_{PH}$ provided the basis, as in so many other cases, for the rationalization of the calculated and experimental data.

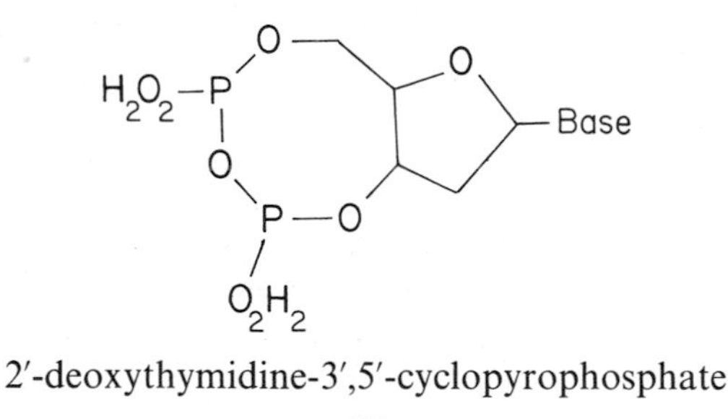

2′-deoxythymidine-3′,5′-cyclopyrophosphate

54

5.4. Rings with Nine Members (Phosphonanes)

Little information is available at present on the conformational properties of this ring system; unlike the seven- and eight-membered rings, conformational studies (some rather fragmentary) have been performed only on systems with carbon and phosphorus in the ring, but in every case two or more of the ring carbons are sp^2 hybridized. Thus, nothing is known about the shape or conformational preferences of the parent saturated phosphonane, even though this system has been recently synthesized.[64]

The first conformational consideration given to a nine-membered ring appears to be that presented by Waite and Tebby[65] on the fully unsaturated phosphonin derivative **55**. This unusual compound arose from a rearrangement reaction and its structure became evident from NMR (^{1}H, ^{31}P), IR, and UV spectral studies. It has received no further study, although it is of special interest as the only reported example of the potentially aromatic (10 π-electron) monocyclic phosphonin system. The NMR spectra required interpretation on the basis of a nonplanar ring, and one of the double bonds was deduced to have *trans* geometry. A conformation such as **55** (or its P invertomer) was proposed from model studies as being of minimized strain and crowding, and capable of giving the observed spectral data. The proposal remains tentative, however.

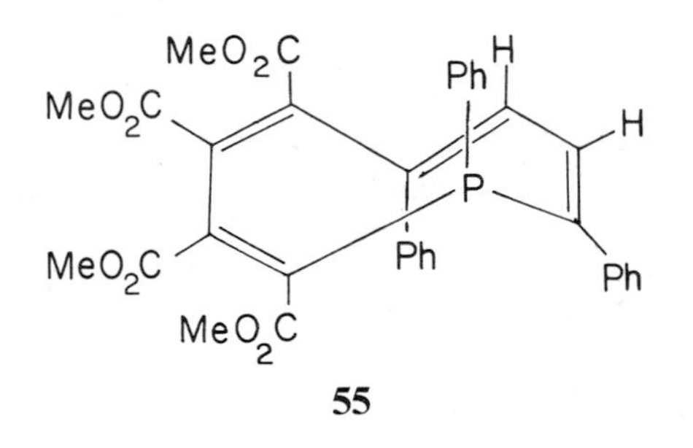

55

The next conformational study of a nine-membered ring was that of Hellwinkel and Lindner[66] on the tetrabenzo-phosphonin system (**56**). A proton NMR study revealed that the four methyl substituents on **56** were in different environments, and remained so up to 120°. This clearly means that the ring is not planar, nor would it have been expected to be planar from the considerable ortho–hydrogen interaction present in such a structure. A puckered conformation would have minimized H—H interactions and would account for the four different methyl signals. The presence of a conformational equilibration was suggested by variable-temperature studies. On raising the temperature further, the number of methyl signals was reduced to two, in a process having T_c of 130° and $\Delta G^{\ddagger}$ of 22 kcal/mol. This could be explained by a ring inversion or a pyramidal inversion of phosphorus, although the latter process is usually found to have

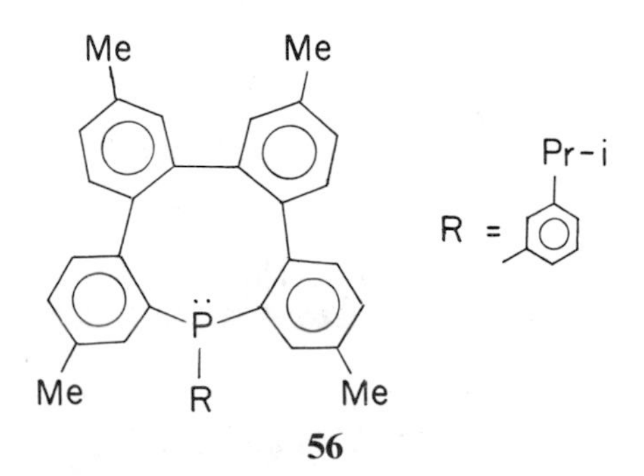

56

a higher energy barrier. An explanation for the NMR results resting on a ring inversion between conformers **57a** and **57b** was favored by the authors.

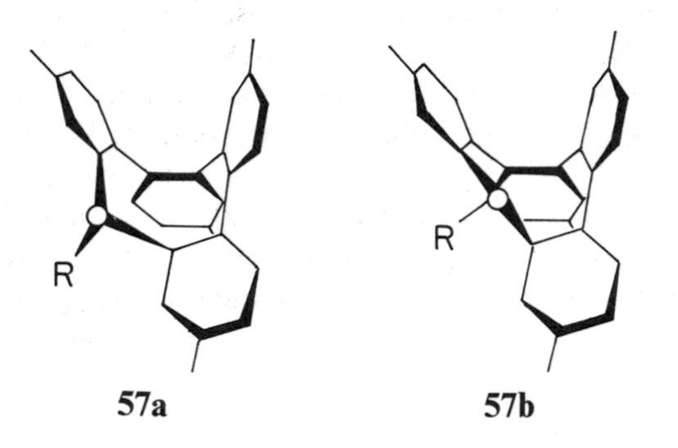

57a **57b**

A nonplanar ring was also required to explain the NMR spectral data for **58a,b**.[64] One double bond in both compounds was clearly *cis* and the other *trans*, from the $^3J_{HH}$ coupling differences. In the phosphine **58b**, the twisting of the ring caused the $^2J_{PH}$ values to be quite different. One value (35 Hz) can be associated with proximity of the proton to the lone pair; the smaller value (11 Hz) suggests a more distant relation.

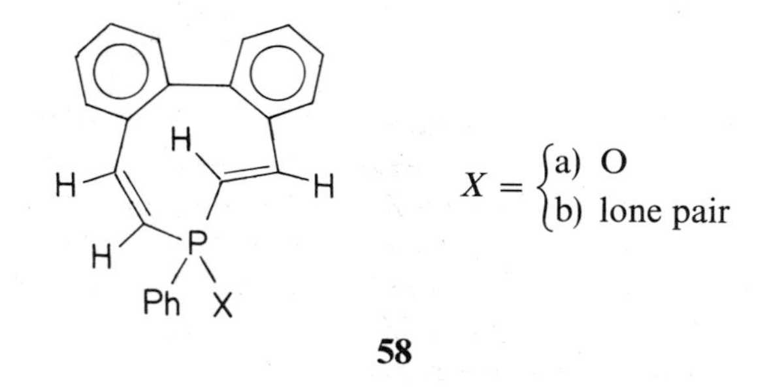

58

A symmetrical conformation seems to be present in the rather unstable bis-trimethylsiloxy phosphonin derivative **59**.[67] This was indicated by the ^{13}C NMR spectrum, which showed spectral equivalence of comparable carbons on either side of the ring. The shape of the ring is not known, but it is unlikely to be planar.

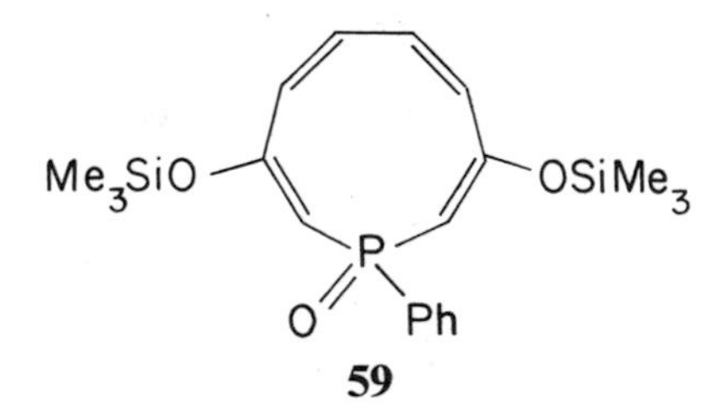

59

3,8-Diketo derivatives of the parent monocyclic phosphonane ring recently became available and proved to be crystalline solids.[68] The X-ray analysis of two such compounds (**60** and **61**) has been accomplished and has provided the only structural data on nine-membered phosphorus ring systems.

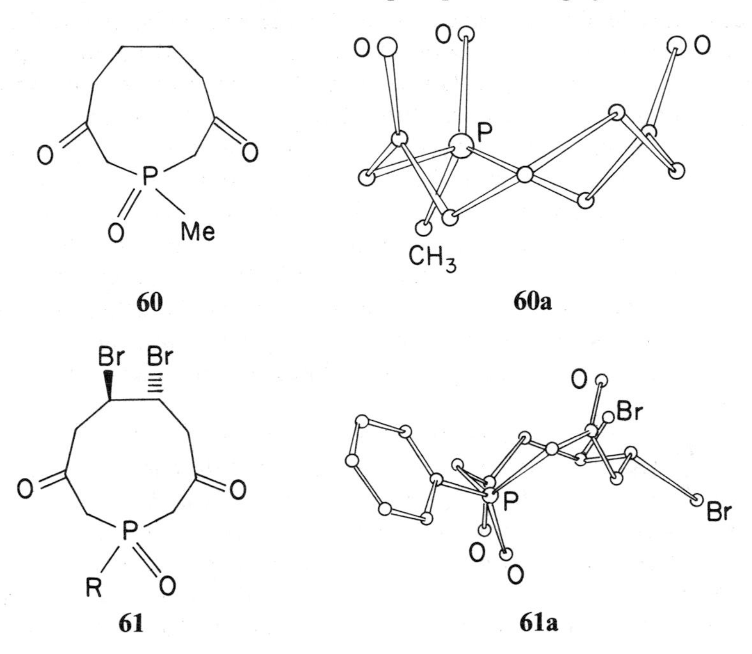

In each, the ring approximated the twist–chair–chair shape (**62**) calculated[69] to be one of the low-energy forms of cyclononane. The longer C—P bond and planarity at the carbonyl carbons caused only moderate deformation from the cyclononane shape. Structures **60a** and **61a** are seen to be different invertomeric forms of the twist–chair–chair conformation. The most striking difference is in the location of the carbonyl oxygens; in **60**, these oxygens, as well as the phosphoryl oxygen, are on the same face of the ring, whereas in **61** the carbonyl oxygens are anti to each other. It is probable that the presence of the additional substituents on **61** influences the selection of the invertomeric form; that form adopted has the phenyl on phosphorus, as well as both bromine atoms, in pseudoequatorial positions, which minimize nonbonded interactions.

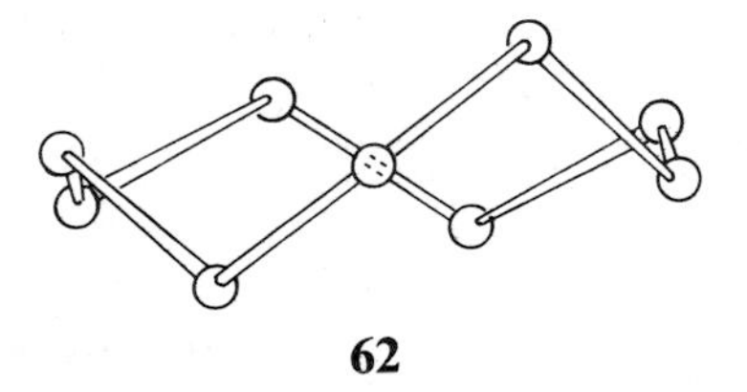

62

When diketone **60** is placed in solution,[68] its ^{13}C NMR spectrum clearly reveals that a conformational change occurs. Comparable carbons on either side of the ring have identical shifts, which would not be the case if the unsymmetrical structure found in the crystal were retained. Either a new and symmetrical conformation has been adopted, or there is rapid equilibration between two (or more) conformations to give time-averaged NMR signals. That the twist–chair–chair conformation has been retained in solution may be deduced from the magnitude of $^3J_{PC}$ to ring carbons 4,7; these are indicated by the X-ray analysis to have a dihedral angle of 90°, and there should be no coupling to ^{31}P. None is, in fact, observed. On lowering the temperature, the signals for P-methyl and C-5,6 show small splitting that suggests the freezing out of two slightly different conformers. An approximate value for T_c of $-84°$ and for $\Delta G^{\ddagger}$ of 9.6 kcal/mol was derived from the data. These results show that the simple spectrum obtained at room temperature is the result of rapid interconversion among invertomers. On the other hand, **61** has a ^{13}C spectrum where all ring carbons have different shifts, as would be true if the solid-state conformation were retained intact. The substituents on the ring may be acting to raise the barrier to ring inversion so that a single species prevails in solution, although a rapid equilibration among unequally populated conformations could give the same result.

The *cis*-5,6-dimethyl derivative **63** has been prepared[68] but not subjected to X-ray analysis. Its ^{13}C NMR spectrum is indicative of the conformational situation, however. Assuming a twist–chair–chair, the two methyls will be in pseudoaxial and pseudoequatorial positions regardless of the invertomeric form. Two typical forms are shown as **63a** and **63b**. Were a rigid conformation present, two signals for the methyls should be seen. In fact, only one is observed, implying that there is a rapid equilibration between invertomers that gives an averaged spectrum.

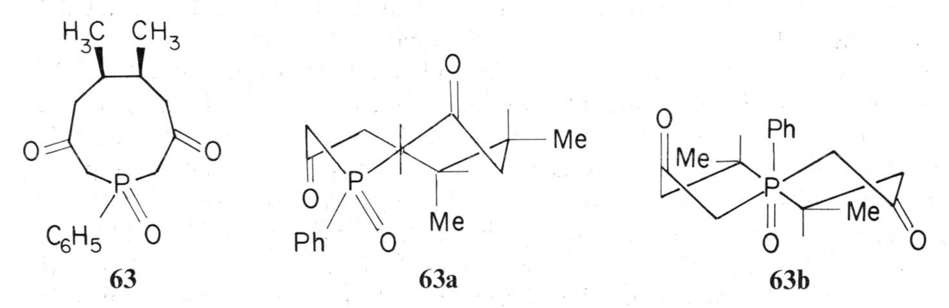

Another situation is presented by the dibenzo derivative **64**[68]; models show that the rotational freedom is severely restricted and the preferred conformation is the highly twisted **64a**. The carbonyl oxygens and methylenes are locked in

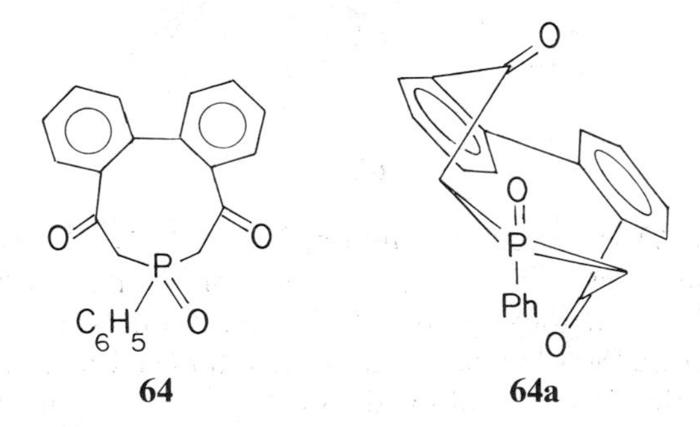

different positions relative to the P-phenyl and to phosphoryl oxygen, and each gives two signals on the ^{13}C NMR spectrum. Reduction of carbonyl to carbinol gives diol **65**, which also must have a twisted and rigid conformation such as **65a**; as a result, the carbinol carbons give different signals.

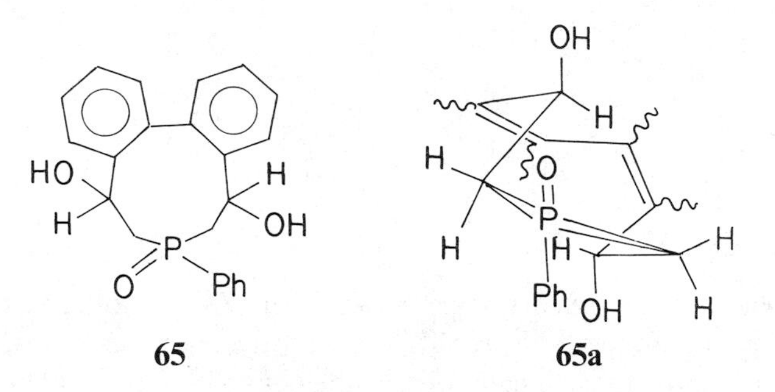

65 **65a**

It may be seen from the immediately preceding discussion that conformational analysis of even a system as complex as that with nine members can proceed on fairly safe ground. Although many different ring shapes, with invertomeric possibilities, are conceivable, the system avoids these complexities and is surprisingly easily studied.

5.5. Rings with 10 or More Members

Conformational information on the larger ring systems is severely limited, and very few of the many possible heterocycles have been studied.

X-ray analysis of the 10-membered ring derivative **66** showed[70] that the conformation in the solid (**66a**) was very similar to that of the most stable conformation of cyclodecane (**67**). This is easily seen on comparing the torsion angles established in the two systems; these are shown in the formulas.

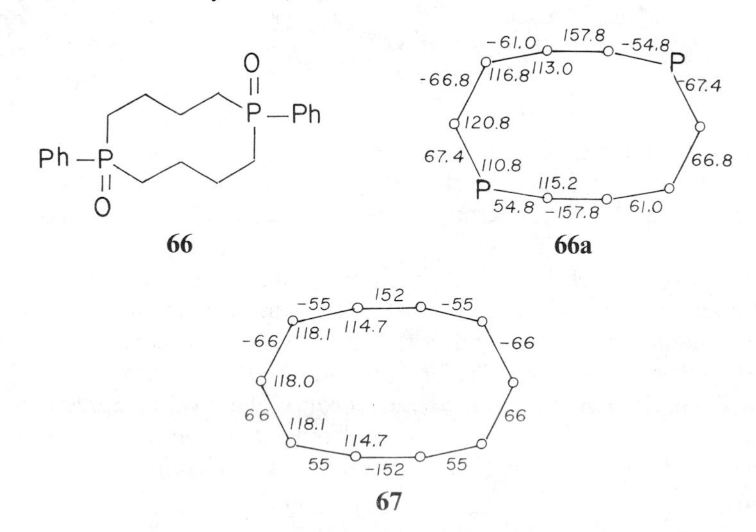

66 **66a**

67

At both phosphorus atoms of **66**, the larger (phenyl) substituent takes up a pseudoequatorial position. No information is available on the solution preference for this compound.

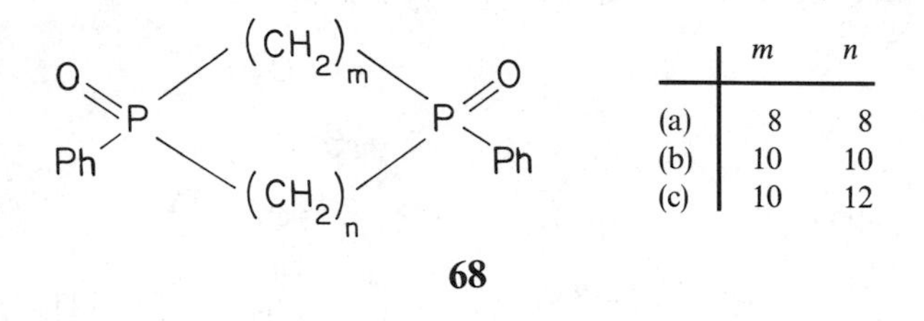

	m	n
(a)	8	8
(b)	10	10
(c)	10	12

68

Bisphosphine oxides (**68**) are also known with 18, 22, and 22 members in the ring.[71] The synthetic method provided a mixture of isomeric (*cis,trans*) forms of these compounds. In each case, one isomer has a significantly smaller dipole moment than the other; ranges of 4.12–5.34 and 6.69–7.10*D* were observed. The more polar isomers appear to have a *cis* structure; their dipole moments are nearly twice those of noncyclic model compounds (R_3PO, 4.35–4.44*D*) and this is suggestive of a conformation in which, by calculation, the dipoles of the two groups have an angle of 75° between them. This is easily accommodated by a *cis*, rather than a *trans*, structure, since the latter would require excessive bending of the ring to achieve this relation. Two conformations are indicated for the *cis* compounds from studies of models; these are shown as **69a** and **69b**, each of which has the angle of 75° with respect to the C_2 axis. Since those *trans* isomers with $m = n$ also have a significant dipole moment, they cannot have a centrosymmetric structure like **69c**, where the two group dipoles cancel each other to give a net value of zero.

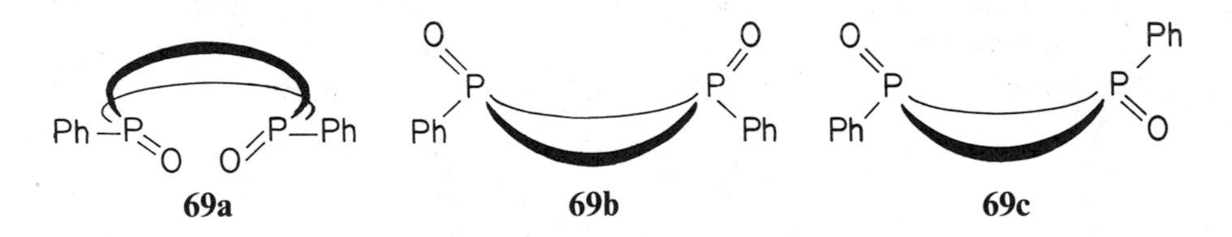

69a **69b** **69c**

The *trans* isomers must possess some distortion from bending of this ring to give the dipole moments observed, which must arise from a nonantiparallel alignment of the phosphoryl groups. The true shape of the ring is not defined by these observations, however.

The conformation of the 12-membered ring compound **70** with *cis* substitution was determined[72] to resemble that of the square shape of cyclododecane, with the phosphorus function and the Me_2C groups at the corners. The torsion angles are given on structure **70**. The proton NMR spectrum had the $^3J_{POCH}$ values of 5.5 and 6.5 Hz for the methylene group. These values agree with those that can be expected from the torsion angles derived from the X-ray analysis, and this is a strong indication that the solid and dissolved forms have the same conformation.

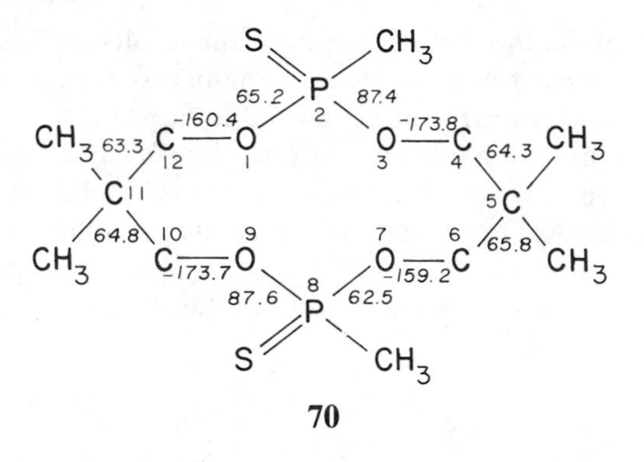

70

A major investigation of phosphorus macrocycles with 11 or 14 members, with fused benzo groups, has resulted in a number of reports[73–75] where X-ray structural data are presented. The compounds for which data are reported are summarized by structures **71**, **72**, and **73**.

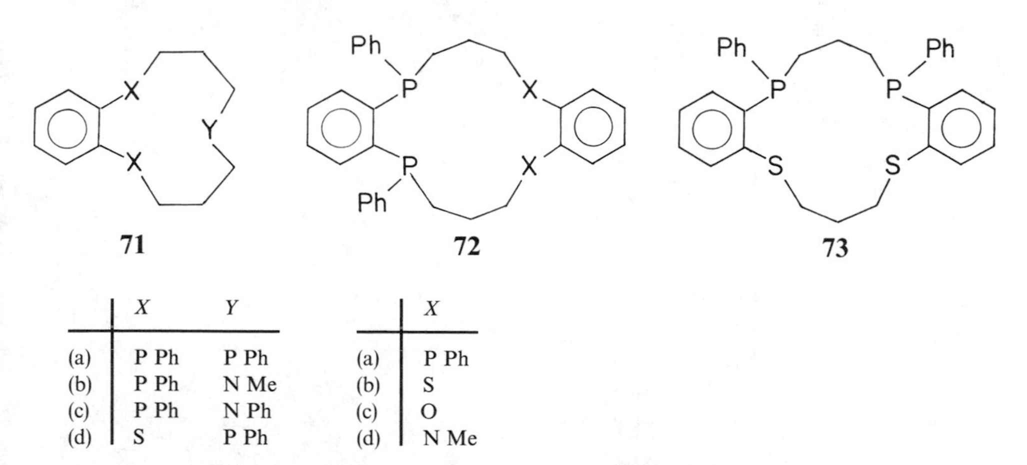

71 **72** **73**

	X	*Y*
(a)	P Ph	P Ph
(b)	P Ph	N Me
(c)	P Ph	N Ph
(d)	S	P Ph

	X
(a)	P Ph
(b)	S
(c)	O
(d)	N Me

To illustrate the shape of the 11-membered ring system, compound **71a** may be considered.[73] The synthetic method produced the *cis* isomer, and the phenyl substituents were found in a pseudoequatorial position.

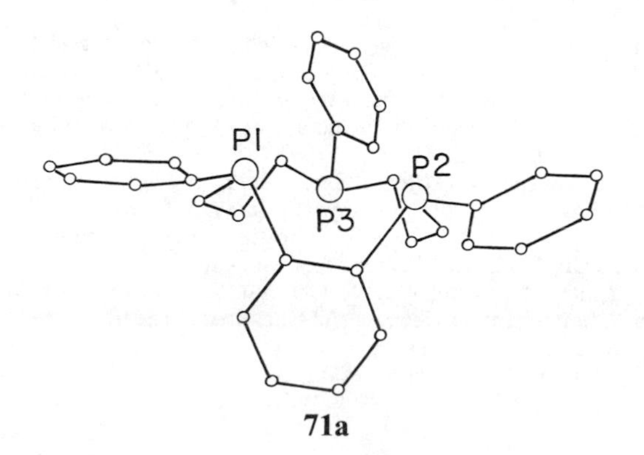

71a

As an example of the 14-membered system, the results for **72a** are typical. The heteroatoms make up a rough plane, with the benzo groups above and below this plane in the *trans* isomers and on the same side in the *cis* isomers. For **73**, the X-ray analysis showed that the *trans* isomer had crystallized in two thermally interconvertible conformations, in the ratio 1:1, Both had pseudoequatorial phenyl substituents; the conformational differences resulted from the adoption of different torsion angles in the —$S(CH_2)_3S$— chains. A simplified description of these complex compounds as proposed by the authors is given as **74**.

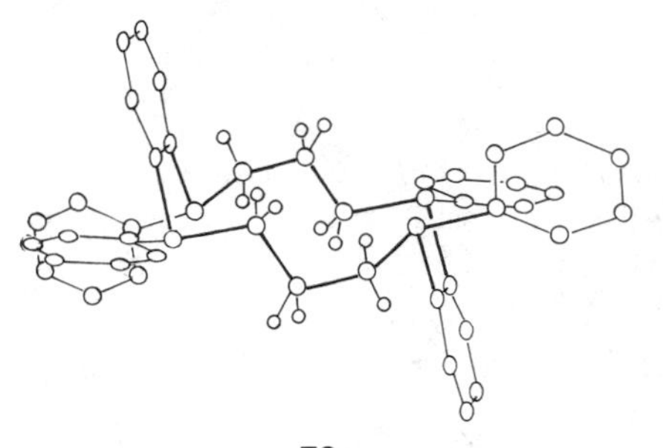

72a

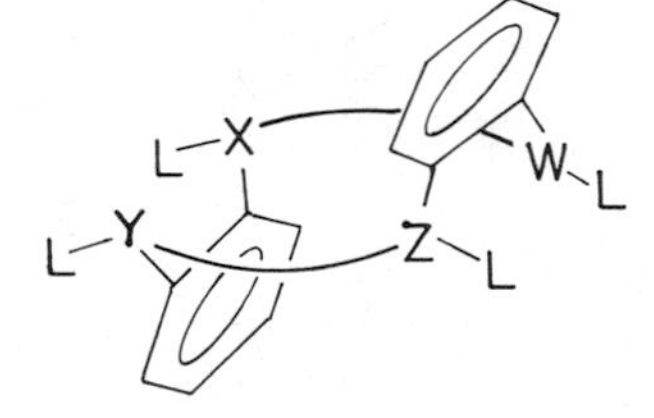

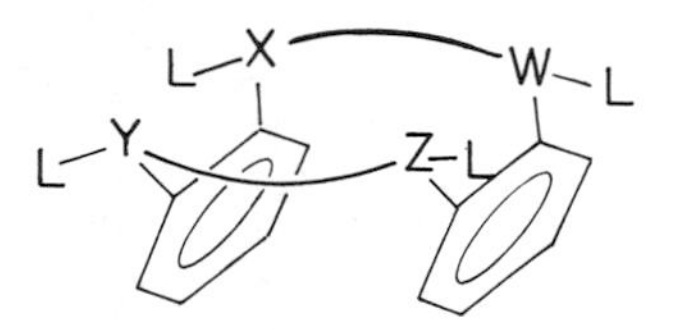

73 **74**

References

1. Gallagher, M. J. In “Stereochemistry of Heterocyclic Compounds,” Pt. II, Armarego, W. L. F., Ed. Wiley: New York, 1977, Chap. 5.
2. Anet, F. A. L. In “Conformational Analysis of Medium-Sized Ring Heterocycles,” Glass, R., Ed. VCH Publishers: New York, 1988, Chap. 2.
3. Corbridge, D. E. C. “The Structural Chemistry of Phosphorus.” Wiley: New York, 1974.
4. Dunitz, J. D. “X-Ray Analysis and the Structure of Organic Molecules.” Cornell University Press: Ithaca, N.Y., 1979, pp 312–318.
5. Bentrude, W. G. In “Phosphorus-31 NMR Spectroscopy in Stereochemical Analysis,” Verkade, J. G.; Quin, L. D., Eds. VCH Publishers: New York, 1986, Chap. 11.
6. Quin, L. D. In “Phosphorus-31 NMR Spectroscopy in Stereochemical Analysis,” Verkade, J. G.; Quin, L. D., Eds. VCH Publishers: New York, 1986, Chap. 12.
7. Quin, L. D.; Gallagher, M. J.; Cunkle, G. T.; Chesnut, D. B. *J. Am. Chem. Soc.* **1980**, *102*, 3136.
8. Dimroth, K. In “Comprehensive Heterocyclic Chemistry,” Katritzky, A. R.; Rees, C. W., Eds. Pergamon: Oxford, 1984, Vol. 1, Chap. 17.
9. Quin, L. D. “The Heterocyclic Chemistry of Phosphorus.” Wiley-Interscience: New York, 1981, Chap. 8.

10. Arbuzov, B. A.; Polezhaeva, N. A.; Arshinova, R. P. In "Chemistry Reviews," Volpin, M. E., Ed., Harwood Academic Publishers: Chur, Switzerland, 1984, Vol. 6, pp 1–123.
11. Maryanoff, B. E.; Hutchins, R. O.; Maryanoff, C. A., In "Topics in Stereochemistry," Allinger, N. L.; Eliel, E. L., Eds. Wiley-Interscience: New York, 1979, Vol. 11, pp 187–326.
12. Coulter, C. L. *J. Am. Chem. Soc.* **1975**, *97*, 4084.
13. Guimaraes, A. C.; Robert, J. B.; Taieb, C.; Tabony, J. *Org. Magn. Reson.* **1978**, *11*, 411.
14. Sato, T.; Goto, K. *J. Chem. Soc. Chem. Commun.* **1973**, 494.
15. Shagidullin, R. R.; Shakirov, I. K.; Musyakaeva, R. K.; Matrosov, E. I. *Izv. Akad. Nauk SSSR* **1981**, 2253.
16. Guimaraes, A. C.; Robert, J. B. *Tetrahedron Lett.* **1976**, 473.
17. Shagidullin, R. R.; Fazliev, D. F.; Musyakaeva, R. K.; Plyamovatyi, O. N.; Nuretdinova, O. N.; Vandyukova, I. I. *Izv. Akad. Nauk SSSR* **1981**, 567.
18. Fazliev, D. F.; Shagidullin, R. R.; Nuretdinova, O. N. *Izv. Akad. Nauk SSSR* **1977**, 2638.
19. Arbuzov, B. A.; Kadyrov, R. A.; Arshinova, R. P.; Sharikov, I. K.; Shagidullin, R. R. *Zhur. Obshch. Khim.* **1985**, *55*, 1975.
20. Arbuzov, B. A.; Kadyrov, R. A.; Klochkov, V. V.; Arshinova, R. P.; Aganov, A. V. *Izv. Akad. Nauk SSSR* **1982**, 588.
21. Shagidullin, R. R.; Shakirov, I. K.; Plyamovatyi, A. K.; Arshinova, R. P.; Kadyrov, R. A.; Arbuzov, B. A. *Izv. Akad. Nauk SSSR* **1984**, 1803.
22. Grand, A.; Robert, J. B. *Acta Cryst.* **1978**, *B34*, 199.
23. Arbuzov, B. A.; Aganov, A. V.; Klochkov, V. V.; Kadyrov, R. A.; Arshinova, R. P. *Izv. Akad. Nauk SSSR* **1982**, 1195.
24. Arbuzov, B. A.; Kadyrov, R. A.; Arshinova, R. P.; Mukmeneva, N. A. *Izv. Akad. Nauk SSSR* **1981**, 784.
25. Pastor, S. D.; Spivack, J. D.; Steinhuebel, L. P.; Matzura, C. *Phosphorus Sulfur* **1983**, *15*, 253.
26. Reddy, C. D.; Rao, C. V. N.; Reddy, D. B.; Thompson, M. D.; Jasinski, J.; Holt, E. M.; Berlin, K. D. *Ind. J. Chem.* **1985**, *24b*, 481.
27. Lampin, J. P.; Mathey, F.; Sheldrick, W. S. *Acta Cryst.* **1974**, *B30*, 1628.
28. Macdonell, G. D.; Berlin, K. D.; Ealick, S. E.; van der Helm, D. *Phosphorus Sulfur* **1978**, *4*, 187.
29. Allen, D. W.; Nowell, I. W.; Walker, P. E. *Z. Naturforsch.* **1980**, *35b*, 133.
30. Segall, Y.; Shirin, E.; Granoth, I. *Phosphorus Sulfur* **1980**, *8*, 243.
31. Winter, W. *Chem. Ber.* **1976**, *109*, 2405.
32. Winter, W. *Chem. Ber.* **1978**, *111*, 2942.
33. Kung, W.-J.; Marsh, R. E.; Kainosho, M. *J. Am. Chem. Soc.* **1977**, *99*, 5471.
34. Litvinov, I. A.; Yufit, D. S.; Struchkov, Y. T.; Arbuzov, B. A.; Gurarii, L. I.; Mukmenev, E. T. *Dokl. Akad. Nauk SSSR* **1982**, *265*, 884.
35. Contreras, R.; Murillo, A.; Klaebe, A. *Heterocycles* **1984**, *22*, 1307.
36. Kalinin, A. E.; Andrianov, V. G.; Struchkov, Y. T. *Zhur. Strukt. Khim.* **1975**, *16*, 1041.
37. Piccinni-Leopardi, C.; Reisse, J.; Germain, G.; Declercq, J. P.; Van Meerssche, M.; Jurkschat, K.; Mügge, C.; Zschunke, A.; Dutasta, J. P.; Robert, J. B. *J. Chem. Soc., Perkin Trans. II*, **1986**, 85.
38. Grand, A. *Cryst. Struct. Commun.* **1982**, *11*, 569.
39. Devillers, J.; Houalla, D.; Bonnet, J. J.; Wolf, R. *Nouv. J. Chim.* **1980**, *4*, 179.
40. Piccinni-Leopardi, C.; Germain, G.; Declercq, J. P.; Van Meerssche, M.; Robert, J. B.; Jurkschat, K. *Acta Cryst.* **1982**, *B38*, 2197.
41. Martin, J.; Robert, J. B. *Acta Cryst.* **1979**, *B35*, 1623.
42. Arbuzov, B. A.; Erastov, O. A.; Nikonov, G. N.; Litvinov, I. A.; Yufit, D. S.; Struchkov, Y. T. *Dokl. Akad. Nauk SSSR* **1981**, *257*, 127.
43. Arbuzov, B. A.; Erastov, O. A.; Nikonov, G. N.; Litvinov, I. A.; Yufit, D. S.; Struchkov, Y. T. *Izv. Akad. Nauk SSSR, Ser. Khim.* **1981**, 2279.
44. Arbuzov, B. A.; Erastov, O. A.; Nikonov, G. N.; Yufit, D. S.; Struchkov, Y. T. *Dokl. Akad. Nauk SSSR* **1982**, *267*, 650.
45. Dutasta, J. P.; Robert, J. B. *J. Am. Chem. Soc.* **1978**, *100*, 1925.
46. Patsanovskii, I. I.; Ishamaeva, E. A.; Dyakov, V. M.; Remizov, A. B.; Kuznetsova, G. A.; Lazarev, I. M.; Voronkov, M. G.; Pudovik, A. N. *Zhur. Obshch. Khim.* **1981**, *51*, 980.
47. Sharma, R. K.; Sampath, K.; Vaidyanathaswamy, R. *J. Chem. Res. (S)* **1980**, 12.
48. Märkl, G.; Yu Jin, G.; Schoerner, C. *Tetrahedron Lett.* **1980**, 1409.
49. Arbuzov, B. A.; Erastov, O. A.; Nikonov, G. N.; Arshinova, R. P.; Romanova, I. P.; Kadyrov, R. A. *Izv. Akad. Nauk SSSR* **1983**, 1846.
50. Dutasta, J. P.; Robert, J. B.; Vincens, M. *Tetrahedron Lett.* **1979**, 933.
51. Dutasta, J. P.; Martin, J.; Robert, J. B. *Heterocycles* **1980**, *14*, 1631.

52. Litvinov, I. A.; Struchkov, Y. T.; Arbuzov, B. A.; Arshinova, R. P.; Ovodova, O. V. *Zhur. Strukt. Khim.* **1984**, *25*, 118.
53. Pätoprsty, V.; Malik, L.; Goljer, I.; Göghova, M.; Karvas, M.; Durmis, J. *Magn. Reson. Chem.* **1985**, *23*, 122.
54. Arshinova, R. P.; Danilova, O. I.; Ovodova, P. V. *Dokl. Akad. Nauk SSSR* **1986**, *287*, 852.
55. Odorisio, P. A.; Pastor, S. D.; Spivack, J. D.; Steinhuebel, L. *Phosphorus Sulfur* **1983**, *15*, 9.
56. Odorisio, P. A.; Pastor, S. D.; Spivack, J. D. *Phosphorus Sulfur* **1984**, *19*, 285.
57. Pastor, S. D.; Spivack, J. D. *J. Heterocyclic Chem.* **1983**, *20*, 1311.
58. Abdou, W. M.; Denney, D. B.; Denney, D. Z.; Pastor, S. D. *Phosphorus Sulfur* **1985**, *22*, 99.
59. Arshinova, R. P.; Danilova, O. I.; Ovodova, O. V. *Dokl. Akad. Nauk SSSR* **1986**, *287*, 1135.
60. (a) Cheng, C. Y.; Shaw, R. A.; Cameron, T. S.; Prout, C. K. *Chem. Commun.* **1968**, 616. (b) Cameron, T. S. *J. Chem. Soc. Perkin Trans. II* **1972**, 591.
61. Cameron, T. S.; Prout, C. K.; Howlett, K. D. *Acta Cryst.* **1975**, *B31*, 2331.
62. Winter, W. *Z. Naturforsch.* **1977**, *32b*, 1335.
63. Vorobev, Y. N.; Badashkeeva, A. G.; Lebedev, A. V. *Zhur. Strukt. Khim.* **1982**, *23*, 29.
64. Quin, L. D.; Middlemas, E. D.; Rao, N. S. *J. Org. Chem.* **1982**, *47*, 905.
65. Waite, N. E.; Tebby, J. C. *J. Chem. Soc. C* **1970**, 386.
66. Hellwinkel, D.; Lindner, W. *Chem. Ber.* **1976**, *109*, 1497.
67. Rao, N. S.; Quin, L. D. *J. Org. Chem.* **1984**, *49*, 3157.
68. Quin, L. D.; Middlemas, E. D.; Rao, N. S.; Miller, R. W.; McPhail, A. T. *J. Am. Chem. Soc.* **1982**, *104*, 1893.
69. Hendrickson, J. B. *J. Am. Chem. Soc.* **1964**, *86*, 4854.
70. Dräger, M. *Chem. Ber.* **1974**, *107*, 3246.
71. Chan, T. S.; Ong, B. S. *J. Org. Chem.* **1974**, *39*, 1748.
72. Dutasta, J. P.; Grand, A.; Robert, J. B. *Acta Cryst.* **1978**, *B34*, 3820.
73. Kyba, E. P.; John, A. M.; Brown, S. B.; Hudson, C. W.; McPhaul, M. J.; Harding, A.; Larsen, K.; Niedzwiecki, S.; Davis, R. E. *J. Am. Chem. Soc.* **1980**, *102*, 139.
74. Kyba, E. P.; Davis, R. E.; Hudson, C. W.; John, A. M.; Brown, S. B.; McPhaul, M. J.; Liu, L.-K.; Glover, A. C. *J. Am. Chem. Soc.* **1981**, *103*, 3868.
75. Kyba, E. P.; Clubb, C. N.; Larson, S. B.; Schueler, V. J.; Davis, R. E. *J. Am. Chem. Soc.* **1985**, *107*, 2141.

Conformational Analysis of Peptides Containing Medium-sized Heterocycles

Victor J. Hruby
University of Arizona, Tucson, AZ

Patricia S. Hill
Millersville University, Millersville, PA

6.1. Introduction

For decades there has been much interest in the structure of small peptides, particularly with respect to how their structure affects their biological function. Peptides are known to regulate and modulate such diverse biological functions as sexual maturation and reproduction, blood pressure regulation, satiety, analgesia, glucose metabolism and storage, enzyme inhibition, and learning and memory.[1] Because peptides play these important roles, there is great interest in developing an understanding of the underlying physical and chemical basis for information transfer by these molecules. Conformational analysis of peptides in terms of both three-dimensional space and dynamic motion is essential to the determination of these underlying principles.

The fact is that many peptides are small, linear, conformationally flexible molecules whose conformations are highly environment dependent. Determining predominant conformations, if they exist, and deciding which of them may have biological significance are difficult tasks. One approach to overcoming these problems is to reduce the number of possible conformations to one or several conformational models through the use of conformational restrictions.[2–9] A simple method of conformational restriction is through cyclization. Cyclic peptides occur naturally and they have proved to be interesting models for conformational studies.[6,10–17] Such studies, particularly those involving small cyclic peptides, indicate that many of these molecules have relatively restricted conformations in solution, some being quite well defined.

Detailed investigation of these restricted molecules has relied upon a wide variety of physical techniques, each adding corroborating evidence to support the contention that small cyclic peptides do exist in preferred, if not rigid, conformations. Most of these techniques, such as NMR, X-ray crystallography, and IR and Raman spectroscopy, are quite common in conformational studies of organic molecules in general. Others, such as circular dichroism (CD) and optical rotary dispersion (ORD), are more common in the study of optically active macromolecules such as peptides.

This review will focus on the conformational analysis of small cyclic peptides. These are arbitrarily defined here as peptides containing a heteronuclear ring of between eight and 14 atoms as part of their structure. The major classes include the cyclic tripeptides (nine-membered rings), cyclic tetrapeptides (12-membered rings), several classes of antibiotics (eight-, 12–14-membered rings), and peptides containing small disulfide loops (eight- to 14-membered rings). Larger cyclic peptides, such as penta-, hexa-, and octapeptides where ring size varies from 14 to 24 atoms, are currently under intense conformational investigation and many publications are available.[6,11,18–34] Even in such large rings, there is some precedent for the observance of "preferred conformations," especially in the case of the ion-binding octapeptides.[35–38] These phenomena will be briefly pointed out in this review.

6.1.1. Definitions

Throughout this chapter, the standard nomenclature and conventions are those adopted by the IUPAC-IUB Commission on Biochemical Nomenclature.[39] Unless otherwise noted, all amino acids refer to α-amino acids of the *L*-configuration.

Peptides consist of amino acid residues linked together through amide bonds. If one disregards side-chain groups of amino acids, a peptide chain is reduced to a series of repeating "peptide units." The "peptide backbone," therefore, consists exclusively of N, C—α, and C′ (carbonyl carbon) atoms. From X-ray studies, Pauling and Cory[40] found that bond lengths and bond angles in the peptide backbone depend very little on the nature or size of the side-chain groups and have relatively fixed values. The amide bond of the peptide unit has partial double-bond character and is considered planar in either a *cis* or, more stable, *trans* configuration. The barrier to rotation about the amide bond is of the order of 14–20 kcal/mol.[41] Since the bond lengths and bond angles in peptides are relatively fixed, a peptide's backbone conformation can be described by the dihedral angles about the covalent bonds linking the peptide units. These dihedral angles, defined as ϕ, ψ, and ω, refer to rotation about the C^{α}—N, C′—C^{α}, and C′—N bonds, respectively, and are illustrated in Fig. 6-1.[42]

Any dihedral angle may be defined by the position of four atoms. For example, the torsional angle ϕ is defined by the position of the atoms C′—N—C^{α}—C′ as shown in Fig. 6-2. If all four atoms are in a planar *cis* configuration, then by

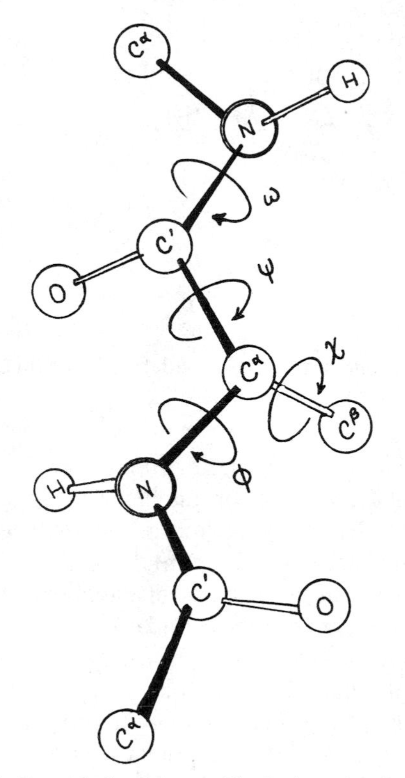

Figure 6-1. Peptide backbone dihedral angles Φ, Ψ, and ω.

definition the torsional angle is 0°. If they are planar *trans*, then the angle is ±180°. Positive values are described by a clockwise rotation about the central bond while negative values are described by counterclockwise rotation (Fig. 6-2). As mentioned previously, the peptide bond prefers the planar *trans* form. Therefore, in the extended form of the peptide chain, the dihedral angles $\phi = \psi = \omega = 180°$.

Conformation of side-chain groups in peptides can also be defined in terms of rotational angles, $\chi_1, \chi_2, \ldots, \chi$ ($\chi_1 = C_\alpha - C_\beta$; $\chi_2 = C_\beta - C_\gamma$) (Fig. 6-1). The angle χ is defined as 0° when the substituent groups on the carbons of interest are *cis* (as defined by Prelog's rules). Again angles are measured from 0° to ±180°. Side-chain χ angles are generally found to be in the vicinity of one of

Figure 6-2. *Cis* and *trans* torsional Φ angles.

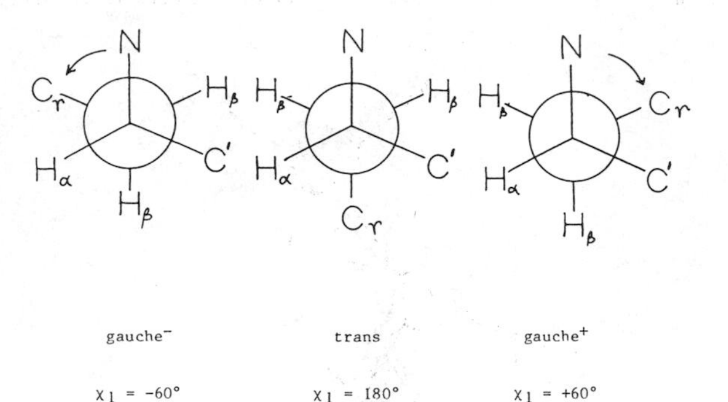

Figure 6-3. Classical rotamers for amino acid side chains; gauche(−), *trans* and gauche(+).

three rotamer states. These predominant rotamers are $\chi = -60^\circ$ or gauche(−); $\chi = \pm 180^\circ$ or *trans*; and $\chi = +60^\circ$ or gauche(+) (Fig. 6-3).

Peptides can be classified as either homodetic or heterodetic.[43] Homodetic peptides have a peptide backbone that contains only the usual amide linkages. Heterodetic peptides contain other types of backbone linkages in addition to amides. The most common are disulfides and ester linkages. Peptides containing ester groups are called depsipeptides. Naturally occurring depsipeptides are commonly of microbial origin. Disulfides occur in many peptides and contribute to the secondary, tertiary, and quaternary structure of proteins. Bioactivity of peptides is often directly linked to disulfide structure.

6.1.2. Structural Elements in Peptides

Covalent forces determine the primary structure of peptides; however, it is noncovalent forces that greatly influence their secondary and tertiary structure. One of the best known is the intramolecular hydrogen bond. Most commonly, a hydrogen bond occurs between the amide NH of one residue and the carbonyl of another residue separated by a certain number of amino acid residues on the peptide chain. These intramolecular hydrogen bonds contribute to the stability of peptide and protein conformations in solution and allow peptides to form the familiar α-helical and β-sheet structures as well as a variety of turns. A turn or bend in a peptide is merely the site where the peptide chain reverses its direction.[11,31,34] Various patterns of hydrogen bonding occur where the H bonds span three, four, and five residues respectively. This leads to different types of turns. The $3 \rightarrow 1$ hydrogen bonds, although not common, are known as C_7 or γ-turns for turns utilizing three residues. A β-turn involves four amino acid residues: i, $i + 1$, $i + 2$, and $i + 3$. A hydrogen bond between the NH of the $i + 3$ residue and the CO of the i residue is often present, but is not a necessary feature of such turns.[44] Turns are characterized by the dihedral angles (ϕ, ψ) $i + 1$ and (ϕ, ψ) $i + 2$ of the two corner residues. β-turns have been classified[177,178] into 10 types based on the values of these ϕ, ψ angles. The major categories are types I, II, and III (Fig. 6-4*a,b,c*). Table 6-1 lists the character-

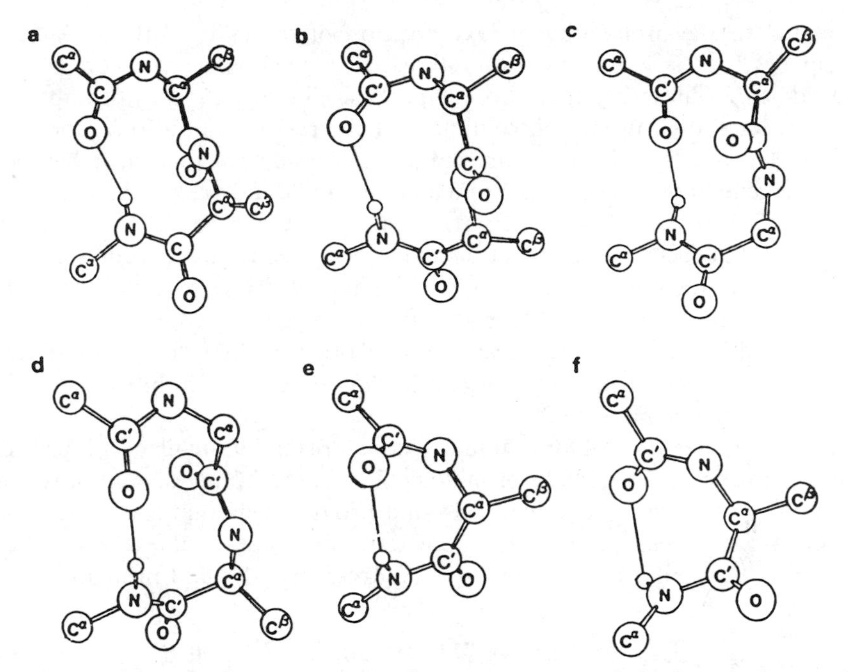

Figure 6-4. Diagrams of structures of various reverse turns. Only those hydrogens involved in hydrogen bonding are shown. Side-chain (C′) carbons are shown in those positions that are considered to be preferred for L-residues according to Venkatachlam[18] (β-turns). (*a*) Type I β-turn. (*b*) Type I′ β-turn. (*c*) Type II β-turn. (*d*) Type II′ β-turn. (*e*) γ-turn. (*f*) Inverse-γ-turn. Types III and III′ β-turns, which are parts of 3_{10} helices, are not illustrated because they closely resemble types I and I′ respectively.

TABLE 6-1. Dihedral Angles of β-Turn Types

Type	$i+1$	$i+1$	$i+1$	$i+2$
I	−60	−30	−90	0
I′	60	30	90	0
II	−60	120	80	0
II′	60	−120	−80	0
III[a]	−60	−30	−60	−30
III′	60	30	60	30

[a] Actually a 3_{10}-helix.

istic ϕ, ψ angles for these three β-turn types. The bend character can be maintained despite deviations from these values and the lack of a $4 \rightarrow 1$ H bond.[44,45]

An important point to note is that turns in peptides are stabilized by the formation of noncovalent medium-sized rings. For example, the β-turn generally results in a 10-membered H-bonded ring, and for this reason is sometimes referred to as a $^{3}10$ or C_{10} turn. Recently there has been an increased interest in studying turn structures.[31,34] Reversal of peptide chain direction is a key feature of the closure of rings in cyclic structures as the rate of cyclization is related to the probability of juxtaposition of the ends of the open-chain precursor.[46,47] Since turns tend to optimize both backbone chain compactness and side-chain clustering, it has been hypothesized that turns may be important not only for stabilization of three-dimensional structure, but also for the interaction of peptides with macromolecules (e.g., receptors). In fact, it has been found in general that peptides contain a large number of turns in their bioactive sequences. Further evidence supporting this hypothesis comes from the synthesis of small cyclic fragments of bioactive peptides. Highly potent analogues of the tetradecapeptide somatostatin have been made by head-to-tail cyclization of the active tetrapeptide sequence as part of a cyclic hexapeptide.[4,48] Likewise, a shortened sequence of the linear tridecapeptide α-MSH, forced to adopt a turn structure by cyclization through side-chain to side-chain links, resulted in superactive cyclic analogues.[49,50]

Understanding the conformational characteristics of small cyclic peptides may aid in the understanding of biological activity. It may also improve our ability to synthesize desired analogues and provide compounds that can serve as models for correlating spectral parameters with topological features that can be used in the conformational analysis of larger peptides and proteins.

6.2. Physical Methods for Peptide Conformational Analysis

6.2.1. Introduction

The experimental methods currently available for determination of peptide conformation rely on a combination of solid-state and solution techniques. The earliest efforts to determine conformational properties of peptides involved the use of X-ray crystallography.[51] The importance of solution conformation was noted at about the same time by Schwyzer et al.[46,52] in his work with cyclic hexapeptides. Since then the study of peptides in solution has grown tremendously, involving the use of a series of physicochemical measurements—NMR, CD, IR, UV, dipole moments, etc.—as well as theoretical calculations. With such a "composite" approach, it is possible to obtain a reliable structural picture of a molecule, whereas the use of only one or two methods often gives equivocal results.[13]

Foremost among the methods available is nuclear magnetic resonance (NMR) spectroscopy, which can provide the most detailed conformational information. Another important technique particularly suited to peptide study is circular dichroism (CD) spectroscopy. Although it does not possess the high-resolution conformational analysis capabilities of NMR spectroscopy, it provides a way to determine secondary structure such as α-helix and β-sheet, in conformationally flexible structures. It is extremely useful in examining and comparing the relative conformation or rigidity in a series of similar analogues. Other less frequently used but important methods include fluorescence, infrared (IR), and Raman spectroscopies. These methods are often used to complement other biophysical measures and to test conformational models.

6.2.2. Solid-state Studies

Of all the physical–chemical methods of analysis, X-ray diffraction studies can provide the most exact information about a peptide's three-dimensional structure. However, a limiting factor in X-ray studies is the availability of suitable crystals, and this has been a particularly serious problem in peptide chemistry. However, many of the medium-size ring-containing peptides have been crystallized. Interestingly, cyclic peptides are often easier to crystallize than linear peptides, and this has been ascribed to the reduced number of conformations in solution.[53]

A second, more fundamental problem of solid-state studies of peptides is that there is no reason to assume that small flexible molecules will have the same conformation in solution or, for that matter, at a biological receptor as found in a crystal. In fact, recent studies have shown that peptides can exist in several conformational states in a single crystal.[54–56] Furthermore, X-ray studies generally do not provide insight into the dynamic properties of peptides in solution, such as conformational equilibria, averaging, and interconversion dynamics. Crystal structures are dominated by intermolecular forces, while intramolecular forces are more important for solution conformations. Therefore, the crystal conformation may have little to do with biologically relevant structures. The primary usefulness of X-ray analysis lies in its ability to provide a starting point in terms of bond lengths and bond and torsional angles for theoretical and dynamic properties studies that seek to determine low-energy conformational minima of peptides. When used in combination with structure–biological activity relationships and other conformational studies, such analysis often can provide useful insights into possible relationships between peptide conformation and biological activity.

A second technique that may help bridge the gap between solution studies and crystal studies is solid-state NMR. High-resolution spectra from solids can now be obtained through the use of cross-polarization magic angle spinning (CP-MAS) techniques.[57–60] Solid-state NMR is particularly valuable when no suitable crystals are available for X-ray studies. This technique has also been applied to the study of crystalline peptides where a comparison of the peptide's solution and solid-state conformation is desired. Conformational changes upon dissolution have been followed using this technique.[6,61–63]

6.2.3. NMR Spectroscopy

Nuclear magnetic resonance spectroscopy is a valuable technique for the study of solution conformations of peptides. Information can be extracted from a peptide's NMR spectrum that can answer questions about (1) hydrogen bonding in the molecule, (2) proximity of atoms or residues, (3) dihedral angles of the peptide backbone and side chains, (4) *cis-trans* isomerization of amide bonds, (5) molecular dynamics phenomena, and (6) in favorable cases, the overall conformation and topology of the peptide.

The usefulness of NMR depends upon being able to assign all resonances in a peptide's spectrum. Until recently, this involved extensive decoupling experiments, and often the use of selectively deuterated derivatives. Application of two-dimensional (2-D NMR) techniques[64–72] has revolutionized assignment and extraction of relevant spectral parameters for peptides. Among the more widely used techniques are homonuclear,[73–78] and heteronuclear,[79,80] 2-D *J*-resolved spectroscopy,[81] and heteronuclear 2-D shift-correlated[79,80,82,83] and homonuclear 2-D shift-correlated spectroscopy.[77,80,82–85] New techniques that often can give critical insight into intramolecular distances and other conformational and dynamic properties are being introduced and exploited at an increasingly accelerated rate.

6.2.3.1. Hydrogen Bonds. Hydrogen bonding is one of the major noncovalent interactions that provides conformational stability to small peptides in solution. The orientation of peptide bonds is defined by hydrogen bonding either intramolecularly or with the solvent. NMR can be used to determine the presence of intramolecular hydrogen bonds through hydrogen–deuterium exchange studies, by examining amide proton chemical shifts as a function of solvent, and by following the temperature dependence of amide proton chemical shifts. The amide protons of intramolecular H bonds will exchange more slowly and will be less affected by solvent and temperature than will less shielded or exposed amide protons. In DMSO, for example, temperature gradients for the chemical shift of the NH signal greater than 4×10^{-3} ppm/°K generally are considered evidence of external NH orientation.[11] Values under 2×10^{-3} ppm/°K are considered to indicate solvent shielding or intramolecular H bonding.[20] Newer techniques of N^{15} NMR can also detect hydrogen bonding in peptides. The amide nitrogen will exhibit a down-field shift when its neighboring carbonyl is involved in a hydrogen bond.[86–89]

6.2.3.2. Proximity. Several measures of proximity can be derived from NMR studies. It is well known that the chemical shift for the C proton in free amino acids is sensitive to the ionization state of the amine and carboxylic acid functions.[90–92] Alpha carbon protons are also sensitive to the orientation of the nearby carbonyl group. Anisotropy effects from the carbonyl will shift the alpha carbon proton to lower fields if it is located in the same plane as the carbonyl group.[77,84,93–95] Likewise, anisotropy effects from side-chain aromatic groups will influence the chemical shifts of nearby protons, shifting them to a higher field if they are above the aromatic plane in the shielding region and to a lower field if they are in the deshielding region of the benzene ring. Implications of spatial arrangement can be drawn from these shifts.[96]

Another measure of proximity comes from nuclear Overhauser enhancement (NOE) studies.[97] In one-dimensional NMR, these effects are seen as a change in signal intensity following irradiation at the resonance frequency of a nearby nucleus. Homonuclear intramolecular NOEs between protons are caused by a dipolar relaxation mechanism and there is a correlation between them and the interproton distance. However, NOEs can usually only be observed if the distance between atoms is no greater than 3–4.5 Å.[98] Both NOE-difference spectra[99,100] and 2-D NOE spectroscopy[101–106] have been used successfully to sequence peptides without the need for deuterated derivatives. NOEs will be observable between the CH and NH of adjacent residues due to the *trans* nature of the peptide bond, which places these protons within 3 Å of each other. NOE measurements have also been used successfully to determine the presence and type of turns in both linear and cyclic peptides.[107]

6.2.3.3. Dihedral Angles. The potentially most useful information for the conformational analysis of peptides that can be obtained from the NMR spectrum are the spin–spin coupling constants (J) for the NH—C_αH protons and for the C_αH—C_βH protons. The vicinal $^3J_{NHC\alpha H}$ constant depends on the dihedral angle θ between the H—N—C_α and the N—C_α—H planes and J and θ can be related by a Karplus-like relationship. Equations relating J and θ have been both theoretically and semiempirically derived.[108–110] Once derived, the dihedral angle can be directly related to the torsional or rotational angle by $\phi = |\theta - 60°|$. Unambiguous results can only be obtained for J values less than 3 Hz or greater than 9 Hz. Measured coupling constants between 3 and 9 Hz can correspond to four different bond angles. The NMR time scale is slow when compared with the speed of many conformational interconversions that occur in microseconds or faster times. This conformational averaging or equilibrium between various conformers can lead to the observation of averaged coupling constants. In these cases, the conformational parameters obtained may not represent any of the true low-energy conformations actually assumed by the peptide. In many cases, especially for the C_αH—$C_\beta H_2$ systems, due to second-order coupling, accurate J values can only be derived from simulation spectra where one assumes various chemical shifts and coupling constants, obtains theoretical spectra, and by an interactive process compares them with the experimentally derived ones until a proper “match” is obtained.

Coupling between the protons on the alpha and beta carbons can be used to determine the dihedral angle χ by using the Karplus relationship, $J = A \cos^2 \theta + B \cos \theta + C \sin^2$.[111–114] In this case, it is assumed that a rapid equilibrium exists between several side-chain rotamers. If the electronegativity of substituents is accounted for and reliable J values obtained, this method can be used to acquire relative residence times and free-energy differences for the rotational isomers.[115,116]

With the improvement of high-field instruments and advent of modern techniques such as INEPT[117] and heteronuclear double-quantum spectroscopy, both C-13 and N-15 nuclei can be used to further enhance the ability to determine Φ and Ψ angles.

6.2.3.4. *Cis-trans* Isomers. Due to its partial double-bond character, the peptide bond can exist in either a *cis* or a *trans* configuration. The barrier to rotation is of the order of 17–20 kcal/mol and the *trans* (Z)-conformation is energetically favored. The local interactions between C_α carbons of two adjoining residues in the *cis* form are believed to be primarily responsible for its higher energy rather than the *trans* form. The *trans* conformation can be deduced from heteronuclear coupling constants between the amide N and its directly bonded proton using the INEPT technique.[117] In linear and macrocyclic peptides, *trans*-peptide linkages have been observed in nearly every case except for X-Pro bonds. However, in peptides containing N-alkyl amino acids, the difference in energy between *cis*- and *trans*-peptide bonds is decreased and the possibility exists for the presence of both configurations. Much work has been done with the amino acid proline. Carbon-13 NMR can be used with reliability to determine *cis-trans* isomerization in proline peptide bonds. In particular, the chemical shifts of the C_β and C_γ atoms will give information on the stereochemistry at this bond.[17,118]

6.2.3.5. Molecular Dynamics. Conclusions about the mobility of a peptide's backbone or side-chain groups are usually made on the basis of C-13 spin-relaxation times T_1.[11,70] The exchange of NH protons with solvent or the transition between several conformations often results in line broadening or signal coalescence when the energy barriers involved are in the range of 5–25 kcal/mol.[119–121] Unfortunately, only limited conclusions can be drawn by following exchange with deuterons or by saturation transfer techniques unless one is confident of a rather rigid conformation. Fortunately, this is often the case with small ring peptides.

6.2.4. Optical Rotary Dispersion and Circular Dichroism

Many types of spectroscopy (NMR, IR) primarily measure properties of individual groups, with conformational effects providing minor perturbations of the basic spectrum. Optical activity depends upon molecular geometry. For peptide chromophores, optical activity is the result of an asymmetric array of interacting chromophores and perturbing polar groups. Optical rotary dispersion (ORD) and circular dichroism (CD) can be powerful probes into the conformation of peptides in solution. ORD measures a molecule's ability to rotate a plane of linearly polarized light as a function of wavelength while CD evaluates the unequal absorption of right- and left-handed circularly polarized light by a molecule.[122,123]

Amide groups are the most abundant chromophore present in peptides, and the parameters of the ORD and CD curves of peptides are determined by the mutual orientation of these chromophores. All optically active peptides display chiroptical effects at 180–205 nm and 210–250 nm corresponding to $\pi \rightarrow \pi^*$ and $n \rightarrow \pi^*$ transitions respectively. The peptide chromophore region is composed of broad overlapping bands arising from this ensemble of transitions. Therefore, this spectrapolarimetric method is not good for obtaining detailed structural data but is particularly sensitive to changes in the overall conformational states of molecules. Early studies using optical methods relied on ORD. With the appearance of commercial CD instruments in the mid-1960s, CD became the preferred technique because each optically active electronic transition gives rise to only one CD band instead of a (+) and (−) signal as with ORD.

The earliest conformation properties of peptides studied were the onset of helicity and the presence of α-helix and β-structure in homologous polypeptides. The α-helical content of peptides and proteins can be quite accurately quantitated by CD because of its characteristic spectrum. These studies were based on model peptides and they examined the CD bands attributable to the $\pi \rightarrow \pi^*$ and the $n \rightarrow \pi^*$ transitions of the amide bonds that dominate the CD spectrum below 250 nm.

More recently, CD has been used to study turns, polyproline helices, and disulfide conformations in biologically active linear and cyclic peptides.[123,129] Interpretation of CD spectra is often fraught with problems due to interfering chromophores or to the fact that several different conformational features can yield similar CD spectra. Nonbackbone chromophores that can interfere with the interpretation of amide bands include aromatics residues (Phe, Tyr, Trp), which absorb in the range of 240–300 nm; cystine disulfide and histidine, which also absorb in the 185–240-nm range; and the disulfide in the 250–320-nm peptide region.[130]

Although CD and ORD have limited value for the direct determination of three-dimensional peptide structures, they can give unequivocal evidence as to the persistence of alteration of conformational states or of the environment of the peptide.

6.2.5. *Vibrational Spectroscopy*

Vibrational spectroscopy (IR and Raman) has been used extensively as a complementary method of peptide conformational analysis. Vibrational spectroscopy has a fast time scale (10^{-15} seconds), and vibrational bands for all species are seen. However, it is the amide bands and disulfide vibrations that have been the most thoroughly investigated.

IR spectroscopy has been most reliable in the studies of amide bands. Amide vibrations have been found to be sensitive to hydrogen bonding.[19,131–135] The presence of NH (amide A) bands at 3430–3480 cm^{-1} corresponds to free NH groups while bands in the 3300–3380-cm^{-1} region are evidence for hydrogen bonding of the amide protons.[136] Amide A frequencies have also been used to discriminate between *cis* and *trans* secondary amide bonds. The amide I region (carbonyl bands) is also useful in that band position can be correlated with the presence of α-helix, random coil, and β-structure.

IR can be used to study peptides in a crystalline state, in nonpolar solvents and in argon matrices. However, the study of peptides in aqueous media is difficult because water is opaque to IR below about 1500 cm^{-1}.

Raman spectroscopy, however, is well suited to the study of peptides in aqueous solutions since water is not a Raman scatterer. Raman has the potential to yield substantial structural information, but it has not been used extensively. Its most common use has been in the attempt to define disulfide dihedral angles in cyclic peptides. Its application to the determination of turn conformations has been discussed by Fox et al.,[137] Williams,[138] and Williams and Dunker.[139]

One disadvantage of Raman spectroscopy is its low sensitivity, which often necessitates concentrations in excess of 1 mM, though enhancements using modern computer systems are now possible, thus making selected studies possible in the micromolar range in favorable cases.

6.3. Eight-Membered-Ring-Containing Peptides

Cyclic dipeptides, or diketopiperazines, make up the simplest group of cyclic peptides. They contain two amino acids linked through *cis* amide bonds to form a six-membered ring. They will not be the subject of this review, although extensive conformational studies have been performed on them.[140–145] However, several cyclic dipeptides do exist that contain medium-sized rings. These are cysteinyl cysteine disulfide (**1**), cyclo cystine (**2**), and cyclo di-β-alanine (**3**), which all possess eight-membered rings (Fig. 6-5).

6.3.1. Disulfide Loops

Disulfide bridges between pairs of cysteine residues are the most important covalent cross-linkages found in peptides and proteins. Therefore, it is not surprising that simpler disulfide-containing compounds have been sought as models for conformational studies of the disulfide moiety. Greenstein[146,147] was the first to synthesize *L*-cysteinylcysteine and then to study its oxidation to the disulfide.[148,149] The yield of cyclic monomer formed was dependent upon the pH, giving a 70% yield at pH 6.5 but only a 35% yield at pH 8.5. With the decrease in monomer, concomitant increase in cyclic dimer formation occurred.[150] The cyclization of a series of disulfide peptides of the general formula Cys(Gly)$_n$Cys ($n = 0$–6) was examined, and it was found that the proportion of monomeric cyclic disulfide obtained increased with increasing ring size.[151–153] The most difficult ring size to form seemed to be the 11-membered cycles of the formula Cys-Gly-Cys.

The question of the configuration of the peptide bond in *L*-cysteinyl-*L*-cysteine disulfide has been investigated by theoretical studies[154,155] and by IR on closely related analogues.[156] These studies suggested the presence of a *cis*-amide

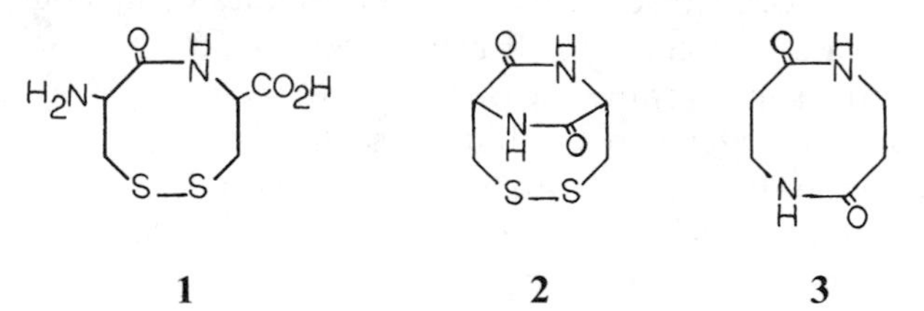

Figure 6-5. Structures of cysteinyl-cysteine disulfide (**1**) cyclo cysteine (**2**) and cyclo-di-β-alanine (**3**).

bond. This was later confirmed by X-ray crystallography on analogues of the dipeptide.[157,158]

More extensive conformational work has been done on cyclo-*L*-cystine (referred to as CC). CC is a bicyclic compound containing a diketopiperazine ring, which is bridged by a disulfide at the β-carbons to form an eight-membered ring. It is easily prepared by amide cyclization of *L*-cysteinyl-*L*-cysteine disulfide (**1**). The ease of cyclization lends further evidence to the presence of a *cis*-amide bond in cysteinylcysteine disulfide.[159] CC was investigated as a simple model for applying and testing the theory of chiroptical properties of disulfides by Donzel et al.[159] From UV, ^{1}H NMR, and CD data, Donzel et al. determined that the diketopiperazine ring was in a boat conformation and the disulfide dihedral angle was 90°. Donzel et al. pointed out that CC could exist in conformations with either a right-handed (P) or left-handed (M) disulfide with $[\chi_{s-s}] = 90°$. They concluded from chemical-shift arguments that only one conformer, the P conformer, was present. Applying the quadrant rule, they also concluded that the P form should have a significantly more negative $n \rightarrow \sigma^*$ rotational strength than the M form. Jung and Ottnad[160] observed splitting of C-13 signals for cyclo cystine and noted their temperature dependence. They also noted a slight temperature dependence of the CD spectrum. They concluded that the conformation of CC was a mixture of P-helical and M-helical (15–20%) in solution. Ottnad et al.[161] studied the dissolution of cyclo cystine in ethanol and noted a time-dependent decrease in ellipticity of the negative Cotton effect at 225 nm, and interpreted this as an indication of conformational change in going from the crystal to solution. All these studies maintained that the predominant conformer was the P-helical. A semiempirical energy calculation for cyclo cystine was done by Mitra and Chandrasekaran.[162] The M-helical conformer was found to be of slightly lower energy and to have a greater number of low-energy conformations than the P model. Extended Huckel-type calculations by Gregory and Przybylska[163] also conclude that the M helix is preferred over the P helix. The crystal structures of both the CC-acetic acid complex[164] and cyclo-cystine[165] show the disulfide in an M conformation. The diketopiperazine ring is in a twisted-boat form and the C—S—S—C bridge has an angle of $-94°$. The two peptide bonds are significantly nonplanar with ω values of $-10°$ and $-16°$. The earlier spectroscopic results warrant reinterpretation in light of the observed crystal conformation. Woody[130] argued that crystals studied by Ottnad et al.[161] may be different from those studied by Varguhese et al.,[165] and that a thorough theoretical CD study is needed to ascertain the contribution of the disulfide and amide transitions to the observed spectrum.

6.3.2. *Malformin*

Malformin A, a cyclic pentapeptide known to induce curvature in corn roots and malformation of stems in bean plants,[166] contains the eight-membered-ring —Cys—Cys— fragment[167,168] (see Fig. 6-6). Ptak's NMR and CD studies,

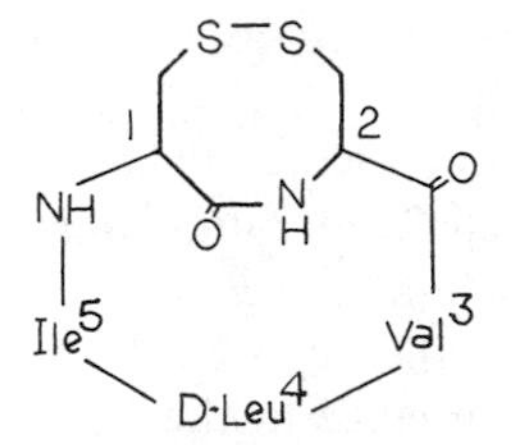

Figure 6-6. Structure of malformin A.

based on an incorrect primary sequence, were reinterpreted by Tonelli.[168] He concluded that there were two structures of malformin possible that were consistent with the experimental data. Both possessed an internally buried *D*-Cys amide proton, a gamma turn between the NH of *D*-Leu and CO of *D*-Cys, and a positive chirality of the —S—S— bond with an angle of 80°. Tonelli also proposed only *trans*-peptide bonds in malformin despite the fact that cysteinylcysteine disulfide is known to have a *cis*-amide bond in its eight-membered ring. Mitra and Chandrasekaran[169] have questioned Tonelli's results and have examined the conformational flexibilities available to malformin A with respect to the *cis-trans* nature of the peptide bond across the S—S bridge and the helicity of this bridge. They studied ring closure by a linked-atom least-squares (LALS) technique.[170] Energy calculations were done for the resulting conformations to assess their relative stabilities. Their study resulted in six different possible conformational states of malformin A, two with *cis*-peptide bonds and four with *trans* bonds. Each of these six conformers is fairly rigid, and barriers to the transition between them would be expected to be high.

6.3.3. Cyclo Di-β-Alanine

Cyclo di-β-alanine, or β-alanine diketopiperazine, first synthesized by Rothe,[171] also forms an eight-membered ring. X-ray diffraction analysis[172] has shown it to possess two *cis*-amide groups. The amide bonds appear to be planar and the molecule adopts a "flexible" C symmetric boat conformation analogous to the boat form of cyclohexane. There is a slight torsional angle of ≈27° in the N—CH_2—CH_2—C′ bonds, giving the overall structure a slightly twisted

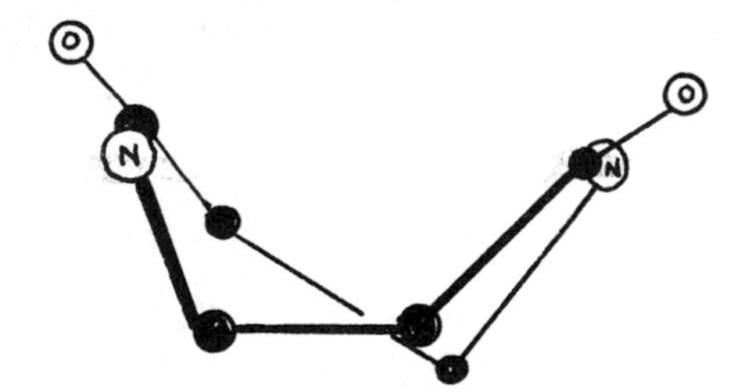

Figure 6-7. Twisted-boat formation of cyclo-di-β-alanine.

appearance (Fig. 6-7). Its conformation has also been shown to be similar to the cyclic alkene, *cis, cis, cis*-cycloocta-1,1-diene.[173,174]

6.4. Nine-membered-ring-containing Peptides

Perhaps the most extensively studied and best characterized medium-size ring-containing peptides are those possessing nine atoms. These are the cyclotripeptides, and although there are no known naturally occurring cyclic tripeptides, these compounds have served as model compounds for spectroscopic studies of peptide conformation.

6.4.1. Synthesis

Until 1965 no synthesis of a cyclic tripeptide had been achieved. Upon cyclization, linear tripeptides with secondary CONH groups undergo dimerization to yield cyclic hexapeptides with unstrained 18-membered rings without even traces of cyclotripeptides.[175] The first successful synthesis of a nine-membered cyclotripeptide, cyclotri-L-prolyl, was done by Rothe et al.[176] Venkatachalam[177,178] predicted that all peptide bonds in cyclic tripeptides would be of a *cis* configuration. Cyclization of linear tripeptides, therefore, requires an appreciable population of *cis*-peptide bonds in the precursor molecule. This only occurrs with N-alkylated amino acids. Cyclic tripeptides have been synthesized which incorporate Pro,[113,176,179–182] Hyp,[113,179,181,182] Sar,[93,183] Aze,[184] Bzl-Gly,[185] and o-nitrobenzyl-Gly. Until recently no cyclic tripeptide possessing a "free" NH group in an amide bond had been observed.[186]

Synthetic studies of the above-reported cyclo-tripeptides indicate various factors that affect the yield of cyclization and occurrence of higher cyclic products. Cyclization of the all *L*-Pro-containing tripeptide results exclusively in the nine-membered ring in greater than 90% yields when cyclization is carried out at "infinite dilution." Even when the concentration is 0.1 M, a 24% yield of c(*L*-Pro)$_3$ can be obtained.[175] This is believed to be due to rigid helical structures that begin to develop at small chain length ($n \geqslant 3$) for polyproline.[187] Boni et al.,[184] in studying tripeptide cyclization of the proline homolog, *L*-azetidine carboxylic acid (Aze), stated that three factors come into play: (1) the number of sites of possible *cis-trans* isomerism of the linear precursor; (2) the deviation from planarity of the peptide groups in the forming cyclopeptide; and (3) and the side-chain rigidity. Furthermore, Rothe and Mastle[175] found in the study of cyclization of *L, D*-triproline derivatives that drastic differences in cyclization tendencies were observed depending upon the configurational sequence of the linear precursors. The *LLL* and the *DLL* precursors were found to form cyclotripeptides in high yield with little higher cycles while the *LDL* and *LLD* sequences led to poor yields of cyclotripeptides and increased yields of cyclic hexapeptides. They proposed a quasi-"cisoid" disposition of end groups with respect to the peptide backbone in the former and a quasi-"transoid" conformation in the latter to explain these cyclization tendencies.

6.4.2. Conformational Studies

Conformational analysis of the cyclotripeptides has involved both solution and solid-state studies in an attempt to define the backbone conformation of the nine-membered rings as well as the side-chain flexibility, particularly in the side-chain pyrrolidine ring.

Using the analogy between the peptide bond and a carbon-carbon double bond, one can compare the backbone conformation of cyclotripeptides with that of the *cis, cis, cis*-cyclonona-1,4,7-trienes.[186] The hydrocarbon's most stable conformation is a "crown" with C_3 symmetry.[188] By substituting a CH_2 group for the olefinic *cis*-double bonds, the nonatriene can be compared with the cyclohexane chair and boat conformations.[189] Cyclotripeptides can adopt either of two backbone conformations—a "crown" or a "boat." In the crown conformation, the three carbonyl groups are all on the same side of the molecule and the three alpha protons point inward on the opposite side in an orientation that is comparable to the axial orientation of the 1,3,5-hydrogens of cyclohexane (Fig. 6-8). The boat conformation differs by the inversion of one α-carbon to the opposite side of the ring, and in unsymmetrical molecules, the possibility exists for the presence of six different boat conformations. These would be degenerate in the case of homologous achiral amino acids (e.g., c[Sar_3]).

Whether the molecule prefers one conformation exclusively or exists as an equilibrium mixture of both conformers depends upon the chirality and relative configuration of the three amino acids in the ring.[77] If all the amino acids are chiral, then the preferred conformation is determined by their chirality. For homochiral tripeptides (i.e., c[*L*-Pro_3]), only the C_3 symmetric crown is allowed and the crown assignment is made from the observation of the equivalence of the three amino acids in their ^1H and ^{13}C NMR spectra.[186] Dipole moment measurements also support the assignment of crown conformations. The orienta-

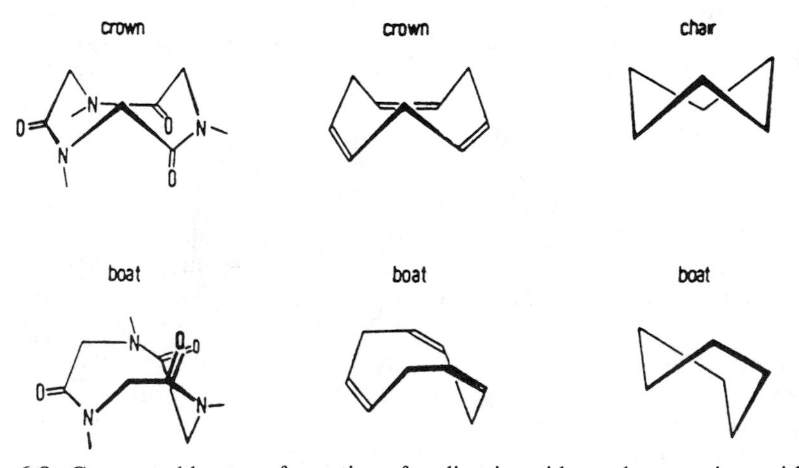

Figure 6-8. Crown and boat conformation of cyclic tripeptides and comparison with the conformational analogues cyclonoatriene-(1,4,7) and cyclohexane. (By permission of H. Kessler. In "Stereodynamics of Molecular Systems," R. H. Sarma, Ed. Pergamon Press: New York, 1979.)

tion of the three C=O groups on one side of the ring results in higher dipole moments for the crown than for the boat. For example, the dipole moment of c(Sar_3) is 4.66 in benzene, which conforms to the three carbonyl groups being on the same side of the ring.[190]

For cyclotripeptides of mixed chirality (e.g., c[*L*-Pro_2-*D*-Pro]), the boat or twisted-boat form is the only possible conformation.[191] This is due to strong steric interactions occurring between substituents at the inwardly oriented α-positions, which dramatically destabilize the crown conformation. When cyclotripeptides contain at least one achiral amino acid (e.g., Sar, N-Bzl-Gly), the crown and boat conformations coexist in equilibrium. Assignment of crown or boat conformation relies on observation of NMR data and comparison of C-13 chemical shifts with cyclotripeptides known to exist exclusively in the crown (e.g., cPro_3) or boat (e.g., c[Pro_2-*D*-Pro]) conformation.

In the case of cyclic tripeptides containing only achiral amino acids (e.g., c[Sar_3], c[N-Bzl-Gly_3]), the crown is again the preferred conformation, although small amounts (6–10%) of boat conformation have been observed by ^1H NMR in relatively nonpolar solvents.[186] Complete ring inversion—crown ⇄ boat ⇄ crown—is only possible in the case of achiral cyclotripeptides with no C^α substituents. The barrier for the process of crown ⇄ boat can be deduced from the high signal coalescence temperatures observed in the NMR spectra for these cyclic tripeptides and is of the order of 18–20 kcal/mol. However, the thermodynamic stability of the flexible boat conformation is similar to that of the crown.[186] On the other hand, cyclo-[*L*-Pro-Bzl-Gly-*D*-Pro] was found to possess two interconverting boat conformations, with one strongly favored.[192]

Cyclo[*L*-Pro_3] has been extensively studied by X-ray crystallography[180,181]; by 220-MHz, 270-MHz, and 500-MHz ^1H NMR[84,113,115,193,194]; and by C-13 NMR.[26,84,179,194] It has been found to exist in the crown form both in the crystal and in solution. Although the backbone conformation appears to be quite rigid, the pyrrolidine ring of the Pro side chain maintains substantial flexibility as observed by X-ray,[180,181,195] ^1H NMR coupling constants,[193] and C-13 relaxation data.[84,194] Interestingly, the X-ray structures of the cyclodepsipeptides cyclo(HyIV-Pro-Pro), and cyclo(D-HyIV-Pro-Pro) show that all peptide bonds are *cis* in these compounds, but the lactone bonds are *trans*.[196]

The only cyclotripeptide known from X-ray analysis and complete NMR studies to exist exclusively in a boat conformation is c[*L*-Pro_2-*D*-Pro].[26,77,86,194,197] In fact, X-ray has found that this boat form is not the "ideal boat" but somewhere in between the ideal and a "twisted" boat conformation. Extensive 2-D NMR studies have confirmed this assignment for the solution conformation as well.[77] Furthermore, the proline rings in c[*L*-Pro_2-*D*-Pro] are not conformationally homogeneous. In the unit cell of the crystal, two molecules of different conformation have been observed by X-ray and solid-state C-13 NMR.[26,191] Kessler[186] has summed up the cyclotripeptides synthesized and the types of backbone conformation they assume. Peptides synthesized since then are included in the summary given in Table 6-2.

Not much CD work has been done with cyclotripeptides of this type. Deber et al.[198] reported that cyclo$(Pro)_3$ has a strong CD spectrum. Another CD study

TABLE 6-2. Possible Cyclotripeptide Conformations Based on Chirality of the Component Amino Acids

Chirality	Conformation	Tripeptide
SSS/RRR	Crown	c[Pro_3] c[Pro_2Hyp]
SSR/RRS	Boat	c[Pro_2D-Pro]
SSO/RRO[a]	Crown + boat	c[Pro_2Bzl·Gly] c[Pro_2Sar]
SRO	Two boats	c[ProBzl·GlyD-Pro]
SOO/ROO	Crown + three boats	c[ProBzl·Gly_2] c[ProSar_2]
OOO	Two crowns + six boats	c[Bz·Gly_3] c[Sar_3]

[a] O means an achiral amino acid.

of cyclo(Pro)$_3$, as well as its lower homologue, cyclo(D-Aze)$_3$, was conducted by Vicar et al.[199] The problem with these studies is in the ambiguous assignment of the CD band to various transitions in the molecule. Woody[123] maintains that detailed theoretical calculations must be done on these systems in order to resolve the questions that have arisen.

6.4.3. Nine-membered Cyclols

As we saw earlier, ring closure of tripeptides with secondary CONH amide groups to yield nine-membered peptide rings is generally not successful. The usual products are cyclic dimers and oligomers. In 1965, Shemyakin and co-workers[202] noted that incorporation of hydroxyl or aminoacyl groups into cyclic amides takes place with ease only when the size of the resulting ring exceeds a certain critical value—that value being greater than 10 atoms. In 1971, Rothe and co-workers[200] postulated the formation of cyclols (Fig. 6-9) as tetrahedral intermediates during tripeptide cyclizations. They are presumed to form as a result of transannular interaction between spacially close amide groups in the highly strained nine-membered rings. Although cyclols do not possess medium-sized rings, they are bicyclic isomers that are tautomeric with medium-sized cyclopeptides,[201] cyclodepsipeptides,[202] and cyclothiodepsipeptides.[203]

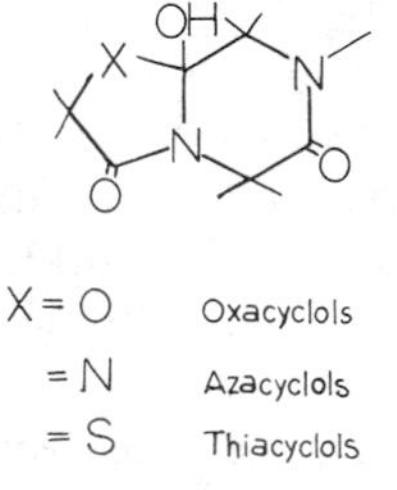

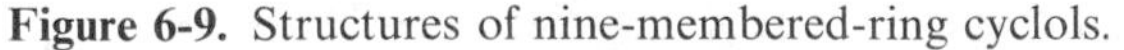

Figure 6-9. Structures of nine-membered-ring cyclols.

Rothe et al.[201] prepared azacyclols via an aminoacyl insertion reaction on diketopiperazines in order to determine their stabilities and properties. Investigations of the cyclization of a variety of tripeptides containing secondary amide bonds have led to a complete picture of possible reactions and side products formed by isomerization, dehydration, oxidation, and hydrolysis of azacyclols-cyclotripeptides. These possible products are illustrated in Fig. 6-10.

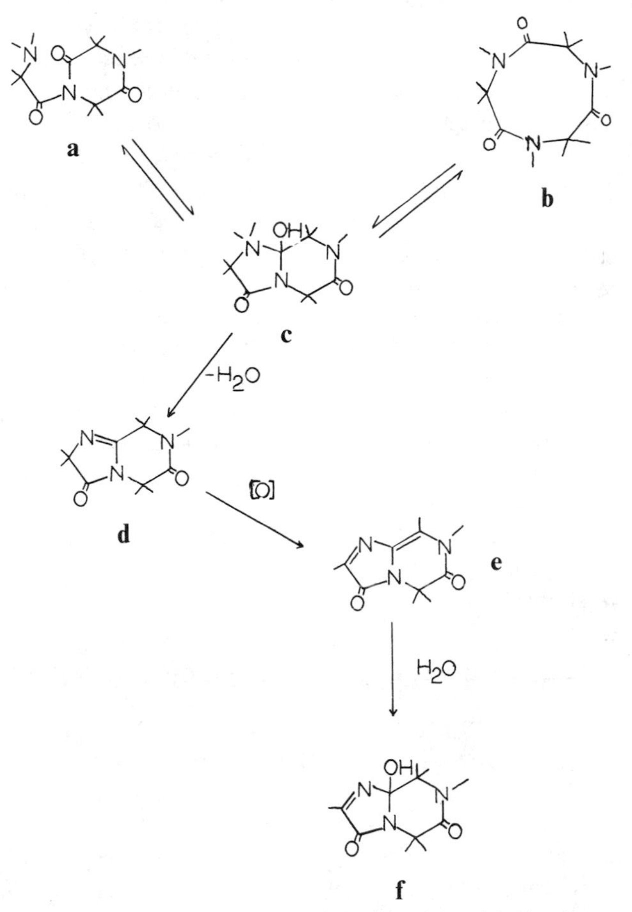

Figure 6-10. Reaction products formed by isomerization, dehydration, oxidation, and hydrolysis of azacyclols-cyclotripeptides.

a = aminoacyl diketopiperazine

b = cyclic tripeptide

c = free azacyclol

d = acylamidine

e = dehydroacylamidine

f = dehydrocyclol

The first proof of formation of a cyclol as an intermediate during peptide cyclization came from the isolation of an acylamidine as one of the products in the cyclization of Ala-Phe-Pro-p-nitrophenyl ester.[201] This product could only result from the dehydration of the cyclol. The first X-ray structure of a tripeptidic cyclol was reported by Lucente et al.,[204,205] who more recently[206] examined the formation of tripeptidic azacyclols from Ala-Phe-Pro-ONp·TFA and from phenylacetyl-Ala-Phe-Pro-ONp. Rothe et al.[207] studied the cyclization of 12 secondary amide-containing tripeptides. In the case of c(Pro-Phe-Pro), the cyclol was isolated in a 20% yield. The stability of peptidic cyclols isomeric with homodetic and heterodetic nine-membered cyclotripeptides depends on the nature, configuration, and sequence of the three amino acids involved. An interesting recent example is that reported by Zanotti et al.[208] with tripeptides containing cysteine as the N-terminal residue.

The first synthesis of a nine-membered cyclotripeptide containing a secondary amide bond was also accomplished by Rothe and Mästle.[175] Cyclo(D-Phe-Pro-Pro) was obtained in a 22% yield and found to undergo slow cyclolization in protic solvents. Investigations into the mechanism of cyclol formation show that tripeptides of the general formula X-Pro-Pro form cyclols via transannular interactions between two peptide groups, while those of structures Pro-X-Pro and X-Y-Pro form cyclols via aminoacyl incorporation into diketopiperazine intermediates.[209] Rothe and co-workers[209] have found c(Pro-Val-Pro) to be the most stable secondary amide-containing cyclic tripeptide obtained so far. It is the first cyclotripeptide to be observed with a *cis-trans-cis* amide bond configuration and the isomerization between *cis* and *trans* forms can be observed by the amide II band at 1515 cm^{-1} in the IR spectrum. The *trans*-amide bond has also been confirmed by X-ray analysis and the entire molecule assumes an unusual twist conformation.[209]

6.5. Ten-membered Cyclic Peptides and Cyclols

Relatively little attention has been devoted to the study of 10-membered cyclic peptides. Rothe and co-workers[175] found that c(β-Ala-Gly-Pro) could not be obtained from direct cyclization of the linear tripeptide precursor, but rather required aminoacyl incorporation from β-alanyl diketopiperazine. Rothe hypothesized the intermediate formation of an unstable azacyclol containing two fused six-membered rings. Relative to nine-membered cyclotripeptides, the 10-membered analogues should possess less ring strain and thus have a reduced tendency toward transannular interactions between amide groups. Pinnen and co-workers[210,211] have investigated these ring systems in the hope of isolating the azacyclols tautomeric with 10-membered cyclotripeptides. By incorporating N-methyl anthranylic acid (N-Me-Ant) as the N-terminal residue, they were able to isolate in 33% yield an azacyclol containing two fused six-membered rings upon cyclization of N-Me-Ant-Phe-Pro-ONp. The 10-membered cyclotripeptide was also obtained in 17% yield. Both compounds were stable in the solid state and in chloroform. However, when dissolved in hydroxylated solvents, the cyclol slowly tautomerized to the cyclic peptide.

When the tripeptides c(β-Ala-Gly-Pro) or Boc-Ser-Phe-Pro-ONp were cyclized, 10-membered rings formed. In these cases, there is less strain in the ring and, therefore, the cyclotripeptide is preferred to the tautomeric azacyclol, which would contain two fused six-membered rings.

6.6. Twelve-Membered-Ring-Containing Peptides

The largest group of 12-membered-ring-containing peptides is the cyclic tetrapeptides. Unlike the cyclic tripeptides, the cyclotetrapeptides include naturally occurring biologically active compounds such as chlamydocin, tentoxin, Cyl-2, and the AM-Toxins. Cyclotetrapeptides particularly those consisting of imino acids, are also noted for their ion-binding capacities.

6.6.1. Synthetic Cyclic Tetrapeptides

The first synthetically produced cyclotetrapeptide was $c(Gly)_4$.[212] Early theoretical studies on $c(Gly)_4$ and $c(Ala)_4$[213] based on contact criteria and potential energy calculations concluded that nonplanar *trans*-peptide bonds were necessary for ring closure. These bonds deviated from planarity by 10–15°. Furthermore, the cyclic structure would possess fourfold symmetry with ϕ, ψ angles similar to those of the α-helix. In fact, Balasubramanian and Wetlaufer[214] had found the ORD of c(Ala) to be similar to the ORD of α-helical structures. Later, Grathwohl and co-workers[215] performed ^{13}C and ^{1}H NMR studies on $c(Gly)_4$ and concluded that the four peptide groups were magnetically equivalent on the NMR time scale, but that they were not in a standard *trans* or *cis* form. They termed this form "*transoid*." The proton NMR shows an A_2X pattern for the $CH_\alpha NH$ signals and indications of solvent-exposed amide protons. The C-13 shifts do not correspond with either *cis* or *trans* amide models.

When Groth[216] examined the crystal structure of cyclotetrasarcosine, $c(Sar)_4$, he found that the four amide bonds were not all the same but alternated their configurations in a *cis-trans-cis-trans* manner. Again, none were strictly planar. The two *trans* groups were found to be close enough (3.08 Å) to exert a stabilizing effect on the ring through transannular attractions. Similar conclusions were drawn from solution NMR studies of five sarcosine-containing cyclic tetrapeptides.[94] $Cyclo(Sar)_4$ exhibited a characteristic splitting pattern of the four $—CH_2—$ groups, indicating a rigid centrosymmetric conformation. The coalescence temperature for the NMR signals was quite high (180–200°C), indicating a high barrier to ring inversion. This is presumably due to the strong transannular interactions across the ring. When secondary amide bonds occur in the cycle (substituting Gly or Ala), the same *cis-trans-cis-trans* pattern is observed; however, the coalescence temperatures drop dramatically. Dale and Titlestad[183] also studied all combinations of Gly-Sar and Ala-Sar in cyclotetrapeptides and found that chirality of the amino acids were crucial to cyclization yields. Deber and co-workers[198] examined the cyclic tetrapeptide $cyclo(L\text{-}Pro\text{-}Gly)_2$ and found that it too has a *cis-trans-cis-trans* sequence of peptide bonds. However, it also showed evidence for slow rotation about the proline ψ angles adopting *trans'* and *cis'* forms. These are characterized by values of 120° and −60°, respectively, for the Pro C_α—CO bonds (ψ angles). Evidence for this motion came from ^{13}C studies at varying temperatures. The C_α of Pro and Gly exhibited splitting at 0°C while the C_β signal of Pro did not shift, indicating the persistence of the *cis* Pro peptide bond. Neither c(D-Pro-Gly-*L*-Pro-Gly) nor c(*L*-Pro-Sar)$_2$ showed coalescence behavior in their NMR.

The all-*cis* cyclic tetrapeptide structure was shown by Sarathy[217] not to be stereochemically favorable. However, the crystal structure and NMR studies of the cyclic tetrapeptide, cyclo(*L*-Pro-Sar)$_2$, indicated that it possesses four *cis* peptide bonds with angles that deviate by 2–10° from the ideal value of 0°.[218–220] A minor asymmetric conformation, characterized by *cis-cis-cis-trans* peptide bonds, was also found. This conformation was seen to increase with decreasing solvent polarity in the NMR experiments. However, cyclo(*L*-Pro-Val)$_2$ and cyclo(*L*-Pro-*D*-Val)$_2$ have *cis-trans-cis-trans* conformations in $CDCl_3$ and DMSO-d_6, respectively, but the latter compound appears to have a *trans* conformation in trifluoroethanol-d_3.

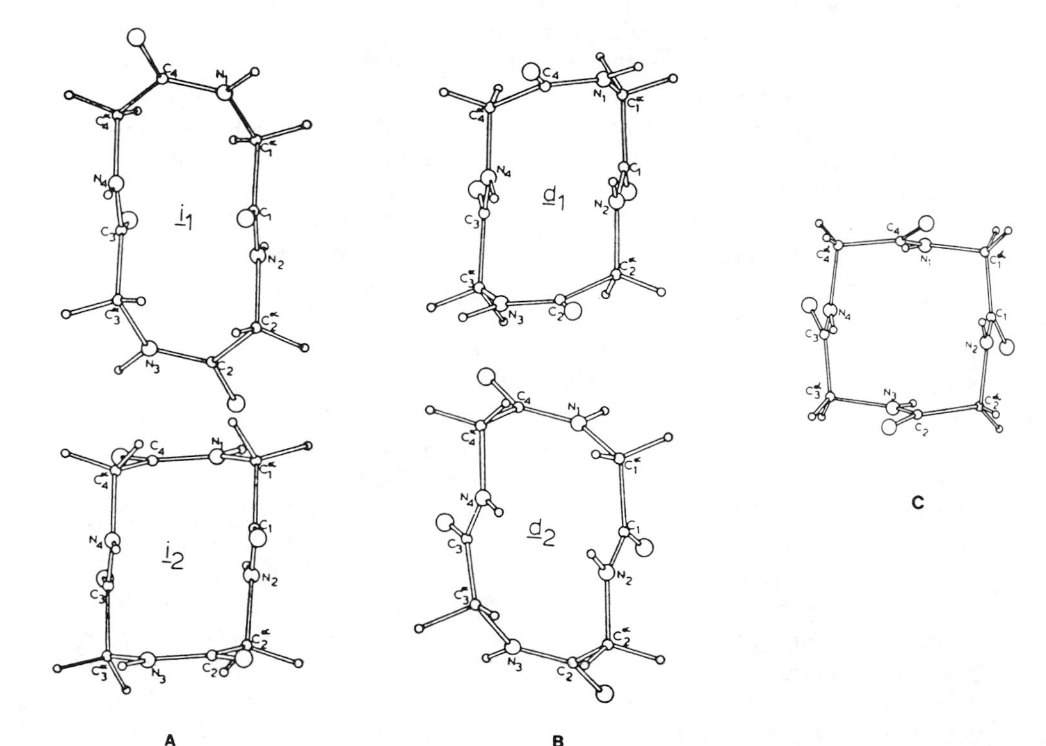

Figure 6-11. Three major conformational forms of cyclic tetrapeptides. (By permission of Kawai, M.; Jasensky, R.; Rich, D. H. JACS, **1983**, *105*, 4456–4462.)

Theoretical examinations of the 12-membered rings of cyclic tetrapeptides have shown that these peptides can adopt at least three distinct conformations (Fig. 6-11). These conformations depend upon the sequence of chiral amino acids and the presence, number, and sequence of tertiary amide bonds,[216,221,222] The *d* types (d_1 and d_2) possess a two-fold axis and the *i* types (i_1 and i_2) have a center of inversion. The S_4 type possesses all-transoid amide bonds. The "*i* symmetry" conformer was found first[216] and the *i* types include all the sarcosine-containing tetrapeptides. Approximate d_2 symmetry was found in the crystal structure of c(Phe-Pro-Ala-Pro)[223,224] and its epimer c(*D*-Phe-Pro-Ala-Pro) was found to be an asymmetric version of d_2. Although the S_4 symmetric conformation has not yet been unambiguously observed, a related all-transoid cyclic tetrapeptide conformation has been established for the solution conformation of chlamydocin peptides.[225,226]

Although it is a tripeptide, cyclo tri-β-analyl (CTBA) contains a 12-membered ring. Theoretically, it can accommodate all *trans* amide bonds without excessive strain. A comparison study involving energy minimization of CTBA and its cyclic alkene analogue, t,t,t-cyclodeca-1,5,9-triene (CDDT), showed no conformational relationships.[227] NMR and X-ray analysis were also attempted on CTBA. However, the ^1H NMR results were ambiguous, indicating the presence of several conformations. After a two-year effort, X-ray studies produced no interpretable electron-density map indicating high thermal motion and multiple conformations in the crystal. Energy minimization likewise determined 12 possible unique low-energy conformations. Apparently this 12-membered ring with no side-chain substituents is quite floppy, indicating that side-chain groups contribute substantially to rigidification of medium-sized cyclic systems.

6.6.2. *Chlamydocin*

Chlamydocin (Fig. 6-12) is a cytostatic cyclic tetrapeptide isolated from *Diheterospora chlamydosporia*. An X-ray study of a closely related analogue,

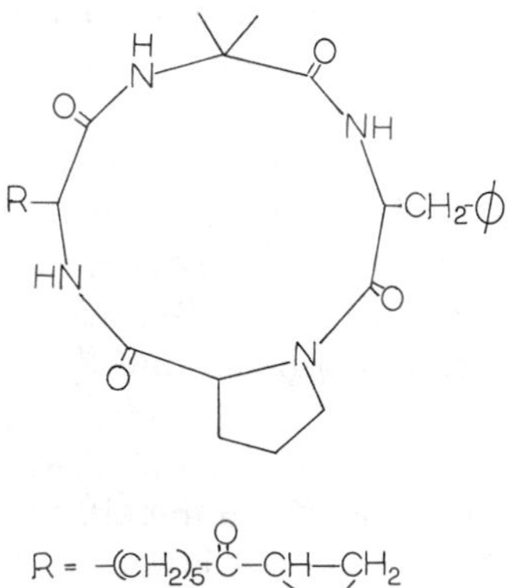

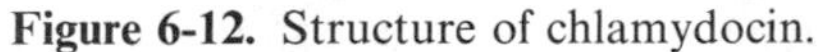

Figure 6-12. Structure of chlamydocin.

dihydrochlamydocin,[228] showed the peptide to contain four transoid amide bonds. The ω angles varied from 14 to 24° from the ideal value of 0°.

A number of related cyclic tetrapeptides have been synthesized and extensively studied by both NMR and CD. Rich and Jasensky[229] synthesized c(Aib-Phe-*D*-Pro-Ala) or Ala-chlamydocin—a model for the chlamydocin ring system in which Ala replaces the epoxy ketone amino acid, *L*-2-amino-8-oxo-9,10-epoxidecanoic acid (Aoc). ^{13}C resonances for the Pro C_β and C_γ indicated a *trans X*-Pro amide bond, while temperature-dependency studies for the amide protons indicated the presence of two solvent-shielded or H-bonded amide protons. Both the CD spectrum and the NMR high-field shift of the Pro C_α indicated the possibility of a γ-turn with two $3 \rightarrow 1$ intramolecular hydrogen bonds, one between the Ala-NH and the Phe-CO and the other between the Phe-NH and the Ala-CO.

In a search for optimal conditions for the preparation of [Ala^4]chlamydocin, Pastuszak and co-workers[230] found that of four possible sequences of the linear tetrapeptide precursors, only one resulted in a favorable cyclization. Ala-Aib-Phe-*D*-Pro-OSu could be cyclized to give a 44% yield of desired cyclic tetrapeptide while the other three sequences resulted in only 2–3% yields. The authors could offer no explanation for this difference but suggested studies of the solution conformation of the linear precursors to answer these questions. Unfortunately, no such studies have been made to date.

Further conformational analysis of [Ala^4]chlamydocin by Kawai et al.[231] has turned up evidence for the existence of two conformers in solution. Conformer I, which is preferred in pure chloroform or dimethyl sulfoxide, possesses four transoid amide bonds and the two bis γ-turn hydrogen bonds. Conformer II is only observed in mixed solvents with high concentrations of DMSO. It possesses a *cis-trans-trans-trans* sequence of amide bonds, the result of *cis-trans* isomerization of the Phe-Pro peptide bond. This isomerization was established on the basis of the Pro C_β and C_γ resonances.[15] The Ala-Aib peptide bond was determined to be *trans* as a result of low-temperature NOE studies, which showed an NOE between the Ala $C_\alpha H$ and the Aib NH. The authors suggested that chlamydocin adopts this *cis-trans-trans-trans* conformation in preference to the alternating *cis-trans* conformation in order to relieve transannular steric interactions. This has been somewhat supported by studies of [Gly^1-Ala^4]-chlamydocin,[226] which indicate a mixture of all four known cyclic tetrapeptide ring conformations: the bis γ-turn; the *cis-trans-cis-trans*, i_1 and d_1; and the *cis-trans-trans-trans* conformation. Replacement of Aib with Gly relieves the destabilizing steric interactions. Further evidence for the *cis-trans-trans-trans* conformations comes from recent[232] NMR studies of WF-3161, cyclo(-*L*-Leu-*L*-Pip-*L*-Aoe-*D*-Phe), a more constrained analogue with apparently only one conformation in solution.

Similar behavior has been observed for the chlamydocin analogue [Ada^4]-chlamydocin (Ade = ethylene ketal of 2-amino-10-ethoxy-8-oxo-decanoic acid). Haslinger and Kalchhauser[233] employed homo- and heteronuclear 2-D NMR techniques to assign the carbon and proton spectrum of this cyclic tetrapeptide. The thermodynamic characteristics of the *cis-trans* isomerization of the Phe-Pro peptide bond was studied by saturation transfer NMR experiments,[234] and $\Delta G^{\ddagger}$ was found to be of the order of 70 kJ/mol, which corresponds well with established values for *cis-trans* isomerization of *X*-Pro bonds.[235,236]

6.6.3. Tentoxin

Meyer et al.[237] correctly assigned the structure cyclo(N-Me-Ala-Leu-N-Me-dehydroPhe-Gly) to the phytotoxic metabolite of the fungus *Alternaria tenuis*, tentoxin. Reduction of the dehydrophenylalanine residue to give the dihydro derivative of tentoxin yielded crystals suitable for X-ray analysis.[238] The X-ray structure corresponded to that deduced from ^{1}H NMR studies.[237] Meyer and co-workers[239] used extensive proton NMR work to determine the sequence and configuration of tentoxin. They proposed a structure, which agrees with that commonly found in cyclic tetrapeptides; that is, the alternating *cis-trans* amide bonds i_1-type conformation.[93,94,183]

Tentoxin was subsequently synthesized,[240] which further confirmed the proposed structure. Proton NMR and CD studies[240] on the synthetic tentoxin and its linear tetrapeptide precursor indicated similar conformations. The CD spectra for both showed comparable positive ellipticities in the 260–320-nm region, indicating similar environments for the α,β-dehydro chromophore. Likewise, the NMR chemical shifts for both the C_α and C_β protons of Leu and the methyl and vinyl protons of N-methyl-Z-Δ-Phe were almost identical in both precursor and cyclic tetrapeptide. NMR and CD studies of Leu-Me-Δ-Phe-OMe[241] indicated that this dipeptide ester can exist as a mixture of *cis* and *trans* isomers. Hence it is believed that the linear tetrapeptide precursor of tentoxin adopts a folded conformation with a *cis* amide bond between the Leu and Me-Δ-Phe residues. This folded conformation facilitates ring closure by placing the reactive ends in close proximity.

During the course of structure-activity studies on tentoxin, Rich and co-workers[242] noted that [Pro^1]tentoxin exhibited the same biological activity and CD and NMR parameters as tentoxin. This called into question the conformation proposed earlier,[237,239,241] because this Pro-containing analogue cannot adopt the proposed conformation due to the proline ring system. C-13 NMR established a *cis* Gly-Pro bond and UV data established a *cis* Leu-Me-Δ-Phe peptide bond. The $^3J_{NHC\alpha H}$ coupling constants for the Gly residue were interpreted as giving a 170° ϕ angle for Gly, and the chemical shift of the vinyl protons indicate an *s-cis* CH=C—C=O configuration. This led to the proposal of conformation shown in Fig. 6-13 for tentoxin.

The analogue [*D*-MeAla^1]tentoxin also was synthesized, and it was found to exist in multiple conformations.[95] Two conformers could be separated by preparative TLC at 4°C, and both were examined by ^{1}H and ^{13}C NMR, UV, and CD spectroscopies. The two conformers, labeled 2U and 2L for the faster and slower moving conformer, respectively, were shown to have an activation energy of interconversion of 23 kcal/mol. Low-temperature NMR studies indicated that 2L is a mixture of two conformers that interconvert with an activation energy

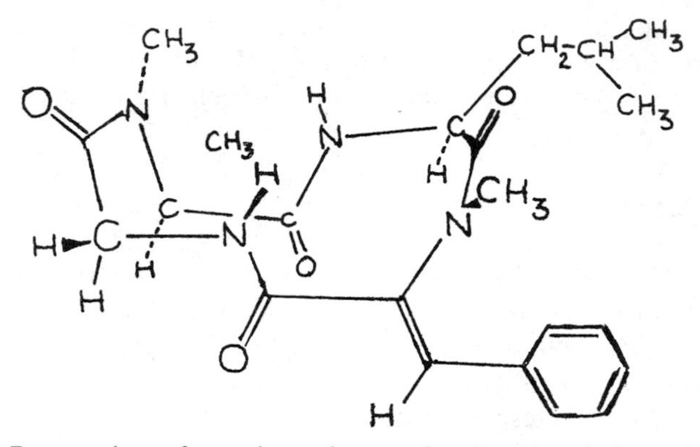

Figure 6-13. Proposed conformation of tentoxin. (By permission of Rich, D. H.; Bhatnagar, P. In "Peptides: Proceedings of the Fifth American Peptide Symposium," Goodman, M.; Meienhofer, J., Eds. Academic Press: New York, 1977, p. 342.)

of 13 kcal/mol. The two conformers 2U and 2L have also been examined for their ability to inhibit coupled electron transport in isolated chloroplasts. Both are less potent than tentoxin, but each conformer shows distinctly different biological activity.[243] This is the only example so far of conformers of the same structure being able to be separated and shown to have different biological activities.

6.6.4. Tuftsin

Tuftsin, a linear tetrapeptide that stimulates phagocytosis, was shown to have the sequence Thr-Lys-Pro-Arg.[244] As a result of physicochemical studies[199,245–249] and energy calculations,[250,251] two different models were proposed for the backbone conformation of tuftsin. One model[245] proposed proximity of the α-amino and C-terminal residues. The second predicted proximity between the ϵ-amino group of lysine and the C-terminal carboxyl group.[251] Chipens et al.[252] synthesized two cyclic tuftsin analogues in order to test these proposed models. One analogue was the 12-membered cyclic tetrapeptide c(Thr-Lys-Pro-Arg) and the other contained the 12-membered cycle formed by reaction of the lysine ϵ-amino group with the arginine C-terminal carboxyl giving Thr-Lys-Pro-Arg. Since the latter compound showed biological activity comparable with tuftsin itself, conformational studies were performed upon it using energy calculations, CD, and NMR spectroscopies.[253]

Energy calculations indicated that the *trans* conformation of the Lys-Pro bond is necessary for all low-energy structures. This is supported by ^{13}C NMR data, which showed proline C_β and C_γ resonances at 29.65 and 25.52 ppm, respectively, corresponding to a *trans* conformation. Compared with tuftsin, the cyclic analogue is predicted to be conformationally less mobile. CD studies show vastly different spectra between tuftsin and the cyclic compound, indicating that the "averaged" conformation of tuftsin is different than that important for biological activity. Further conformational work on this peptide may provide further insights into the link between conformation, structure, and activity.

6.6.5. AM Toxins and Cyl-2

Studies of the AM toxins (I, II, III) and Cyl-2, naturally occurring phytotoxins with the cyclic tetrapeptide structures shown in Fig. 6-14, have been focused on elucidating the correct sequences, determining the peptide backbone conformation, and establishing the configuration of the Aoe residue in Cyl-2.

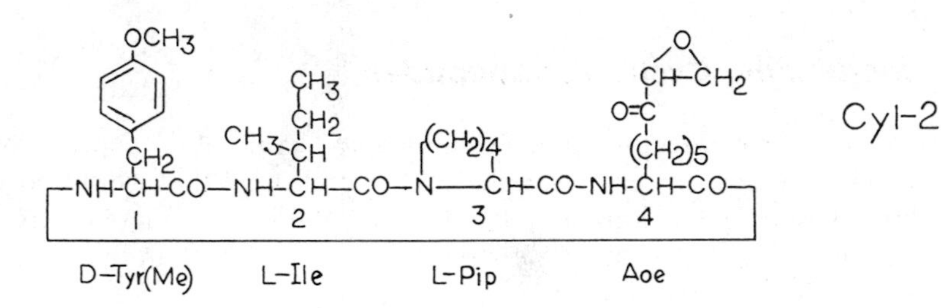

R
CH2 CH2 CH3 CH3 CH3
-NH-CH-CO-NH-C-CO-NH-CH-CO-O-CH-CO-
1 2 3 4
L-X Δ-Ala L-Ala L-Hmb

	R	X
AM-toxin I	OCH_3	Amp
AM-toxin II	H	App
AM-toxin III	OH	Ahp

Figure 6-14. Structure of the AM-toxins and Cyl-2.

Aoe = 2-amino-8-oxo-9,10-epoxidecanoic acid; Pip = pipecolic acid

Amp = 2-amino-5-p-methoxyphenylpentanoic acid

App = 2-amino-5-phenylpentanoic acid

Ahp = 2-amino-5-p-hydroxyphenylpentanoic acid

Hmb = 2-hydroxy-3-methylbutanoic acid

ΔAla = dehydroalanine

Structure-activity studies on analogues of the AM toxins have indicated the importance of the Ala and arylpentanoic residues and the backbone conformation for biological activity.[254] Conformational analysis of analogues of AM toxin I containing *D*- or *L*-Ala in place of Ala suggested a possible backbone conformation for the AM toxins in which all amide bonds are *trans* and three of the four carbonyl groups are on the same side of the ring.

Comparison of ^{1}H and ^{13}C NMR of simplified analogues of Cyl-2 with natural Cyl-2 have suggested that this phytotoxin contains a unique *trans-trans-cis-trans* backbone conformation, with a *cis* Ile-Pip bond.[254,255] The configuration of the Aoe residue, however, has not been firmly established, although based on the similarity among ^{1}H NMR spectra of [*L*-Leu]Cyl-2, [*L*-Pro-*L*-Leu]Cyl-2, and natural Cyl-2, the configuration appears to be of the *L*-type.[256]

6.6.6. Ion-binding Cyclic Tetrapeptides

Interest in the dodecadepsipeptide valinomycin, a microbial peptide ionophore, led investigators to seek a simpler cyclic compound that would possess the ability to bind metal ions and be readily soluble in nonpolar organic solvents, in the hope of producing more efficient ionophores than the naturally occurring one. When Dale and Titlestad[183] investigated sarcosine-containing cyclopeptides, they noted that these cyclic peptides had the ability to bind metal ions. However, Shimizu and Fujishige[37] studied cyclo(Pro-Sar)$_2$ and concluded that efficient ion–dipole interaction was unlikely to take place due to a lack of conformational flexibility on the part of the peptide. They did note, however, that small cyclic peptides tended to form 2:1 peptide–cation complexes with some metal ions. This led to the synthesis, conformational, and ion-binding studies on the bis(cyclic tetrapeptide) S,S′-bis[cyclo(Gly-*L*-hemiCys-Sar-*L*-Pro)] (BCGCSP) shown in Fig. 6-15.

Proton NMR studies in DMSO indicated that the conformations in the two cyclic portions of BCGCSP are identical as those observed in the single tetrapeptide [Gly-*L*-Cys(Bzl(OMe))-Sar-*L*-Pro].[257] Dihedral angles in the di-

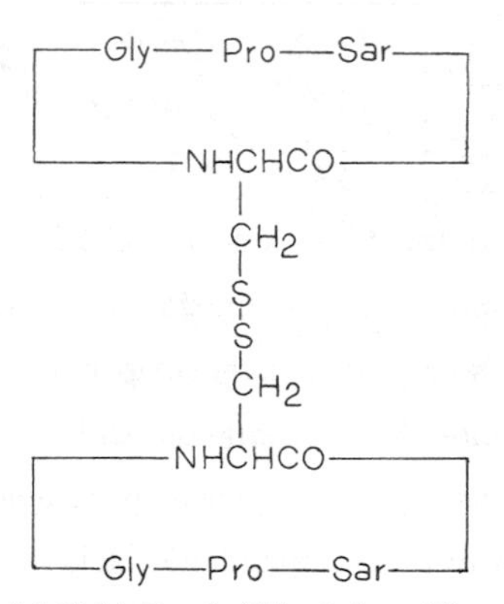

Figure 6-15. Structure of S,S′-bis[cyclo(Gly-*L*-hemiCys-Ser-*L*-Pro)] (BCGCSP).

sulfide bridge linking the two cyclic portions, determined by Raman spectroscopy and CD, were suggested to allow the molecule to adopt a "castanet-type" structure that would allow metal ion binding. Conformational changes upon complexation of the cyclic tetrapeptide with cations in acetonitrile were readily observed by ^{1}H and ^{13}C NMR.[258] The cyclic tetrapeptide was found capable of binding Li^{+}, Mg^{++}, Ca^{++}, and Ba^{++}. Conformational aspects of the complexation of the bis(cyclic tetrapeptide) BCGCSP with metal ions were studied using CD.[219,220] A red shift of the CD disulfide (S—S) band was seen upon complex formation, suggesting that a conformational fitting of peptide takes place upon complexation. Among monovalent cations, BCGCSP was found to selectively bind Rb in preference to Li^{+}, Na^{+}, and K^{+}. It also exhibited marked selectivity in the binding of Ba^{++} over Ca^{++}.

6.7. Thirteen- and 14-Membered-Ring-Containing Peptides

The largest medium-ring-containing peptides to be discussed in this review are those containing 13 and 14 members. These are composed of a group of naturally occurring peptide alkaloids, some cyclic enkephalins, several small disulfide-loop peptides, and peptides designed to mimic β-bends.

6.7.1. *Models for β-Bends*

As mentioned earlier, β-bends in both naturally occurring and model peptides have been the focus of intense investigation. Several dipeptides cyclized with ϵ-aminocaproic acid have been designed as models for various types of β-bends. Deslauriers et al.[259] reported the synthesis of cyclo(Ala-Gly-Aca), where Aca designates the residue ϵ-aminocaproic acid coupled within the ring. Conformational properties of this model β-bend peptide have been examined by energy calculations, NMR, IR, Raman, and CD. The goal of these studies has been to characterize the spectroscopic properties of β-bends so they can be applied to detect β-bends in conformational studies of other peptides in solution.

Nemethy and co-workers[260] have carried out an analysis of the conformational space of c(Ala-Gly-Aca) using energy calculations. Low-energy conformations only occur when all three peptide units are *trans*. Ten low-energy conformations were determined; the lowest-energy structure possessed a type II β-turn. Type I and III bends were also seen with only slightly higher energies. Therefore, based on these calculations, Nemethy suggested that the type II turn is the predominant conformation for this 14-membered cyclic dipeptide. Experimental measurements[261,312] utilizing NMR, Raman, and CD also support the energy-calculation studies, indicating that the predominant conformation in solution is a type II β-turn, with minor contribution from type I and type III bends. Carbon-13 spin-lattice relaxation times (NT_1) for the backbone CH_2 and CH groups did not show any gradient of motion within the peptide. This indicates either little overall flexibility or considerable flexibility throughout the structure. The large chemical shift difference between the

Gly $C_\alpha H$ protons (0.45 ppm) suggests one major conformation. This is also supported by the coupling patterns of the $C_\alpha H$ and $C_\epsilon H_2$ protons of the Aca residue. The pattern seen is typical of that for a conformationally locked cyclic alkane rather than a mobile hydrocarbon chain.[261,313] Chemical shift and temperature studies of the amides indicate the Aca NH to be the most shielded in the molecule. NOE data indicate a proximity of the Ala $C_\alpha H$ and Gly·NH (~2.95 Å). Such short distances can occur only for positive values of Ψ Ala, and hence only for type II bends. The conformation consistent with the coupling constants, NOE values, and all *trans* peptide bonds has the dihedral angles (ϕ, Ψ)Ala = $(-80°, 70°)$ and (ϕ, Ψ)Gly = $(120°, 20°)$, and is a type II β-bend.

The CD spectra of the cyclized model dipeptide and a linear N- and C-terminal blocked dipeptide show negative bands at 225–230 nm and positive bands at 205 nm. Both the position and intensity of the bands vary with solvent composition, which indicates some flexibility in the molecule. These spectra are similar to class B types of CD spectra computed by Woody[124] for β-bends. Theoretically calculated CD spectra for c(Ala-Gly-Aca) for type II bends are similar to the experimental CD spectra.

Two other model cyclized dipeptides, c(Ala-Ala-Aca) and c(Ala-*D*-Ala-Aca), have been synthesized and studied by Bandekar et al.[262] Energy calculations and NMR and CD spectroscopic investigations indicated that c(Ala-Ala-Aca) is constrained to form a type I + III (these types are very similar) bend while c(Ala-*D*-Ala-Aca) forms a type II′ bend. This is consistent with the observation that *D* amino acids tend to stabilize β-turns in peptides. Raman and IR were also employed to determine the predominant conformations. Calculated frequencies for the amide I, II, III, and V modes of the computed low-energy conformations were compared with experimental spectra. Amide frequencies were found to be sensitive to conformation, in particular to bend structures, and indicated predominant conformations consistent with the NMR and CD data for both cyclic peptides.

6.7.2. *Disulfide Loop Peptides*

A second type of model turn containing peptides are those involving medium-ring peptide disulfides (i.e., Cys-X_n-Cys). An example of a naturally occurring $n = 2$ loop (14-membered ring) is the active site of the redox protein thioredoxin,[263] which consists of the segment -Cys-Gly-Pro-Cys- as a protrusion from the core of the protein. Very little conformational work has been done with this system within the native protein, although it is believed that reduction of the disulfide causes a conformational change.

It has been observed that peptides with a Pro-X sequence have a high probability for peptide chain reversal.[31,264–266] Therefore, Venkatachalapathi, Venkataram Prasad, and Balaram[267] combined the Pro-X sequence with ring formation through medium-sized disulfide loops to design model compounds to study turn characteristics. They reported the synthesis and conformational analysis of Boc-Cys-Pro-X-Cys-NHMe where X = Aib, Ala, *D*-Ala, Gly, and Val.[268,269] They concluded that all five cyclic peptides favor β-turn conformations with a transannular $4 \rightarrow 1$ H bond. Furthermore, they attempted to distinguish the type of β-turn favored by each peptide by using NOE studies. They claimed that positive NOEs between the $C_\alpha H$ of the $i+1$ and NH of the $i+2$ residues in a turn can be used to distinguish type II β-turns and γ-turns. Only the Pro-Aib and Pro-*D*-Ala peptides exhibited NOEs of this type. Thus, they concluded that these two peptides possessed Type II β-turns. However, the X-ray study of Pro-Aib disulfide indicated a type III β-turn. This discrepancy may be due to the fact that these crystals were prepared from a mixture of $CDCl_3$ and DMSO while the NMR work was done in $CDCl_3$.[270,271] The presence of the $4 \rightarrow 1$ H bond was supported by the low-temperature dependence, solvent dependence of the chemical shift, and H—D exchange of the Cys^4—NH groups in all peptides. Supporting evidence from IR studies showed a broad, intense band below 3350 cm^{-1}, indicative of an H-bonded amide.

Disulfide conformations were investigated by laser Raman[272] and CD studies. These results indicated a right-handed chirality with either a gauche-gauche-*trans* (*ggt*) or *trans*-gauche-gauche (*tgg*) arrangement for the C_α—C_β—S—S—C_β—C_α atoms. CD suggested a χ_{SS} angle of $+60°$ or $+120°$ consistent with the Raman band at 522 cm^{-1}. The authors conclude that cyclic peptide disulfides are relatively rigid and can serve as useful model systems for the simultaneous spectroscopic characterization of β turns and disulfide linkages.

6.7.3. Cyclopeptide Alkaloids

Another important group of 13- and 14-membered-ring-containing peptides is the cyclopeptide alkaloids.[273] These are a class of heterodetic cyclic peptides of plant origin that contain aryl-alkyl ether linkages. Found in leaves, bark, roots, and seeds of plants in the *Rhamnaceae* family, they possess antibiotic and antifungal properties. A general structure for these compounds, as well as examples of the four major classes of cyclopeptide alkaloids, is shown in Fig. 6-16. Of particular interest are the 14-membered-ring phencyclopeptines.[274] Structurally these are both cyclophanes and cyclodipeptides consisting of β-hydroxyamino acid, p-hydroxystyrylamine, and α-amino acid residues within the ring. In frangulanine, the β-hydroxyamino acid is commonly β-hydroxyleucine; however, other natural phencyclopeptine alkaloids may contain different β-hydroxy residues. Due to their low natural availability and the difficulty in synthetically preparing the 14-membered macrolide ring, thorough studies of the biological, pharmacological, and conformational properties of the compounds have been lacking.

Studies have focused predominantly on the synthetic challenge presented by these molecules—the creation of a 14-membered cycle. Ring closure at various positions of the cycle has been extensively investigated.[275–277] More success has been achieved with the synthesis of 13- and 15-membered cyclopeptides than with the 14-membered ring systems.

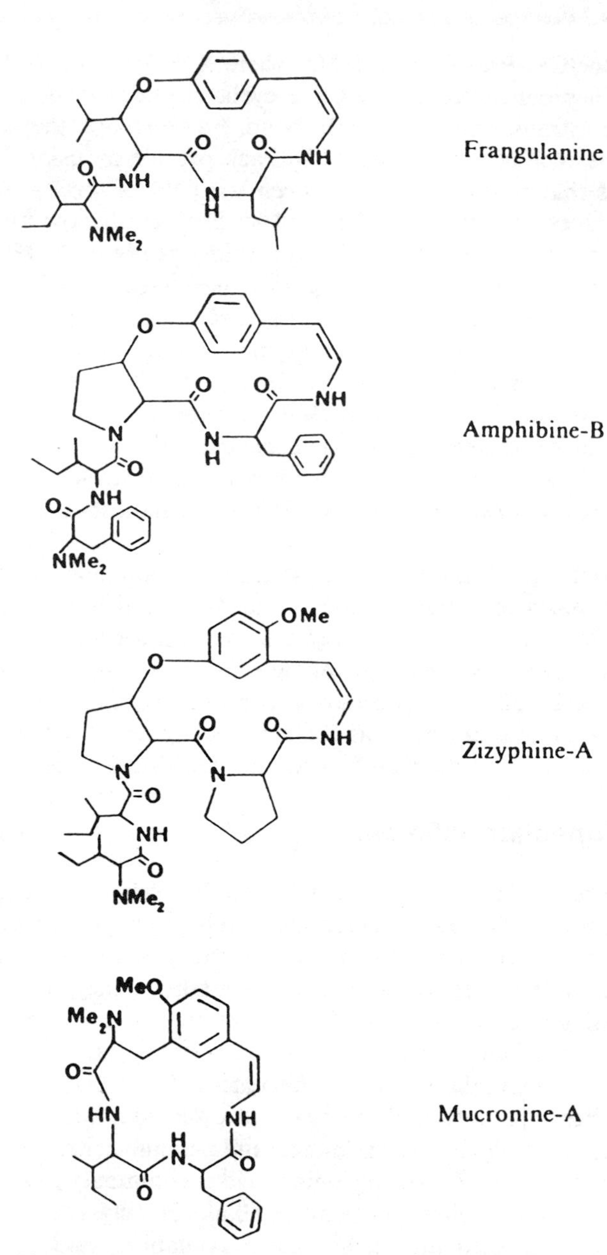

Figure 6-16. Four major classes of cyclopeptide alkaloids.

Structural studies have focused primarily upon structural determination via chemical degradation and spectroscopic methods. Because these compounds tend to crystallize easily, X-ray analysis has been carried out on two cyclopeptide alkaloids: mauritine-A[278,279] and a derivative of frangulanine.[280,281] X-ray studies have also been performed upon simplified synthetic models of the 14-membered dihydro macrocycle.[274] These studies have confirmed a lack of planarity between the aromatic ring and the eneamido chromophore in the phenethylamine moiety. This had been suspected from earlier UV studies. The UV spectra of the 14-membered cyclopeptide alkaloids do not exhibit an absorption maximum for the conjugated styrylamine moiety (272 nm, $\epsilon = 4.35$). In contrast, the UV for 13-membered ring alkaloids had shown absorption bands characteristic for this chromophore. All this is considered evidence for ring strain in the 14-membered system that prohibits π-orbital overlap between the aromatic ring and eneamido group, and contributes to the difficulty of synthesis of these compounds.

Detailed NMR investigations of the hydroxyleucine-containing alkaloids[282,283] and the 14-membered ring alkaloids discarine-B[284] and frangulanine[285–288] also have indicated conformational rigidity in these cyclopeptide alkaloids. From ^{1}H and ^{13}C NMR, NOE measurements, and temperature studies, the amide NH of the Leu residue in frangulanine appears to be involved in an intramolecular H bond forming a γ-turn conformation in $CDCl_3$. The constancy of coupling constants of protons on vicinal carbons in the 14-membered ring of discarine-B in various solvents and at various temperatures also is consistent with conformational rigidity of the ring.

Recent C-13 NMR studies of cyclopeptide alkaloids[289,290] have shown restricted motion of the aromatic ring, with four separate aromatic methine resonances observed. The two styrene carbons also exhibit almost equivalent chemical shifts (124.0 and 125.4 ppm in discarine-B), which are consistent with a nonconjugated double-bond system. These NMR parameters all support a picture of rigidity and macrocyclic ring strain in these cyclopeptide alkaloids.

Finally, X-ray diffraction studies of mauritine-A provided clear evidence of ring strain, with the benzene ring slightly bent and the attached atoms considerably out of the benzene plane. As noted earlier, the strain prevents π-orbital overlap in the styrylamide system. X-ray also established the *trans*-stereochemistry of the β-hydroxyproline residue in mauritine-A and the erythro-stereochemistry of the hydroxyleucine in frangulanine.[288]

6.7.4. Cyclic Enkephalins

The enkephalins are flexible linear peptapeptides, existing in multiple conformations in solution and solid state, which are the natural endogenous ligands for the opioid receptors and to which the rigid and semirigid nonpeptidic opiates also bind. Investigators have sought an explanation for this, based upon putative structural similarities between the enkephalins and opiates, but as yet no agreement exists regarding the biologically active conformation(s) of the enkephalins.[291] There are many reasons for this, not the least of which is the recent recognition that there are several different opioid receptors. These receptor types, which include the μ, δ, and κ receptors, and perhaps others, have been recognized as important and distinct receptor systems.

Based upon suggestions from energy calculations,[292] X-ray studies,[53,293] solutions studies using NMR, Raman, and fluorescence, and a variety of conformational considerations,[5] cyclic analogues of the enkephalins have been prepared in order to examine the conformational requirements for the opioid receptors. A number of these cyclic enkephalins contain 13- and 14-membered rings formed by cyclization between side chains at positions 2 and 5 via either an amide or disulfide linkage (see below).

Schiller and Di Maio[9,294] found that cyclization between the side-chain amino group of a *D*-amino acid residue in position 2 with the terminal carboxyl group in enkephalin pentapeptides led to enhanced μ-receptor selectivity. Interestingly, the smaller 13- and 14-membered lactam rings with *D*-α,β-diaminopropionic acid and *D*-α,γ-diaminobutyric acid, respectively, in position 2 (**1** and **2** in Fig. 6-17) showed only minor preference for μ versus δ receptor and relatively low potency at both receptors. On the one hand, the larger, more flexible 15- and 16-membered lactam ring analogues with *D*-ornithine and *D*-lysine, respectively (**3** and **4**, Fig. 6-17), showed much greater selectivity, and in the latter case higher potency.[9,295] The conformational and structural factor responsible for these differences is not yet determined, though in most cases tested, it has been shown that ring closure leads to higher receptor selectivity.[291]

In a different approach, cyclic lactams between side groups of *D*-amino acids at position 2 and *L* or *D* amino acids in position 4 of ring contracted cyclic enkephalin analogues have led to medium-sized ring compounds of high receptor potency, and very good selectivities for μ versus δ opioid receptors.[291,296] For example, the *D*-Orn^2, Asp^4 analogue **5** (Fig. 6-17), which has a 13-membered lactam ring, has high μ versus δ opioid receptor selectivity, as does the reverse sense cyclic peptide lactam **6** (Fig. 6-17), and both have moderate binding potencies at the μ receptor. Interestingly, the corresponding analogues with *D*-amino acids in positions 2 and 4, such as compound **7**, were much less receptor selective, as were the larger-ring more conformationally flexible analogues such as the 15-membered lactam analogue (Fig. 6-17). In the latter case, the compounds actually bind more strongly to the μ-opioid receptor than compounds **5** and **6** but have much reduced selectivity because they bind much more strongly to δ-opioid receptors than **5** and **6**.

Side-chain to side-chain cyclization via a disulfide ring closure[5] also has been examined. Sarantakis[297] and Schiller et al.[9,298] prepared the *D*-Cys^2, *D*- or

H-Tyr-D-A$_2$pr-Gly-Phe-Leu **1**

H-Tyr-D-A$_2$bu-Gly-Phe-Leu **2**

H-Tyr-D-Orn-Gly-Phe-Leu **3**

H-Tyr-D-Lys-Gly-Phe-Leu **4**

H-Tyr-D-Orn-Phe-Asp-NH$_2$ **5**

H-Tyr-D-Asp-Phe-Orn-NH$_2$ **6**

H-Tyr-D-Orn-Phe-D-Asp-NH$_2$ **7**

H-Tyr-D-Lys-Phe-Glu-NH$_2$ **8**

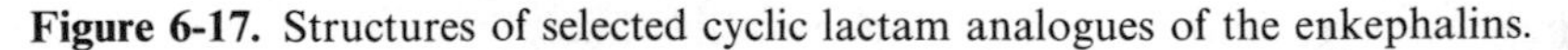

Figure 6-17. Structures of selected cyclic lactam analogues of the enkephalins.

H-Try-D-Cys-Gly-Phe-Cys-NH2	**1**
H-Tyr-D-Cys-Gly-Phe-D-Cys-NH2	**2**
H-Tyr-D-Pen-Gly-Phe-Cys-OH	**3**
H-Tyr-D-Pen-Gly-Phe-D-Cys-OH	**4**
H-Tyr-D-Pen-Gly-Phe-Pen-OH	**5**
H-Tyr-D-Pen-Gly-Phe-D-Pen-OH	**6**

Figure 6-18. Structures of some selective disulfide-containing enkephalin analogues.

L-Cys^5 enkephalin amide analogues **1** and **2** (Fig. 6-18). These compounds and related analogues showed moderate to high binding potencies at opioid receptor, but little or no receptor selectivity. However, in a series of papers, we have shown, using the conformationally restrictive amino acid *D*-penicillamine (β,β-dimethylcysteine, Pen) at position 2 and/or *D*(*L*)-cysteine or *D*(or *L*)-penicillamine at position 5, that highly potent and delta receptor-selective compounds such as **3**, **4**, **5**, and **6** could be obtained.[299–303] In this case, it should be noted that the most δ-receptor selective compounds have a C-terminal carboxylate. [We have shown (Hruby et al., unpublished results) that the *D*-Cys^2, *D*(*L*) Cys^5 enkephalin acid analogues also are more δ-receptor selective than the corresponding amides.] Indeed, the compound [*D*-Pen^2, *D*-Pen^5]enkephalin (**6**, Fig. 6-18, DPDPE) is the most δ-receptor compound currently available.[301,304–307] An interesting structural feature is that the bis-penicillamine analogues such as **5** and **6** (Fig. 6-18) have 14-membered rings that have within them two geminal dimethyl-containing carbon atoms, which, of course, places severe conformational restrictions on this 14-membered ring system.

Conformational studies on the cyclic amide enkephalins have indicated high rigidity and the presence of turn structures.[308] Even in certain 16-membered cyclic lactam analogues, Kessler, Holzemann, and Geiger[309] have observed a nonequivalence of the Gly^3 $C^{\alpha}H_2$ protons, and the temperature dependence and coupling constants indicate rigidity and the presence of β-bend structure. Conformational studies using ^{1}H NMR comparing the cysteine- and penicillamine-containing analogues[310] indicated similar overall conformations, but with some differences in conformation and flexibility in the C-terminal regions that may be related to the differences in μ- and δ-receptor selectivity.

In our laboratory, we have performed extensive 1-D and 2-D NMR spectroscopic studies in conjunction with extensive molecular mechanics calculations[311] to examine the low-energy conformations of DPDPE and related analogues. The presence of the two geminal dimethyl groups in the 14-membered ring suggests that a favored conformation should be obtained for the bis-penicillamine compounds. Indeed, our studies indicate that DPDPE assumes a favored low-energy conformation (Fig. 6-19) in which the two aromatic rings of the Tyr^1 and Phe^4 residues interact primarily with the Pen^2 geminal dimethyl groups to stabilize an amphiphilic conformation with a highly lipophilic surface containing the aromatic rings and the penicillamine residues, and a hydrophilic

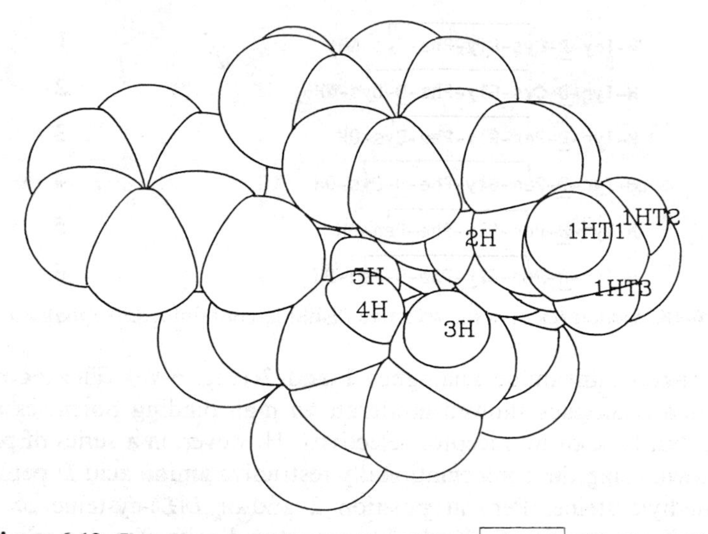

Figure 6-19. Low-energy conformation of [*D*-Pen^2, *D*-Pen^5]enkephalin.

surface in which the three amide carbonyl groups within the 14-membered ring are all pointing out into the aqueous or dimethylsulfoxide environment in solutions of DPDPE. Analysis of the factors that stabilize this indicates that the modified type IV β-turnlike conformations apparently are favored as a result of the van der Waal stabilizations provided by the interactions of the various lipophilic groups within DPDPE. These stabilizations apparently account for the very high yields that were obtained in the synthesis of these compounds.[300,301] Furthermore, the topology created by these interactions is highly favorable for the δ-opioid receptor, but not the μ- or κ-opioid receptors. Further structural and topological studies are in progress to determine further the conformational requirements of the δ-opioid receptor (receptor mapping).

Acknowledgments

The work in Arizona was supported by the National Science Foundation and U.S. Public Health Service Grant AM-17420, NS-19972, and DA-04248. We thank Karen Burgan for her typing and editing of the manuscript.

References

1. Hruby, V. J. In "The Peptides: Conformation in Biology and Drug Design," Hruby, V. J., Ed. Academic Press: New York, 1985, Vol. 7, pp 1–14.
2. Marshall, G. R.; Gorin, F. A.; Moore, M. L. *Annu. Rep. Med. Chem.* **1978**, *13*, 227–238.
3. Hruby, V. J. In "Perspectives in Peptide Chemistry," Eberle, A.; Geiger, R.; Wieland, T., Eds. S. Karger: Basel, 1981, pp 207–220.
4. Veber, D. F. In "Peptides: Synthesis, Structure and Function," Rich, D. H.; Gross, E., Eds. Pierce Chemical Co.: Rockford, Ill., 1981, pp 685–694.

5. Hruby, V. J. *Life Sci.* **1982**, *31*, 189–199. Hruby, V. J. In "Conformationally Directed Drug Design," Vida, J. A; Gordon, M., Eds. ACS Symposium Series 251: Washington, D.C., 1984, pp 9–22.
6. Kessler, H. *Angew. Chem. Int. Ed. Engl.* **1982**, *21*, 512–523.
7. Hruby, V. J.; Hadley, M. E. In "Design and Synthesis of Organic Molecules Based on Molecular Recognition," G. van Binst, Ed. Springer-Verlag: Heidelberg, 1986, pp 269–289.
8. Hruby, V. J.; Mosberg, H. I.; Sawyer, T. K.; Knittle, J. J.; Rockway, T. W.; Ormberg, J.; Darman, P. *Biopolymers* **1983**, *22*, 517–530.
9. Schiller, P. W.; Di Maio, J. In "Peptides: Structure and Function," Hruby, V. J.; Rich, D. H., Eds. Pierce Chemical Co.: Rockford, Ill., 1983, pp 269–278.
10. Hassall, C. H.; Thomas, W. A. *Chem. Brit.* **1971**, *7*, 145–153.
11. Hruby, V. J. In "Chemistry and Biochemistry of Amino Acids, Peptides and Proteins," Weinstein, B. Ed. Marcel Dekker: New York, 1974, pp 1–188.
12. Ovchinnikov, U. A. In "The Chemistry of Polypeptides," Katsoyannis, P. G., Ed. Plenum Press: New York, 1973, pp 169–204.
13. Ovchinnikov, Y. A.; Ivanov, V. T. *Tetrahedron* **1975**, *31*, 2177–2209.
14. Blout, E. R.; Deber, C. M.; Pease, L. G. In "Peptides, Polypeptides and Proteins," Blout, E. R; Bovey, F. A.; Goodman, M.; Lotan, N., Eds. John Wiley & Sons: New York, 1974, pp 266–281.
15. Bovey, F. A.; Brewster, A. I.; Dinshaw, J.; Tonelli, A. E.; Torchia, D. A. *Acc. Chem. Res.* **1972**, *5*, 193–200.
16. Bovey, F. A. In "Peptides, Polypeptides and Proteins," Blout, E. R.; Bovey, F. A.; Goodman, M.; Lotan, N., Eds. John Wiley & Sons: New York, 1974, pp 248–265.
17. Deber, C. M.; Madison, V.; Blout, E. R. *Acc. Chem. Res.* **1975**, *9*, 106–113.
18. Demel, D.; Kessler, H. *Tetrahedron Lett.* **1976**, 2801–2804.
19. Pease, L. G. In "Peptides: Structure and Biological Function," Gross, E.; Meienhofer, J., Eds. Pierce Chemical Co.: Rockford, Ill., 1979, pp 197–200.
20. Bara, Y. A.; Friedrich, A.; Kessler, H.; Molter, M. *Chem. Ber.* **1978**, *111*, 1045–1057.
21. Kondor, P.; Kessler, H. In "Peptides: Structure and Biological Function," Gross, E.; Meienhofer, J., Eds. Pierce Chemical Co.: Rockford, Ill., 1979, pp 181–184.
22. Kessler, H.; Kondor, P. *Chem Ber.* **1979**, *112*, 3536–3551.
23. Khaled, M. A.; Urry, D. W.; Okamoto, K. *Biochem. Biophys. Res. Commun.* **1979**, *72*, 162–169.
24. Pease, L. G. In "Peptides: Structure and Biological Function," Gross, E.; Meienhofer, J., Eds. Pierce Chemical Co.: Rockford, Ill., 1979, pp 197–200.
25. Pease, L. G.; Niu, C. H.; Zimmerman, G. *J. Am. Chem. Soc.* **1979**, *101*, 184–191.
26. Kessler, H.; Bermel, W.; Förster, H. *Angew. Chem. Int. Ed. Engl.* **1982**, *21*, 689.
27. Karle, I. L. *J. Am. Chem. Soc.* **1978**, *100*, 1286–1289.
28. Karle, I. L. *J. Am. Chem. Soc.* **1979**, *101*, 181–184.
29. Karle, I. L. In "Perspectives in Peptide Chemistry," Eberle, A.; Geiger, R.; Wieland, T., Eds. Karger: Basel, 1981, pp 261–271.
30. Einspahr, H.; Cook, W. J.; Bugg, C. E. *Am. Cryst. Assoc. Abstr.* **1980**, *7*, 14.
31. Smith J. A.; Pease, L. G. *CRC Crit. Rev. Biochem.* **1980**, *4*, 315–400.
32. Mauger, A. B.; Stuart, O. A.; Highet, R. J.; Silverton, J. V. *J. Am. Chem. Soc.* **1982**, *104*, 174–180.
33. Ovchinnikov, Y. A.; Ivanov, V. T. In "The Proteins," Neurath, H.; Hill, R. L., Eds. Academic Press: New York, 1982, pp 307–642.
34. Rose, G. D.; Gierasch, L. M.; Smith, J. A. In "Advances in Protein Chemistry," Anfinson, C.; Edsall, J.; Richards, F., Eds. Academic Press: New York, 1985, Vol. 37, pp 1–129.
35. Sugihara, T.; Imanishi, Y.; Higashimura, T. *Biopolymers* **1975**, *14*, 723–731.
36. Sugihara, T.; Imanishi, Y.; Higashimura, T. *Biopolymers* **1976**, *15*, 1529–1542.
37. Shimizu, T.; Fujishige, S. *Biopolymers* **1980**, *19*, 2247–2265.
38. Deber, C. M.; Drobnies, A. E.; Hughes, D. W.; Lanningan, D. A. In "Peptides: Syntheses—Structure—Function," Rich, D. H.; Gross, E., Eds. Pierce Chemical Co.: Rockford, Ill., 1981, pp 331–334.
39. Kendrew, J. C.; Klyne, W.; Lifson, S.; Miyazawa, T.; Nemethy, G.; Phillips, D. C.; Ramachandran, G. N.; Scheraga, H. A. *Biochemistry* **1970**, *9*, 3471–3479.
40. Pauling, L.; Cory, R. B. *Proc. Natl. Acad. Sci. USA* **1952**, *38*, 86–93.
41. LaPlanche, L. A.; Rogers, M. T. *J. Am. Chem. Soc.* **1964**, *86*, 337–341.
42. Ramachandan, G. N.; Sasisekharan, V. In "Advances in Protein Chemistry," Anfinsen, C. B. Jr.; Anson, M. L.; Edsall, J. T.; Richards, F. M., Eds. Academic Press: New York, 1986, Vol. 23, pp 283–438.
43. Kopple, K. J. *Pharmaceut. Sci.* **1972**, *61*, 1345–1346.
44. Lewis, P. N.; Momany, F. A.; Scheraga, H. A. *Biochem. Biophys. Acta* **1973**, *303*, 211–229.

45. Toma, F.; Lam-Thanh, H.; Piriou, F.; Heindl, M. C.; Lintner, K.; Fermandjian, S. *Biopolymers* **1980**, *19*, 781–804.
46. Schwyzer, R.; Sieber, P.; Gorup, B. *Chimia* **1958**, *12*, 90–91.
47. Mutter, M. *J. Am. Chem. Soc.* **1977**, *99*, 8307–8314.
48. Veber, D. F.; Freidinger, R. M.; Schewenk, D.; Perlow, D.; Paleveda, W. J. Jr.; Holly, F. W.; Strachan, R. G.; Nutt, R. F.; Arison, B. H.; Homnick, C.; Randall, W.; Slitzer, M. S.; Saperstein, R.; Hirschmann, R. *Nature* **1981**, *292*, 55–58.
49. Sawyer, T. K.; Darman, P. S.; Hruby, V. J.; Hadley, M. E. In "Peptides: Synthesis—Structure—Function," Rich, D. H.; Gross, E., Eds. Pierce Chemical Co.: Rockford, Ill., 1981, pp 387–390.
50. Sawyer, T. K.; Hruby, V. J.; Darman, P. S.; Hadley, M. E. *Proc. Natl. Acad. Sci. USA* **1982**, *79*, 1751–1755.
51. Kendrew, J. C.; Perutz, M. F. *Ann. Rev. Biochem.* **1957**, *26*, 327–372.
52. Schwyzer, R. *Rec. Chem. Progr.* **1959**, *20*, 147–167.
53. Benedetti, E. In "Peptides, Proceeding of the Fifth American Peptide Symposium," Goodman, M.; and Meienhofer, J.; Eds. John Wiley & Sons: New York, 1977, pp 257–273.
54. Karle, I. L.; Karle, J.; Mastropaolo, D.; Camerman, A.; Camerman, N. *Acta Crystallogr., Sect. B.* **1983**, *39*, 625–637.
55. Pitts, J. E.; Wood, S.; Tickles, I. J.; Trehane, A. M.; Mascarenhas, Y.; Li, J. Y.; Husain, J.; Cooper, S.; Blundell, T.; Hruby, V. J.; Wyssbrod, H. R. In "Peptides: Structure and Function," Deber, C. M.; Hruby, V. J.; Kopple, K. D., Eds. Pierce Chemical Co.: Rockford, Ill., 1985, pp 145–150.
56. Wood, S. P.; Tickle, I. J.; Trehane, A. M.; Pitts, J. E.; Mascarenhas, Y.; Li, J. Y.; Husain, J.; Cooper, S.; Blundell, T. L.; Hruby, V. J.; Wyssbrod, H. R.; Buku, A.; Fishman, A. J. *Science* **1986**, *232*, 633–636.
57. Andrew, E. R. *Arch. Sci.* **1959**, *12*, 103–108.
58. Pines, A.; Gibby, M. G.; Waugh, J. S. *J. Chem. Phys.* **1973**, *59*, 569–590.
59. Schaefer, J.; Stejskal, E. O. *J. Am. Chem. Soc.* **1976**, *98*, 1031–1032.
60. Hruby, V. J. In "Opioid Peptides: Medicinal Chemistry," Rapaka, R. S.; Barnett, G.; Hawks, R. L., Eds. NIDA Research Monograph 69: Rockville, Md., 1986, pp 128–147.
61. Pease, L. G.; Frey, M. H.; Opella, S. J. *J. Am. Chem. Soc.* **1981**, *103*, 467–468.
62. Gierasch, L. M.; Deber, C. M.; Madison, V.; Niu, C.; Blout, E. R. *Biochemistry* **1981**, *20*, 4730–4738.
63. Gierasch, L. M.; Freg, M. J.; Hexem, J. G.; Opella, S. J. In "NMR Spectroscopy: New Methods and Applications," Levy, G. C., Ed. American Chemical Society; Washington, D.C., 1982, pp 233–247.
64. Aue, W.; Bartholdi, P.; Ernst, R. R. *J. Chem. Phys.* **1976**, *64*, 2229–2246.
65. Freeman, R.; Morris, G. A. *Bull. Magn. Reson.* **1979**, *1*, 5–26.
66. Bax, A. "Two-Dimensional Nuclear Magnetic Resonance in Liquids." Delft University Press: Reidel, Netherlands, 1982.
67. Bax, A. In "Topics in Carbon-13 NMR Spectroscopy," Levy, G. C., Ed. John Wiley & Sons: New York, 1984, Vol. 4, pp 199–238.
68. Benn, R.; Günther, H. *Angew. Chem. Int. Ed. Eng.* **1983**, *22*, 350–380.
69. Kessler, H.; Bermel, W. In "Methods in Stereochemical Analysis," Marchand, A. P.; Takeuchi, Y., Eds. In press, 1986b.
70. Kessler, H.; Bermel, W.; Müller, A. In "The Peptides: Conformation in Biology and Drug Design," Hruby, V. J., Ed. Academic Press: New York, 1985, Vol. 7, pp 437–473.
71. Wider, G.; Macura, S.; Kumar, A.; Ernst, R. R.; Wüthrich, K. *J. Magn. Reson.* **1984**, *56*, 207–234.
72. Wüthrich, K. "NMR of Proteins and Nucleic Acids." John Wiley & Sons: New York, 1986.
73. Nagayama, K.; Wüthrich, K. *Eur. J. Biochem.* **1981**, *114*, 365–374.
74. Nagayama, K.; Kumar, A.; Wüthrich, K.; Ernst, R. R. *J. Magn. Reson.* **1980**, *40*, 321–334.
75. Nagayama, K.; Wüthrich, K.; Bachmann, P.; Ernst, R. R. *Biochem. Biophys. Res. Commun.* **1977**, *78*, 99–105.
76. Nagayama, K.; Wüthrich, K.; Ernst, R. R. *Biochem. Biophys. Res. Commun.* **1979**, *90*, 305–311.
77. Kessler, H.; Bermel, W.; Friedrich, A.; Krack, G.; Hull, W. E., *J. Am. Chem. Soc.* **1982**, *104*, 6297–6304.
78. Wider, G.; Baumann, R.; Nagayama, K.; Ernst, R. R.; Wüthrich, K. *J. Magn. Reson.* **1981**, *42*, 73–87.
79. Kessler, H.; Ziessow, D. *Nachr. Chem. Tech. Lab.* **1982**, *30*, 488.
80. Schuck, R. Dissertation, Universität Frankfurt, 1982.
81. Müller, L.; Kumar, A.; Ernst, R. R. *J. Chem. Phys.* **1976**, *65*, 389–840.
82. Morris, G. A.; Hull, L. D. *J. Am. Chem. Soc.* **1981**, *103*, 4703–4711.
83. Bax, A.; Morris, G. A. *J. Magn. Reson.* **1981**, *42*, 501–505.

84. Wider, G.; Lee, K. H.; Wüthrich, K. *J. Mol. Biol.* **1982**, *155*, 367–388.
85. Kessler, H.; Hehlein, W.; Schuck, R. *J. Am. Chem. Soc.* **1982**, *104*, 4534–4540.
86. Kessler, H.; Friedrich, A.; Drack, G.; Hull, W. E. In "Peptides: Synthesis—Structure—Function," Rich, D. H.; Gross, E., Eds. Pierce Chemical Co.: Rockford, Ill., 1981, pp 335–338.
87. Kricheldorf, H. R.; Hull, W. E. *Biopolymers* **1980**, *19*, 1103–1122.
88. Kricheldorf, H. R. *Org. Magn. Reson.* **1981**, *15*, 162–177.
89. Hawkes, G. E.; Randall, E. W.; Hull, W. E.; Convert, O. *Biopolymers* **1980**, *19*, 1815–1826.
90. Van Gorkom, M. *Tetrahedron Lett.* **1966**, 5433–5439.
91. Sheinblatt, M. *J. Am. Chem. Soc.* **1966**, *88*, 2845–2848.
92. Mandel, M. *J. Biol. Chem.* **1965**, *240*, 1586–1592.
93. Dale, J.; Titlestad, K. *J. Chem. Soc. Chem. Commun.* **1969**, 656–659.
94. Dale, J.; Titlestad, K. *J. Chem. Soc. Chem. Commun.* **1970**, 1403–1404.
95. Rich, D. H.; Bhatnagar, P. *J. Am. Chem. Soc.* **1978**, *100*, 2218–2224.
96. Sugg, E. E.; Cody, W. L.; Abdel-Malek, Z.; Hadley, M. E.; Hruby, V. J. *Biopolymers* **1986**, *25*, 2029–2042.
97. Noggle, J. H.; Schirmer, R. E. "The Nuclear Overhauser Effect." Academic Press: New York, 1971.
98. Bothner-By, A. A. In "Biological Applications of Magnetic Resonance," Shulman, R. G. Ed. Academic Press: New York, 1979, pp 117–219.
99. Gibbons, W. A.; Creqaux, D.; Delayre, J.; Dunand, J. J.; Hajdukovic, G.; Wyssbrod, H. R. In "Peptides: Chemistry, Structure and Biology," Walter, R.; Meienhofer, J., Eds. Ann Arbor Science: Ann Arbor, Mich. 1975, pp 127–137.
100. Kuo, M.; Gibbons, W. A. *J. Biol. Chem.* **1979**, *254*, 6278–6287.
101. Jeener, J.; Meier, B. H.; Bachmann, P.; Ernst, R. R. *J. Chem. Phys.* **1979**, *71*, 4546–4553.
102. Kumar, A.; Ernst, R. R.; Wüthrich, K. *Biochem. Biophys. Res. Commun.* **1980**, *95*, 1–6.
103. Braun, W.; Bosch, C.; Brown, L. R.; Go, N.; Wüthrich, K. *Biochim. Biophys. Acta* **1981**, *667*, 377–396.
104. Baumann, R.; Kumar, A.; Ernst, R. R.; Wüthrich, K. *J. Magn. Reson.* **1981**, *44*, 76–83.
105. Macura, S.; Huang, Y.; Suter, D.; Ernst, R. R. *J. Magn. Reson.* **1981**, *43*, 259–281.
106. Macura, S.; Wüthrich, K.; Ernst, R. R. *J. Magn. Reson.* **1982**, *46*, 269–282.
107. Howard, D. W.; Lilley, D. M. J. *Prog. Nucl. Magn. Reson. Spectrosc.* **1978**, *12*, 1–40.
108. Barfield, M.; Gearhart, H. L. *J. Am. Chem. Soc.* **1973**, *95*, 641–643.
109. Bystrov, V. F. In "Progress in Nuclear Magnetic Resonance Spectroscopy," Emsley, J. W.; Feeney, J.; Sutcliffe, L. H., Eds. Pergamon Press: Oxford, 1976, Vol. 10, pp 41–80. Bystrov, V. F. *Russian Chem. Rev.* **1972**, *41*, 281–304.
110. Bystrov, V. F.; Arseniev, A. S.; Gavrilov, Yu. D. *J. Magn. Reson.* **1978**, *30*, 151–184. Bystrov, V. F.; Portnova, S. L.; Tsetlin, V. I.; Iranov, V. T.; Ovchinnikov, Y. A. *Tetrahedron* **1969**, *25*, 493–515.
111. Abraham, R. J.; McLauchlan, K. A. *Molec. Phys.* **1962**, *5*, 513–523.
112. Cung, M. T.; Marrand, M. *Biopolymers* **1982**, *21*, 953–967.
113. Deber, C. M.; Torchia, D. A.; Blout, E. R. *J. Am. Chem. Soc.* **1971**, *93*, 4893–4897.
114. Kopple, K. D.; Wiley, G. R.; Tanke, R. *Biopolymers* **1973**, *12*, 627–636.
115. de Leeuw, F. A. A. M.; Altona, C. *Int. J. Peptide Protein Res.* **1982**, *20*, 120–125.
116. Pachler, K. G. R. *Spectrochim. Acta* **1964**, *20*, 581–587.
117. Morris, G. A.; Freeman, R. *J. Am. Chem. Soc.* **1979**, *101*, 760–762. Morris, G. A. *Ibid.* **1980**, *102*, 428–429.
118. Dorman, D. E.; Bovey, F. A. *J. Org. Chem.* **1973**, *38*, 2379–2383.
119. Kessler, H. *Angew. Chem. Int. Ed. Eng.* **1970**, *9*, 219–235.
120. Binsch, G. H.; Kessler, H. *Angew. Chem. Int. Ed. Engl.* **1980**, *19*, 411–428.
121. Jackman, L. M.; Cotton, F. A. "Dynamic Nuclear Magnetic Resonance Spectroscopy." Academic Press: New York, 1975.
122. Adler, A. J.; Greenfield, N.; Fasman, G. *Meth. Enzymol.* **1973**, *27*, 675–735.
123. Woody, R. W. In "The Peptides: Conformation in Biology and Drug Design," Hruby, V. J., Ed. Academic Press: New York, Vol. 7, 1985, pp 16–114.
124. Woody, R. W. In "Peptides, Polypeptides and Proteins," Blout, E. R.; Bovey, F. A.; Goodman, M.; Lotan, N., Eds. John Wiley & Sons: New York, 1974, pp 338–350.
125. Bush, C. A.; Sarkar, S. K.; Kopple, K. D. *Biochemistry* **1978**, *17*, 4951–4954.
126. Chang, C. T.; Wu, C-S. C.; Yang, J. T. *Analyt. Biochem.* **1978**, *91*, 13–31.
127. Hruby, V. J.; Deb, K. K.; Fox, J.; Bjarnason, J.; Tu, A. T. J. *Biol. Chem.* **1978**, *253*, 6060–6067.
128. Hruby, V. J.; Mosberg, H. I.; Fox, J. W.; Tu, A. T. *J. Biol. Chem.* **1982**, *257*, 4916–4924.
129. Madison, V.; Kopple, K. D. *J. Am. Chem. Soc.* **1980**, *102*, 4855–4863.

130. Woody, R. W. *Biopolymers* **1978**, *17*, 1451–1467.
131. Aubry, A.; Protas, J.; Boussard, G.; Marraud, M. *Acta Cryst.* **1979**, *B35*, 694–699.
132. Aubry, A.; Marraud, M. *Biopolymers* **1983**, *22*, 341–345.
133. Boussard, G.; Marraud, M.; Neel, J. *J. Chem. Phys.* **1974**, *71*, 1081–1091.
134. Kopple, K. D.; Go, A.; Pilipauskas, D. R. *J. Am. Chem. Soc.* **1975**, *97*, 6830–6838.
135. Hseu, Th.; Chang, H. *Biochim. Biophys. Acta* **1980**, *624*, 340–345.
136. Ivanov, V. T.; Kogan, G. A.; Meshcheryakova, E. N.; Senyavina, L. B.; Efremov, E. S.; Ovchinnikov, Yu. A. *Tetrahedron Lett.* **1971**, 2841–2844.
137. Fox, J. A.; Tu, A. T.; Hruby, V. J.; Mosberg, H. I. *Arch. Biochem. Biophys.* **1981**, *211*, 628–631.
138. Williams, R. W. *J. Mol. Biol.* **1983**, *166*, 581–603.
139. Williams, R. W.; Dunker, A. K. *J. Mol. Biol.* **1981**, *152*, 783–813.
140. Anteunis, M. J. O. *Bull. Soc. Chim. Belg.* **1978**, *87*, 627–650.
141. Davies, D. B.; Khaled, M. A. *J. Chem. Soc., Perkin Trans. II* **1976**, 1238–1244.
142. Hooker, T. M. Jr.; Bayley, P. M.; Radding, W.; Schellman, J. A. *Biopolymers* **1974**, *13*, 549–566.
143. Imanishi, Y. *Adva. Polymer Sci.* **1976**, *20*, 1–77.
144. Sammes, P. G. *Fortsch. Chem. Org. Naturst.* **1975**, *33*, 51–118.
145. Tanihara, M.; Hiza, T.; Imanishi, Y.; Higashimura, T. *Bull. Chem. Soc. Jpn.* **1983**, *56*, 1155–1160.
146. Greenstein, J. P. *J. Biol. Chem.* **1937**, *118*, 321–329.
147. Greenstein, J. P. *J. Biol. Chem.* **1937**, *121*, 9–17.
148. Greenstein, J. P. *Arch. Biochem. Biophys.* **1954**, *52*, 203–216.
149. Greenstein, J. P. *Arch. Biochem. Biophys.* **1954**, *53*, 501–513.
150. Wade, R.; Winitz, M.; Greenstein, J. P. *J. Am. Chem. Soc.* **1956**, *78*, 373–377.
151. Heaton, G. S.; Rydon, H. N.; Schofield, J. A. *J. Chem. Soc.* **1956**, 3157–3168.
152. Jarvis, D.; Rydon, H. N.; Schofield, J. A. *J. Chem. Soc.* **1961**, 1752–1765.
153. Large, D. G.; Hydon, H. N.; Schofield, J. A. *J. Chem. Soc.* **1961**, 1749–1751.
154. Chandrasekaran, R.; Balasubramanian, R. *Biochem. Biophys. Acta* **1969**, *188*, 1–9.
155. Chandrasekaran, R.; Lakshminarayanan, A. V.; Pandye, U. V.; Ramachandran, G. N. *Biochim. Biophys. Acta* **1973**, *303*, 14–27.
156. Blaha, K.; Smolikova, J.; Vitek, A. *Coll. Czech. Chem. Commun.* **1966**, *31*, 4296–4311.
157. Capasso, S.; Matlia, C.; Mazzarella, L.; Puliti, R. *Acta Cryst.* **1977**, *B33*, 2080–3083.
158. Hata, Y.; Matsura, Y.; Tanaka, N.; Ashida, T.; Kakudo, M. *Acta Crystallog.* **1977**, *B33*, 3561–3563.
159. Donzel, B.; Kamber, B.; Wüthrich, K.; Schwyzer, R. *Helv. Chim. Acta* **1972**, *55*, 947–961.
160. Jung, G.; Ottnad, M. *Angew. Chem. Int. Ed. Eng.* **1974**, *13*, 818–819.
161. Ottnad, M.; Hartter, P.; Jung, G. *Hoppe-Seylers Z. Physiol. Chem.* **1975**, *356*, 1011–1025.
162. Mitra, A. K.; Chandrasekaran, R. *Int. J. Peptide Protein Res.* **1977**, *10*, 235–239.
163. Gregory, A. R.; Przybylska, M. *J. Am. Chem. Soc.* **1978**, *100*, 943–953.
164. Mez, H. C. *Cryst. Struct. Commun.* **1974**, *3*, 657–660.
165. Varughese, K. I.; Lu, C. T.; Kartha, G. *Int. J. Peptide Protein Res.* **1981**, *18*, 88–102.
166. Curtis, R. W. *Science* **1958**, *128*, 661–662.
167. Ptak, M. *Biopolymers* **1973**, *12*, 1575–1589.
168. Tonelli, A. *Biopolymers* **1978**, *17*, 1175–1179.
169. Mitra, A.; Chandrasekaran, R. *Biopolymers* **1984**, *23*, 2513–2524.
170. Smith, P. J. C.; Arnott, S. *Acta Crystallogr.* **1978**, *A34*, 3–11.
171. Rothe, M. *Chem. Ber.* **1962**, *95*, 783–794.
172. White, D. N. J.; Dunitz, J. D. *Isr. J. Chem.* **1972**, *10*, 249–256.
173. White, D. N. J.; Boville, M. J. *J. Chem. Soc., Perkin Trans. II* **1977**, 1610–1623.
174. Guy, M. H. P.; Heimbach, P.; MacNicol, D. D.; White, D. N. J. *J. Mol. Struct.* **1978**, *48*, 143–146.
175. Rothe, M.; Mästle, W. *Angew. Chem. Suppl.* **1982**, 533–541.
176. Rothe, M.; Steffen, K. D.; Rothe, I. *Angew. Chem. Int. Ed. Engl.* **1965**, *4*, 356. Rothe, M.; Steffen, K. D.; Rothe, I. *Angew. Chem. Int. Ed. Engl.* **1964**, *3*, 64.
177. Venkatachalam, C. M. *Biochim. Biophys. Acta* **1968**, *168*, 402–410.
178. Venkatachalam, C. M. *Biopolymers* **1968**, *6*, 1425–1436.
179. Deslauriers, R.; Smith, I. C. P.; Rothe, M. In "Peptides: Chemistry, Structure and Biology," Walter, R.; Meienhofer, J., Eds. Ann Arbor Science: Ann Arbor, Mich., 1975, pp 91–96. Deslauriers, R.; Smith, I. C. P.; Walter, R. *J. Biol. Chem.* **1974**, *249*, 7006–7010.
180. Druyan, M. E.; Coulter, C. L.; Walter, R.; Kartha, G.; Ambady, G. K. *J. Am. Chem. Soc.* **1976**, *98*, 5496–5502.
181. Kartha, G.; Ambady, G.; Shankar, P. V. *Nature (London)* **1974**, *247*, 204–205.
182. Kartha, G.; Ambady, G. *Acta Cryst.* **1975**, *B31*, 2035–2039.

183. Dale, J.; Titlestad, K. J. *Chem. Soc. Chem. Commun.* **1972**, 255–257.
184. Boni, R.; Verdini, A. S.; Deber, C. M.; Blout, E. R. *Biopolymers* **1978**, *17*, 2385–2399.
185. Kessler, H.; Kondor, P.; Krack, G.; Kramer, P. *J. Am. Chem. Soc.* **1978**, *100*, 2548–2550.
186. Kessler, H. In "Proceedings of the Symposium on Stereodynamics of Molecular Systems," Sarma, R. H., Ed. Pergamon Press: Elmsford, N.Y., 1979, pp 187–196.
187. Rothe, M.; Rott, H. *Angew. Chem. Int. Ed. Engl.* **1976**, *15*, 770–771.
188. Dale, J. In "Topics in Stereochemistry," Allinger, N. L.; Eliel, E. L., Eds. John Wiley & Sons: New York, 1976, pp 199–270.
189. Dunitz, J. D.; Waser, J. *J. Am. Chem. Soc.* **1972**, *94*, 5645–5650.
190. Titlestad, K. *Acta Chem. Scand.* **1975**, *B29*, 153–167.
191. Bats, J. W.; Friedrich, A.; Fuess, H.; Kessler, H.; Mästle, W.; Rothe, M. *Angew. Chem. Int. Ed. Eng.* **1979**, *18*, 538–539.
192. Kessler, H.; Bermel, W.; Krack, G.; Bats, J. W.; Faess, H.; Hull, W. E. *Chem. Ber.* **1983**, *116*, 3164–3181.
193. Kessler, H.; Friedrich, A. *J. Org. Chem.* **1981**, *46*, 3892–3895.
194. de Leeuw, F. A. A. M.; Altona, C.; Kessler, H.; Bermel, W.; Friedrich, A.; Krack, G.; Hull, W. *J. Am. Chem. Soc.*, 1983, *105*, 2237–2246.
195. Druyan, M. E.; Coulter, C. L. In "Peptides: Chemistry, Structure and Biology," Walter, R.; Meienhofer, J., Eds. Ann Arbor Science: Ann Arbor, Mich., 1975, pp 85–88.
196. Pinnen, F.; Zanotti, G.; Lucente, G.; Cerrini, S.; Fedeli, W.; Gavuzzo, E. *J. Chem. Soc., Perkin Trans. II* **1985**, 1931–1937.
197. Bats, J. W.; Fuess, H. *J. Am. Chem. Soc.* **1980**, *102*, 2065–2070.
198. Deber, C. M.; Fossel, E. R.; Blout, E. R. *J. Am. Chem. Soc.* **1974**, *96*, 4015–4017.
199. Vicar, J.; Gut, V.; Fric, I.; Blaha, K. *Collect. Czech. Chem. Commun.* **1976**, *41*, 3467–3473.
200. Rothe, M.; Schindler, W.; Pudill, R.; Kostrzewa, M.; Theysohn, R.; Steinberger, R. In "Peptides 1971," Nesvadba, H., Ed. North-Holland: Amsterdam, 1971, pp 388–399. Rothe, M.; Theysohn, R.; Mulhausen, D.; Eisenbeiss, F.; Schindler, W. In "Chemistry and Biology of Peptides," Meienhofer, J., Ed. Ann Arbor Science: Ann Arbor, Mich., 1972, p. 51.
201. Rothe, M.; Fähnle, M.; Pudill, R.; Shindler, W. In "Peptides: Structure and Biological Function," Gross, E.; Meienhofer, J., Eds. Pierce Chemical Co.: Rockford, Ill., 1979, pp 285–288.
202. Shemyakin, M. M.; Antonov, V. K.; Shkrob, A. M.; Shchelokov, V. J.; Agadzhanyan, Z. E. *Tetrahedron* **1965**, *21*, 3537–3572.
203. Rothe, M.; Steinberger, R. *Tetrahedron Lett.* **1970**, *9*, 649–654.
204. Lucente, G.; Pinnen, F.; Zanotti, G.; Cerrini, S.; Fedeli, W.; Mazza, F. In "Peptides: Synthesis—Structure—Function," Rich, D. H.; Gross, E., Eds. Pierce Chemical Co.: Rockford, Ill., 1981, pp 93–96.
205. Lucente, G.; Pinnen, G.; Zanotti, G.; Cerrini, S.; Fedli, W.; Gavuzzo, E. *Tetrahedron Lett.* **1981**, *22*, 3671–3674.
206. Lucente, G.; Pinnen, F.; Romeo, A.; Zanotti, G. *J. Chem. Soc., Perkin Trans. I* **1983**, 1127–1130.
207. Rothe, M.; Fähnle, M.; Mästle, W. In "Peptides: Synthesis—Structure—Function," Rich, D. H.; Gross, E., Eds. Pierce Chemical Co.: Rockford, Ill., 1981, pp 89–92.
208. Zanotti, G.; Pinnen, F.; Lucente, G. *Tetrahedron Lett.* **1985**, *26*, 5481–5484.
209. Rothe, M.; Fähnle, M.; Mästle, W.; Feige, K. In "Peptides: Structure and Function," Deber, C. M.; Hruby, V. J.; Kopple, K. D., Eds. Pierce Chemical Co.: Rockford, Ill., 1985, pp 177–180.
210. Pinnen, F.; Zanotti, G.; Lucente, G. *J. Chem. Soc., Perkin Trans. I* **1982**, 1311–1316.
211. Pinnen, F.; Zanotti, G.; Lucente, G. *Tetrahedron Lett.* **1984**, *25*, 5201–5204.
212. Schwyzer, R.; Iselin, B.; Rittel, W.; Sieber, P. *Helv. Chim. Acta.* **1956**, *39*, 872–883.
213. Ramakrishnan, C.; Sarathy, K. P. *Biochim. Biophys. Acta* **1968**, *168*, 402–410.
214. Balasubramanian, D.; Wetlaufer, D. B. In "Conformation of Biopolymers," Ramachandran, G. N., Ed. Academic Press: New York, Vol. I, 1967, pp 147–156.
215. Grathwohl, C.; Tun-Kyi, A.; Bundi, A.; Schwyzer, R.; Wüthrich, K. *Helv. Chim. Acta* **1975**, *58*, 415–423.
216. Groth, P. *Acta Chemica Scand.* **1970**, *24*, 780–790.
217. Sarathy, K. P. Ph.D. thesis, Madras University, Madras, India, 1970.
218. Ueno, K.; Shimizu, T. *Biopolymers* **1983**, *22*, 633–641.
219. Shimizu, T.; Tanaka, Y.; Tsuda, K. *Int. J. Biol. Macromol.* **1983**, *5*, 179–185.
220. Shimizu, T.; Tanaka, Y.; Tsuda, K. *Biopolymers* **1983**, *22*, 617–632.
221. Manjula, G. Ph.D. thesis, Indian Institute of Science, Bangalora, India, 1977. Manjula, G.; Ramakrishnan, C. In "Peptides: Proceedings of the Fifth American Peptide Symposium," Goodman, M.; Meienhofer, J., Eds. John Wiley & Sons: New York, 1977, pp 296–299.

222. Manjula, G.; Ramakrishnan, C. *Biopolymers* **1979**, *18*, 591–607.
223. Chiang, C. C.; Karle, I. *Int. J. Peptide Protein Res.* **1982**, *20*, 133–138.
224. Faulstich, H.; Trischman, H.; Dabrowsky, J. In "Peptides 1978. Proceedings of the XV European Peptide Symposium," Siemion, I. Z.; Kupryszewski, G., Eds. Wroclaw University Press: Wroclaw, Poland, 1978, pp 305–310.
225. Rich, D. H.; Kawai, M.; Jasensky, R. D. *Int. J. Peptide Protein Res.* **1983**, *21*, 35–42.
226. Rich, D. H.; Jasensky, R. D. *J. Am. Chem. Soc.* **1980**, *102*, 1112–1119.
227. Hruby, V. J.; Kao, L.-F.; Hirning, L. D.; Burks, T. F. In "Peptides: Structure and Function," Deber, C. M.; Hruby, V. J.; Kopple, K. D., Eds. Pierce Chemical Co.: Rockford, Ill., 1985, pp. 487–490.
228. Flippen, J. L.; Karle, I. L. *Biopolymers* 1976, *15*, 1081–1092. Fox, J. A.; Tu, A. T.; Hruby, V. J.; Mosberg, H. I. *Arch. Biochem. Biophys.* **1981**, *211*, 628–631.
229. Rich, D. H.; Jasensky, R. D. In "Peptides: Structure and Biological Function," Gross, E.; Meienhofer, J., Eds. Pierce Chemical Co.: Rockford, Ill., 1979, pp 487–490.
230. Pastuszak, J.; Gardner, J. H.; Singh, J.; Rich, D. H. *J. Org. Chem.* **1982**, *47*, 2982–2987.
231. Kawai, M.; Jasensky, R. D.; Rich, D. H. *J. Am. Chem. Soc.* **1983**, *105*, 4456–4462.
232. Kawai, M.; Pottorf, R. S.; Rich, D. H. *J. Med. Chem.* **1986**, *29*, 2409–2411.
233. Haslinger, E.; Kalchhauser, H. *Tetrahedron Lett.* **1983**, *24*, 2553–2556.
234. Haslinger, E.; Kalchhauser, H.; Wolscharm, P. *Monatsh. Chemie* **1984**, *115*, 779–783.
235. Govil, G.; Hosur, R. V. "Conformation of Biological Molecules, New Results from NMR." Springer: Berlin, 1982.
236. Gerig, J. T. *Biopolymers* **1971**, *10*, 2435–2443.
237. Meyer, W. L.; Kuyper, L. F.; Lewis, R. B.; Templeton, G. E.; Woodhead, S. H. *Biochem. Biophys. Res. Commun.* **1974**, *56*, 234–240.
238. Meyer, W. L.; Kuyper, L. F.; Phelps, D. W.; Cordes, A. W. *J. Chem. Soc. Chem. Commun.* **1974**, 339–340.
239. Meyer, W. L.; Templeton, G. E.; Grable, C. I.; Jones, R.; Kuyper, L. F.; Lewis, R. B.; Sigel, C. W.; Woodhead, S. H. *J. Am. Chem. Soc.* **1975**, *97*, 3802–3809.
240. Rich, D. H.; Mathiaparanam, P. *Tetrahedron Lett.* **1974**, *46*, 4037–4040.
241. Rich, D. H.; Mathiaparanam, P.; Grant, J. A.; Bhatnagar, P. In "Peptides: Chemistry, Structure and Biology," Walter, R.; Meienhofer, J., Eds. Ann Arbor Science: Ann Arbor, Mich., 1975, pp 943–948.
242. Rich, D. H.; Bhatnagar, P. In "Peptides. Proceedings of the Fifth American Peptide Symposium," Goodman, M.; Meienhofer, J., Eds. John Wiley & Sons: New York, 1977, pp 340–342.
243. Rich, D. H.; Bhatnager, P.; Jasensky, R. D.; Steele, J. A.; Uchytil, T. F.; Durbin, R. D. *Bioorg. Chem.* **1978**, *7*, 207–214.
244. Nishioka, K.; Constantoupoulos, A.; Satoh, P. S.; Najjar, V. A. *Biochem. Biophys. Res. Commun.* **1972**, *47*, 172–179.
245. Konopinska, D.; Nawrocka, E.; Siemion, I. Z.; Szymaniec, S.; Slopek, S. In "Peptides 1976, Proceeding of the 14th European Peptide Symposium," Loffet, A., Ed. l'Université de Bruxelles: Brussels, 1976, pp 535–539.
246. Blumenstein, M.; Layne, P. P.; Najjar, V. A. *Biochemistry* **1979**, *18*, 5247–5253.
247. Sekacis, I. P.; Liepins, E. E.; Veretennikova, N. I.; Chipens, G. I. *Bioorgan. Khim.* **1979**, *5*, 1617–1622.
248. Siemion, I. Z.; Lisowski, M.; Knopinska, D.; Nawrocka, E. *Eur. J. Biochem.* **1980**, *112*, 339–343.
249. Sucharda-Sobczyk, A.; Siemion, I. Z.; Knopinska, D. *Eur. J. Biochem.* **1979**, *96*, 131–139.
250. Fitzwater, S.; Hodes, Z. I.; Scheraga, H. A. *Macromolecules* **1978**, *11*, 805–811.
251. Nikiforovich, G. V. *Bioorgan. Khim.* **1978**, *4*, 1427–1430.
252. Chipens, G. I.; Veretennikova, N. I.; Nikiforovich, G. V.; Atare, Z. A. In "Peptides 1980," Brunfeldt, K., Ed. Scriptor: Copenhagen, 1980, pp 445–450.
253. Nikiforovich, G. V.; Liepina, I. T.; Sekacis, I. P.; Liepins, E. E.; Katayev, B. S.; Verentennikova, N. I.; Chipens, G. I. *Int. J. Peptide Protein Res.* **1984**, *23*, 271–275.
254. Izumiya, N.; Kato, T.; Aoyagi, H.; Shimohigashi, Y.; Yasutake, A.; Lee, S.; Noda, K.; Gross, E. In "Peptides: Structure and Biological Function," Goodman, M.; Meienhofer, J. Eds. John Wiley & Sons: New York, 1977, pp 439–444.
255. Yasutake, A.; Aoyagi, H.; Kato, T.; Izumiya, N. *Int. J. Peptide Protein Res.* **1980**, *15*, 113–121.
256. Yasutake, A.; Aoyagi, H.; Sada, I.; Kato, T.; Izumiya, N. *Int. J. Peptide Protein Res.* **1982**, *20*, 241–253.
257. Shimizu, T.; Tanaka, Y.; Tsuda, K. *Bull. Chem. Soc. Jpn.* **1982**, *55*, 3808–3816.
258. Shimizu, T.; Tanaka, Y.; Tsuda, K. *Bull Chem. Soc. Jpn.* **1982**, *55*, 3817–3823.

259. Deslauriers, R.; Leach, S. J.; Maxfield, F. R.; Minasian, E.; McQuie, J. R.; Meinwald, Y. C.; Nemethy, G.; Pottle, M. S.; Rae, I. D.; Scheraga, H. A.; Stimson, E. R.; van Nispen, J. W. *Proc. Natl. Acad. Sci. USA* **1979**, *76*, 2512–2514.
260. Neméthy, G.; McQuie, J. R.; Pottle, M. S.; Scheraga, H. A. *Macromolecules* **1981**, *14*, 975–985.
261. Deslauriers, R.; Evans, D. J.; Leach, S. J.; Meinwald, Y. C.; Minasian, E.; Neméthy, G.; Rae, I. D.; Scheraga, H. A.; Somorjai, R. L.; Stimson, E. R.; van Nispen, J. W.; Woody, R. W. *Macromolecules* **1981**, *14*, 985–996.
262. Bandekar, J.; Evand, D. J.; Krimm, S.; Leach, S. J.; Lee, S.; McQuie, J. R.; Minasian, E.; Nemethy, G.; Pottel, M. S.; Scheraga, H. A.; Stimson, E. R.; Woody, R. W. *Int. J. Peptide Protein Res.* **1982**, *19*, 187–205.
263. Holmgren, A. *Trends in Biochem. Sci.* **1981**, *6*, 26–29.
264. Chou, P. Y.; Fasman, G. D. *Biochemistry* **1974**, *13*, 222–245.
265. Bramachari, S. K.; Rapaka, R. S.; Bhatnagar, R. S.; Ananthanarayanan, V. S. *Biopolymers* **1982**, *21*, 1107–1125.
266. Zimmerman, S. S.; Scheraga, H. A. *Biopolymers* **1977**, *16*, 811–843.
267. Venkatachalapathi, Y. V.; Venkataram Prasad, B. V.; Balaram, P. *Biochemistry* **1982**, *21*, 5502–5509.
268. Rao, B. N.; Kumar, A.; Balaram, H.; Ravi, A.; Balaram, P. *J. Am. Chem. Soc.* **1983**, *105*, 7423–7428.
269. Ravi, A.; Balaram, P. *Tetrahedron* **1984**, *40*, 2577–2583.
270. Ravi, A.; Venkataram Prasad, B. V.; Balaram, P. *J. Am. Chem. Soc.* **1983**, *105*, 105–109.
271. Prasad, B. V. V.; Ravi, A.; Balaram, P. *Biochem. Biophys. Res. Commun.* **1981**, *103*, 1138–1144.
272. Ishizaki, H.; Balaram, P.; Nagaraj, R.; Venkatachalapathi, Y. V.; Tu, A. T. *Biophys. J.* **1981**, *36*, 509–517.
273. Joullie, M. M.; Nutt, R. F. In "Alkaloids: Chemical and Biological Perspectives," Pelletier, S. W., Ed. Wiley-Interscience: New York, 1985, Vol. 3, pp 113–167.
274. Lagarias, J. C.; Yokoyama, W. H.; Bordner, J.; Shih, W. C.; Klein, M. P.; Rapaport, H. *J. Am. Chem. Soc.* **1983**, *105*, 1031–1040.
275. Schmidt, U.; Griesser, H.; Lieberknecht, A.; Talbiersky, J. *Angew. Chem. Int. Ed. Engl.* **1981**, *20*, 280–281.
276. Schmidt, U.; Lieberknecht, A.; Griesser, H.; Hausler, J. *Angew. Chem. Int. Ed. Engl.* **1981**, *20*, 281–282.
277. Goff, D.; Lagarias, J. C.; Shih, W. C.; Klein, M. P.; Rapoport, H. *J. Org. Chem.* **1980**, *45*, 4813–4817.
278. Kirfel, A.; Will, G. *Z. Krist.* **1975**, *142*, 368–383.
279. Kirfel, A.; Will, G.; Tschesche, R.; Wilhelm, H. *Z. Naturforsch.* **1976**, *31b*, 279–280.
280. Takai, M.; Kawai, K.; Ogihara, Y.; Iitaka, Y.; Shibata, S. *J. Chem. Soc. Chem. Commun.* **1974**, 653.
281. Takai, M.; Ogihara, Y.; Iitaka, Y.; Shibata, S. *Chem. Pharm. Bull.* **1976**, *24*, 2181–2184.
282. Tschesche, R.; Kaussmann, E. U. In "The Alkaloids," Manske, R. H. F., Ed. Academic Press: New York, 1975, Vol. 15, Chap. 4.
283. Morel, A. F.; Bravo, R. V. F.; Reis, F. A. M.; Ruveda, E. A. *Phytochemistry* **1979**, *18*, 473–477.
284. Chang, C. J.; Hagaman, E. W.; Wenkert, E.; Sierra, M. G.; Mascaretti, O. A.; Merkuza, U. M.; Ruveda, E. A. *Phytochemistry* **1974**, *13*, 1273–1279.
285. Haslinger, E. *Tetrahedron* **1978**, *34*, 685–688.
286. Haslinger, E. *Monatsh. Chemie* **1978**, *109*, 523–526.
287. Haslinger, E.; Robien, W. *Monatsh. Chemie* **1979**, *110*, 1011–1018.
288. Takai, M.; Ogihara, Y.; Iitaka, Y.; Shibata, S. *Chem. Pharm. Bull.* **1975**, *23*, 2556–2559.
289. Hindenlang, D. M.; Shamma, M.; Miana, G. A.; Shah, A. H.; Cassels, B. K. *Justus Liebig Ann. Chem.* **1980**, 447–450.
290. Pais, M.; Jarreau, F.-X.; Sierra, M. G.; Mascaretti, O. A.; Ruveda, E. A.; Chang, C. J.; Hagaman, E. W.; Wenkert, E. *Phytochemistry* **1979**, *18*, 1869–1872.
291. Schiller, P. W. In "The Peptides: Analysis, Synthesis, Biology," Vol. 5, Udenfriend, S.; Meienhofer, J., Eds. Academic Press: New York, 1985, pp 219–268.
292. Loew, G. H.; Hashimoto, G.; Williamson, L.; Burt, S.; Anderson, W. *Mol. Pharmacol.* **1982**, *22*, 667–677.
293. Smith, G. D.; Griffin, J. F. *Science* **1978**, *199*, 1214–1216.
294. Schiller, P. W.; Di Maio, J. *Nature* **1982**, *279*, 74–76.
295. DiMaio, J.; Nguyen, T. M.-D., Lemieux, C.; Schiller, P. W. *J. Med. Chem.* **1982**, *25*, 1432–1438.
296. Schiller, P. W.; Nguyen, T. M.-D.; Miller, J. *Int. J. Peptide Protein Res.* **1985**, *25*, 171–177.

297. Sarantakis, D. U.S. Patent 4,098,78, 1979.
298. Schiller, P. W.; Eggimann, B.; Di Maio, J.; Lemieux, C.; Nguyen, T. M.-D. *Biochem. Biophys. Res. Commun.* **1981**, *101*, 337–343.
299. Mosberg, H. I.; Hurst, R.; Hruby, V. J.; Galligan, J. J.; Burks, T. F.; Gee, K.; Yamamura, H. I. *Biochem. Biophys. Res. Commun.* **1982**, *106*, 506–512.
300. Mosberg, H. I.; Hurst, R.; Hruby, V. J.; Galligan, J. J.; Burks, T. F.; Gee, K.; Yamamura, H. I. *Life Sci.* **1983**, 2565–2569.
301. Mosberg, H. I.; Hurst, R.; Hruby, V. J.; Galligan, J. J.; Burks, T. F.; Gee, K.; Yamamura, H. I. *Proc. Natl. Acad. Sci. USA* **1983**, *80*, 5871–5874.
302. Hruby, V. J.; Kao, L.-F.; Hirning, L. D.; Burks, T. F. In "Peptides: Structure and Function," Deber, C. M.; Hruby, V. J.; Kopple, K. D., Eds. Pierce Chemical Co.: Rockford, Ill., 1985, pp 487–490.
303. Hruby, V. J. In "Opioid Peptides: Medicinal Chemistry," Rapaka, R. S.; Barnett, G.; Hawks, R. L., Eds. NIDA Research Monograph 69: Rockville, Md., 1986, pp 128–147.
304. Akiyama, K.; Gee, K. W.; Mosberg, H. I.; Hruby, V. J.; Yamamura, H. I. *Proc. Natl. Acad. Sci. USA* **1985**, *82*, 2543–2547.
305. Corbett, A. D.; Gillan, M. G. C.; Kosterlitz, H. W.; McKnight, A. T.; Paterson, S. J.; Robson, L. E. *Br. J. Pharmacol.* **1984**, *83*, 271–279.
306. Clark, J. A.; Itzhak, Y.; Hruby, V. J.; Yamamura, H. I.; Pasternak, G. W. *Eur. J. Pharmacol.* **1986**, *128*, 303–304.
307. James, I. F.; Goldstein, A. *Mol. Pharmacol.* **1984**, *25*, 337–342.
308. Hall, D.; Pavitt, N. *Biopolymers* **1984**, *23*, 1441–1455.
309. Kessler, H.; Holzemann, G.; Geiger, R. In "Peptides: Structure and Function," Hruby, V. J.; Rich, D. H., Eds. Pierce Chemical Co.: Rockford, Ill., 1983, pp 295–298.
310. Mosberg, H. I.; Schiller, P. W. *Int. J. Peptide Protein Res.* **1983**, *23*, 462–466.
311. Hruby, V. J.; Kao, L.-F.; Pettitt, M.; Karplus, M. *J. Am. Chem. Soc.*, in press, 1988.
312. Maxfield, F. R.; Bandekar, J.; Krumm, F.; Evans, D. J.; Leach, S. J.; Nemethy, G.; Scheraga, H. A. *Macromolecules*, **1981**, *14*, 997–1010.
313. Jackman, L. M.; Sternhell, S. "Application of Nuclear Magnetic Resonance Spectroscopy in Organic Chemistry," 2nd ed. Pergamon Press." Oxford, 1969, p. 289.

Index

Boldfaced numbers in parentheses refer to structures illustrated in the book.